ENCYCLOPEDIA OF DRUG AND DRUG DELIVERY

ENCYCLOPEDIA OF DRUG AND DRUG DELIVERY

Vol. 1

DRUG ADDICTION

By

Dr. Kamlesh Dashora
Associate Professor
Institute of Pharmacy
Vikram University
Ujjain (M.P.)
(India)

&

Dr. Darshan Dubey
Assistant Professor
Institute of Pharmacy
Vikram University
Ujjain (M.P.)
(India)

D P H

DISCOVERY PUBLISHING HOUSE PVT. LTD.
NEW DELHI-110 002

Published by:
Tilak Wasan
DISCOVERY PUBLISHING HOUSE PVT. LTD.
4383/4A, Ansari Road, Darya Ganj
New Delhi-110 002 (India)
Phone: +91-11-23279245, 43596064-65
Fax : +91-11-23253475
E-mail: parul.wasan@gmail.com
discoverypublishinghouse@gmail.com
web : www.discoverypublishinggroup.com

***First Edition:* 2012**

ISBN: 978-93-5056-038-9 (Set)
978-93-5056-039-6 (Vol. 1)
978-93-5056-040-2 (Vol. 2)
978-93-5056-041-9 (Vol. 3)
978-93-5056-042-6 (Vol. 4)
978-93-5056-043-3 (Vol. 5)

Encyclopedia of Drug and Drug Delivery

Printed at:
Shree Balaji Art Press
Delhi

Preface

The Present title "**Encyclopedia of Drug and Drug Delivery**" is the amazing advancement of pharmaceutical studies. It provides the various fundamental aspects of drug designing to be understood adequately. It has been compiled for undergraduate, postgraduate students, research scholars, teachers, technicians, industrial microbiologists and scientists involved in pharmaceutical research and development. The basic concepts have been clearly explained and their functions are adequately highlighted. The presentation of text is simple and systematic. The selection of chapters and topics is according to the specified syllabi of several Indian Universities and are so structured as to enable the student to move easily from the fundamental to the complex. It describes the ideas and techniques that are creating or new domain of science and technology. It holds the promise of revolutionizing field ranging from medical science to pharmaceutical industries. It is our earnest hope that this title will be of great value to all our students.

Organizational and economic changes in the pharmaceutical industry have led to a marked reduction in the supply of new chemical entities. This combined with an increase in the expiration of patents and tighter regulatory control has lead many companies to rethink existing product design. The growing trend to improve the efficacy of existing drug products, to both extend parent life and allow competition with opportunistic generic companies, has allowed for a greater understanding of formulation design. This trend is set to continue for some time as economic and political pressure increases.

There have been many improvements in understanding the theoretical and practical aspects of drug discoveries. Numerous therapeutic compounds have been investigated to treat a variety of different illnesses. There are many pharmaceutical methods requiring a variety of formulation options. Advanced formulation design may be thought of as a way to optimize the mode of drug designing to a patient that will ultimately lead to a more effective therapy and approach an optimal therapeutic outcome for a particular drug. This book is designed to take the reader through various cutting-edge areas of advanced formulation design by describing therapeutic categories that may be beneficial. The reader will be introduced to some novel technologies and updated on the state of the art in areas designed to overcome poor efficacy issues.

To make the work more comprehensive and informative, the authors have consulted many authoritative books, research journals, abstracts, monographs etc., so there can be no claim to originality except in the manner of treatment.

The authors express their thanks to their friends and colleagues whose continue inspirations have initiated them to bring out this book.

The authors express their gratitude to Mr. Wasan and staff of M/s Discovery Publishing House Pvt. Ltd., for their whole hearted co-operation in the publications of this book.

The authors will remain sincerely responsible for any shortcomings of the book and be grateful to the readers for their suggestions and constructive criticism for the continuous betterment of the book. They have this opportunity to appeal to the readers to send their suggestions straightaway to this publisher.

–Authors

Contents

1. Introduction **1–37**

Clinical Aspects of Neuropsychology of Addiction, Objectives of Self-medication, Pleasant Aspects of Drugs, Withdrawal Syndromes, Habit Learning, Substitutive Activities, Reinforcement Learning, Nucleus Accumbens in Motivation and Learning, Pleasurable Aspects of Reward, DA and Instrumental Behaviour, Contingency Activities, Psychomotor Stimulants, Development of Stimulus-response Habits, High Propensity to Relapse, Mechanism of Reinstatement, Observing Drug-induced Reinstatement, Observing Stress-induced Reinstatement, Chronic Administration of Drugs, Recreational Use of Drugs, Subjective and Reinforcement, Examples of Conditioned and Unconditioned Responses, Vulnerability to Drug Addiction, Prolonged Maternal Separation, Social and Environmentally Complex Environment, Early Exposure to Drugs, Hereditary Susceptibility to Addiction, Hereditary Neurovulnerability to Addiction, Sex Differences in Vulnerability to Drugs, Deleterious effects of Drugs of Abuse, Addiction of Drugs, Role of Compulsive Craving in Drug Use and Relapse, Neurobehavioural Economics of Addiction, Development of Inelastic Behaviour, Rational Addiction Therapy, Neuroeconomics of Addiction, Future Drugs of Abuse, Neuroscientific advances.

2. Psychostimulants **38–66**

Histological Details, Clinical Study of Receptors, Immunoreactivity, Anatomical Details of Mesocortical Zone, Pharmacological Study of Amygdaloid Complex, Opioid Receptor Activation, Modulatory Effect of Opioid, Systemic Administration of KOR, Intracerebroventricular Application, Influence of Repeated Administration of MOR, Molecular Mechanisms, Hormonal Axis, Steroidal Hormones as Psychostimulant Drugs, Psychostimulants in Movements, Individual Differences in Drug Responding, Relapse Behaviour, Observing Behavioural Responses, Effects of Stress to Psychostimulants, Stress-induced Sensitization, Stress-induced Corticosterone Secretion, Reinstatement of Behavior, Mesolimbic Dopamine System, Glucocorticoids and the Dopamine Dependent Responses, Dopamine Transmission, Interactive Responses.

3. **Problematic Recreational Substances** 67–92

Historical Background, Understanding Neurotransmitters, Clinical Aspects of Neurotransmitter, Estimating the Prevalence of Drug, Pharmaceutical Importance of Psychoactive Substances, Clinical Treatments, Expansion of Technologies, Clinical Characteristics, Management, Synthesis of Related Compounds, Pharmacological Studies, Treatments, Aberrant Learning, GABA Neurotransmitter, Concluding Remarks.

4. **Brain and Addiction** 93–116

Experimental Psychology, Reward and Reinforcement of Addictive Drugs, Role of Orbitofrontal Cortex, Effect on Emotions, Formation of Habits, Neurotransmitter Systems, Psychoactive Substances, Drugs of Recreational Values, Effects of Psychoactive Substances, Addiction of Short Lived Pleasurable Activities, Beneficial Psychoactive Substances, Resilience to Addiction, Pharmacological Survey, New Techniques, Experimental Psychology, New Communications, Cognitive Psychology, Pharmacodynamic Properties, Concluding Remarks.

5. **Clinical Testing of Psychoactive Drug** 117–131

Biochemical Tests, Analyzing Urine, Analyzing Blood, Analyzing Oral Fluid, Analyzing Sweat, Analysis of Drugs in Hair, Routine Techniques, Fuel Cell Technology, Screening Test, Liquid Chromatography-Mass Spectrometry (LC-MS), Gas Chromatography-Mass Spectrometry (GC-MS), Nuclear Magnetic Resonance (NMR) Spectroscopy, Capillary Electrophoresis (CE), Thin-layer Chromatography (TLC), Supercritical Fluid Chromatography (SFC), Electrochemical Detection (ECD), Advanced Methods, Miniature Instrumentation, Time-of-flight (TOF) Mass Spectrometry, Isotope Ratio Mass Spectrometry (IRMS), Coated Microsphere Technology, Chemiluminescent Nitrogen Detector (CLND), Surface Plasmon Resonance, Surface-enhanced Resonance Raman Spectroscopy, Total Internal Reflectance Fluorescence (TIRF), Concluding Remarks.

6. **Economic Values of Addiction** 132–155

Effects of Using Psychoactive Substances, Management of Psychoactive Substances, Models and Theories, Rational Choice, Commercial Values, Dynamics of Drug-taking Behavior, Computerized Data, Emperical Estimation, Calculating Costs, Factors Affecting Illicit Drug Demand, Literature on Supply Behaviour, Future Impact and Interventions, Economic Evaluation, Drug Policies, Treatment and Rehabilitation, Effectiveness of Programmes, Importance of Workplace, Collective Behavior, Drug Adoption, Basic Model, Conventional Behaviour, Network Analysis, Combined Drug Models, Concluding Remarks.

7. **Hereditary Alcoholism** 156–181

Transgenic Technologies, Hereditary Control, Screening and Mapping, Multistage Mapping Strategy, Recent Findings from Locus to Gene, Alcohol-induced Losses, Congenic Strains, Gene Mapping in Human Population, Ethanol Reward, Genetic Influence of Ethanol, Regular and Continuous Availability, Home-cage Self-administration, Pavlovian Conditioning, Ethanol's Rewarding and Aversive Effects, Recombinant Inbred Strains, Artificial Selection, Locating Genes on a Particular Chromosome, Emerging Genetic Technologies, Regulatory Mechanism, Future Progress.

8. **Effect of Addiction on Nervous Activities** 182–187

Locus Coeruleus of Brain, Neurological Function, Neural Substrate, Concluding Remarks.

9. **Endogenous Opioid System** 188–201

Antinociceptive Responses, Cannabinoid Agonistism, Biochemical Studies, Development of Cannabinoid Tolerance, Mechanism of Opioid Participation, Rewarding Effects, Behavioural Model, Intravenously Self-administration Drugs, Bioelectrophysiological Investigations, Concluding Remarks.

10. **Physiological and Subjective State of Alcoholism** 202–213

Longer Intake of Alcohol, Long-Term Alcohol Self-Administration with Repeated Deprivation Phases, Measurement of Crawing, Behavioral Sensitization of Psychoactive Drugs, Clinical Aspects of Craving, Alcohol Deprivation Effects, Reinstatement Alcohol-seeking Behavior, Systematic Studies, Concluding Remarks.

11. **Tobacco Addiction** 214–228

Constituents of Tobacco, In Vivo Drug Taking Paradigms, Neural Mechanism of the Reinforcement, Positive Reinforcing Efficacy, Relevance of Nicotine, Intravenous Nicotine in Experimental Animals, Pharmacological Studies of Nicotine Reward, Chronic Nicotine Administration Produces AchR Upregulation, Reinforcing Effects of Nicotine, Neuroadaptation to Nicotine Exposure, Mesocorticolimbic DA System, Transcriptional Regulation, Neuroadaptation to Nicotine Exposure, Concluding Remarks.

12. **Vaccination Programme** 229–252

Misuse of Drugs, Consenting Addict, Need of Vaccination, Mind Altering Effect, Pharmacogenetic Studies, Advanced Hereditary Information, Individual's Behaviour, Genetic Drift, Neuropharmacological Studies in Man, Brain Imaging Technologies, Targeted Neurochemical interventions, Neural Enhancement, Social Enhancement, Herbal Drugs with Neurological Effects, Social Consequences, Concluding Remarks.

13. **Prospects of Psychoactive Drugs** 253–269

Objectives of Study, Modernization of Cultures, Technical Changes, Professional Influence, Modern Theories, Commercial Values, Spreading Worldwide, Set and Setting Hypothesis, Management of Psychoactive Drugs, Temperance Movement for the Use of Alcohol, Strict Action, Prohibition Policies, Alarming Use of Drugs, Use at Public Places, Substantial Reduction in Tobacco Consumption, Elasticity in the Psychoactive Drugs, Concluding Remarks.

14. **Addiction of Common Drinks** 270–280

Coca, History, Tolerance and Dependence, Caffeine, Alcoholic Addiction, Combined Effects, Concluding Remarks.

Index 281–284

1

Introduction

Communication within the brain depends on the release of neurotransmitter substances. Many agents, including drugs, but also nutrients and transcranial magnetic or deep brain stimulation, exert their effects through chemical neurotransmission. Many psychoactive drugs work on chemical systems that not only control behaviour, but also respond to behavioural change. Many forms of behaviour, ranging from transcendental meditation to compulsive eating or gambling, may regulate the functioning of the chemical systems of the brain. In the last 50 years or so, the list of chemical neurotransmitter substances in the brain has lengthened from two to over 60. The neurotransmitters include amino-acids and monoamines and structurally more complex neuropeptides. The list is likely to be extended, leading to further drug development.

The discovery of new neurotransmitters goes hand in hand with the mapping of the neurons that contain them in the brain. These substances are not distributed homogeneously in the central nervous system, but are contained within defined tracts and clusters of cells, which may be organised in a complex arrangement to form functional, interconnected brain systems. Moreover, at the synapses between neurons, these substances interact as ligands with complex protein molecules called receptors on the neuronal membranes, to which they bind and thus transduce their chemical signals. There are often several distinct receptor subtypes for a given neurotransmitter, and these are also widely distributed within the brain, generally but not always matching the mapping of the neurotransmitter systems themselves. Further complexity is conferred by variations in protein subunits making up the receptors. The many functional effects of drugs such as nicotine and benzodiazepines such as diazepam can be attributed in part to different receptor subtypes operating preferentially in different brain regions.

The discovery of 'orphan' receptors without obvious neurotransmitter ligands in different brain regions indicates the possibility of further discoveries of new psychoactive substances. In the past, the discovery of psychoactive drugs has often predated the discovery of the endogenous neurotransmitter ligand (e.g. endorphin and enkephalins in the case of opiates such as morphine and heroin, and β-carbolines in the case of the benzodiazepines), as well as the brain receptors upon which they act. The biophysical actions of neurotransmitters at specific receptors range from short-lived electrochemical effects at ion channels to slower cellular signalling via receptors linked to biochemical cascades (e.g. '*second messengers*') and gene transcription. The old adage of one neurotransmitter per neuron has long been disproved by discoveries that many of the actions of the 'classical' neurotransmitters, such as acetylcholine (ACh), noradrenaline (NA), serotonin (5-HT) and dopamine (DA), are augmented by coreleased peptides.

The chemical neurotransmitters affect the functioning of dense, but generally highly organised, sets of connections, conveniently referred to as neuronal networks. They do this either by affecting fast signalling, whether excitatory or inhibitory, within the network, or by slower and spatially more diffuse modulations across the nodes of the network. An individual neuron is subject to many different influences from distinct neurotransmitter systems. The activity of a particular cell and thus of entire networks can be adjusted by variations in the syntax of chemical messages impinging on the cell. Several neurological and neuropsychiatric disorders have chemical pathologies for which a strategy of pharmacological replacement of deficient systems has been adopted.

L-DOPA medication in Parkinson's disease is the classic example, where the loss of DA-containing cells of the substantia nigra leads to the characteristic motor symptoms. L-DOPA remediates some of the cognitive deficits associated with Parkinson's disease, but also produces some undesirable cognitive and emotional side-effects, including for some patients a drive to abuse the drug. L-DOPA's therapeutic effects also tend to diminish with long-term treatment, leading to a gamut of other attempts to treat the disorder which includes other dopaminergic drugs, deep-brain stimulation, neurosurgery and the neural transplantation of embryonic nigral cells. Especially in view of the ethical problems posed by the last-named, a future approach will almost certainly involve the use of stem cells engineered to produce DA.

A similar approach to the treatment of Alzheimer's disease with cholinergic drug treatments (including nicotine) has proved less successful, although such treatment does improve some functions, notably attention. The future strategy (as with the treatment of stroke) is likely to hinge on neuroprotection, preventing through drug treatment the neuronal loss occurring as a consequence of the initial pathology. Many other disorders that result in cognitive or mood-related deficits are now treated with drugs. These include depression, treated with monoamine reuptake inhibitors such as the selective 5-HT reuptake inhibitors (SSRIs), schizophrenia, treated by DA receptor blockers, and attention-deficit hyperactivity disorder (ADHD), which is treated effectively with amphetamine-like stimulant drugs such as methylphenidate. A number of cognitive disorders arising from brain dys-function have been treated on an experimental basis: these include Korsakoff's syndrome (arising from alcoholism), which has been treated with drugs affecting noradrenergic transmission, acute brain injury (treated with DA receptor agonists), and stroke (treated with amphetamine). Whilst the molecular bases of these drugs' actions at the cellular level are well defined, the mechanistic basis of any of these therapeutic effects is less clear.

The most effective anti-psychotic drug, clozapine, is also one of the least specific in pharmacological terms. Anti-depressants such as the SSRIs may work via effects on neurogenesis in the hippocampus. Nevertheless, given the therapeutic efficacy of most of these drugs, and strong evidence from animal models of cognitive function, there is optimism that cognitive and mood-related disorders will continue to respond to interventions based on psychoactive substances. One of the most promising developments from experimental neuroscience has arisen from a neuronal model of learning called long-term potentiation (LTP). It can occur in subtly different forms in many forebrain regions, but has been investigated most intensively in the hippocampus. LTP crucially depends on the excitatory amino-acid neurotransmitter glutamate, and its actions at the AMPA-and N-methyl-*d*-aspartate (NMDA) receptor subtypes.

Several agents affecting glutamate transmission have been developed, including some (such as the AMPA-kines) that have been shown to have positive effects on learning in the laboratory for both normal animals and humans, and have been subject to preliminary clinical trials. Some NMDA receptor antagonists, such as ketamine, have recently been shown to have significant abuse potential. In a more speculative vein, drugs enhancing the transcription factor CREB (cAMP response element binding protein) could also emerge from advances in the application of basic neuroscience. The phenomenon of beneficial effects in normal subjects lacking discernible brain dysfunction is not restricted to drugs affecting glutamate receptors. In specific situations, which may include the infusion of drugs locally to specific brain regions in experimental animals and the engagement of particular cognitive functions, many positive drug effects have been reported for compounds acting on the classical cholinergic, noradrenergic, serotoninergic or dopaminergic systems.

Only some of these systems have also been associated with drug dependence. Thus the positive effects of cholinergic drugs on attention and aspects of mnemonic function are not accompanied by mood-altering effects of potential recreational use. The prediction of psychopharmacological efficacy is often confounded by the surprising emergence of new substances which may have initially appeared to be innocuous or which were initially established in some other functional context. The effective anti-narcoleptic modafinil has stimulant-like actions, but does not appear primarily to affect the brain neurotransmitters implicated in the effects of stimulant drugs such as amphetamine and methylphenidate. This drug has mild beneficial effects on tests of short-term memory and planning, as well as an anti-impulsive action, both in normal adults and in patients with ADHD. Its beneficial effects on vigilance and other aspects of human performance have led to its well-publicised use by the military. These effects indicate again the possibility of cognitive enhancement in intact individuals.

Increasing evidence of individual variability in intellectual function in normal subjects that occurs as a function of genotypical variation and in association with factors such as fatigue or under-arousal in the work-place, may promote self-medication. But although modafinil emerged from a scientific programme of drug development, we do not know how it works. The fact that modafinil is not widely abused indicates that it is feasible to dissociate stimulant from reinforcing actions of drugs of abuse. Whether this dissociation arises from the pharmacokinetic actions of the drug, which is relatively slow-acting, or its distinct neurochemical actions, is a theoretically, as well as practically, important issue. Self-medication prompted by a perceived need to elevate the activation of particular brain neurochemical systems might also affect other domains of forebrain functioning.

The need to enhance activation of the dopaminergic reinforcement ('reward') system might explain why individuals use cocaine and other psychomotor stimulants that operate primarily through this system. Such a view is consistent with evidence that euphoria produced by drugs such as cocaine and methylphenidate may depend on initially low levels of striatal DA (D2) receptors, as revealed by positron emission tomography, which may be indicative of low basal mood states. It may also be relevant to the identification of individuals who indulge in certain behavioural addictions such as gambling. Whether these individual differences arise from genetic or environmental influences is still to be determined.

CLINICAL ASPECTS OF NEUROPSYCHOLOGY OF ADDICTION

Motivated action can be examined by studying instrumental conditioning, the process by which animals alter their behaviour when there is a contingency between their behaviour and a reinforcing outcome. Reinforcement learning has been studied for a long time. At its most basic level, it is the ability to learn to act on the basis of important outcomes such as reward and punishment. Events that strengthen preceding responses are called positive reinforcers, while events whose removal strengthens preceding responses are called negative reinforcers. If reinforcers are defined by their effect on behaviour, then, to avoid a circular argument, behaviour cannot be said to have altered as a consequence of reinforcement. However, to explain behaviour rather than merely describe it, internal processes such as motivation must also be accounted for.

Central motivational states, such as hunger and thirst, account parsimoniously for a great deal of behavioural variability. For example, water deprivation, eating dry food, hypertonic saline injection, and the hormone angiotensin II all induce a common state - thirst - that has multiple effects. Thirsty animals drink more water, drink water faster, perform more of an arbitrary response to gain water, and so on. The ideas of motivational state entered early theories of reinforcement. For example, it was suggested that events that reduce 'drive' states such as thirst are positively reinforcing. However, on its own, this simple model cannot account for many instrumental conditioning phenomena, let alone 'unnatural' reinforcement such as drug addiction. Modern neuropsychological theories recognise that many processes contribute to a simple act such as pressing a lever to receive food or a drug.

Rats and humans exhibit goal-directed action, which is based on knowledge of the contingency between one's actions and their outcomes, and knowledge of the value of those outcomes. These two representations interact so that we work for that which we value. Environmental stimuli provide information about what contingencies may be in force in a given environment. Remarkably, the value system governing goal-directed action is not the brain's only one. This 'cognitive' value system exists alongside the valuation process that determines our reactions when we actually experience a goal such as food, termed 'liking', 'hedonic reactions', or simply 'pleasure'. Under many normal circumstances the two values reflect one another and change together. However, the fact that they are different means that animals must learn which outcomes are valuable in a given motivational state, a process referred to as incentive learning. For example, rats do not know that to eat a particular food while sated is not as valuable as to eat the same food while hungry until they have actually eaten the food while sated. Just as there is more than one value system, there is more than one route to action, and not all action is goal-directed.

With time and training, actions can become habitual, that is, elicited by direct stimulus-response (S-R) associations. S-R habits are less flexible than goal-directed action, because their representation contains no information about what the final outcome will be, and cannot alter quickly if the desirability of a particular outcome changes. But they may help reduce the demands on the cognitive, goal-directed system in familiar settings. Stimuli that predict reward may become conditioned stimuli (CS), associated with the reward (unconditioned stimulus, US) through pavlovian associative learning. Pavlovian CS can influence instrumental behaviour directly (pavlovian–instrumental transfer, PIT) and can serve as the goals of behaviour, termed conditioned reinforcement. Seen in this context, the major neuropsychological theories of drug addiction - none of them mutually exclusive - can be summarised:

Objectives of Self-medication

Drugs are taken for their positive effects. These may include euphoria, enhanced social experiences, enhanced intellectual or attentional performance, and an enhanced effect of other reinforcers such as food or sex. An aspect of this may be that people '*self-medicate*' to achieve a desired level of mood, social performance, and so on, although the extent to which this occurs is debated. Furthermore, the effect of the drug depends upon the user's expectations and prior mood, and varies between people. Tolerance to pleasant drug effects may build up, requiring the user to take more drugs to achieve the same effect. Tolerance can be due to a decrease in drug bioavailability ('*metabolic tolerance*'), a reduction in the number or responsiveness of receptors or intracellular mechanisms ('*pharmacodynamic tolerance*'), or a compensatory mechanism ('*behavioural tolerance*').

Tolerance may develop with chronic use, but in the case of cocaine, can develop in a single session, perhaps explaining cocaine '*bingeing*'. Metabolic tolerance is seen to barbiturates, ethanol and opiates. Pharmacodynamic tolerance is seen to a wide range of drugs including barbiturates, ethanol, opiates, amphetamine, cocaine, nicotine and caffeine. Behavioural or conditioned tolerance has been observed to opiates, ethanol, nicotine, benzodiazepines and other drugs. Since conditioned tolerance may be situation-specific and the lethality of drugs may be increased if the environment changes, the opponent process model of addiction suggests that a key component of addiction is the development of behavioural and neuro-anatomical tolerance, which can counteract the effects of the drug.

Pleasant Aspects of Drugs

Conditioned stimuli associated with the pleasant aspects of drug-taking may act to promote drug-taking. Drug-associated cues including mood states, people, locations, and abuse paraphernalia may induce some of the primary effects of drugs, but can also induce craving in addicts, and trigger relapse. Addicts may also work directly for drug-associated stimuli (conditioned reinforcement), leading them in turn to the drug itself.

Sensitisation ('inverse' or 'reverse' tolerance) occurs when repeated doses of a drug enhance one or more of its effects. Prototypically, moderate, spaced doses of amphetamine enhance the subsequent locomotor response to it. Sensitisation can exhibit environmentally-specific conditioned properties, and changes in drug pharmacodynamics. It has been suggested that the ability of drug-associated CSs to promote drug-seeking or craving also sensitises as a consequence of repeated drug-taking.

Withdrawal Syndromes

Some drugs, notably the opiates and alcohol, produce powerful physical withdrawal syndromes. Thus, it is possible to consider addiction within the framework of both rewarding and aversive consequences. Withdrawal symptoms are improved by the drug, so the drug is taken to escape from withdrawal. However, demonstrations that the neural substrates mediating physical signs of dependence are separate from those of reward support earlier behavioural evidence that physical dependence is not a necessary correlate of opiate addiction. In withdrawal, incentive learning operates for drugs of abuse just as for natural reinforcers.

Just as hunger increases the hedonic impact of food, which teaches the animal that it is more worth working for food when it is hungry, rats learn that heroin has a high value in the state of opiate withdrawal. Hedonic impact may be a 'common currency' for determining the value of widely varying reinforcers. Environmental stimuli may become associated with withdrawal and CSs for withdrawal may then provoke drug-taking just as withdrawal itself does. Drugs such as cocaine that do not produce obvious physical withdrawal syndromes may nonetheless have unpleasant after-effects on mood, which may promote drug-taking in the same way that physical withdrawal does. 'Opponent process' theories use the idea that a long-lasting unpleasant process opposes the euphoric effects of drugs, and that with chronic use, the euphoric effects diminish while the dysphoric process comes to dominate, leading to drug-taking via negative reinforcement.

Habit Learning

Drugs may activate habit-learning systems so that actions that led to the drug are reinforced directly, creating powerful stimulusresonse habits or 'involuntary' responses. A hallmark of habitual responding directly is that it persists stimulus–response habits even if the reinforcer's value is reduced. Habits are sometimes thought of as 'compulsive' responding when they occur at an abnormally high level, since they do not depend on the current value of the goal. Alcohol seeking may primarily reflect habitual responding, and while cocaine-seeking can be goal-directed, under some circumstances responding for cocaine can be more habitual than responding for natural reinforcers. Similarly, soon after acquisition, cocaine-seeking is readily suppressed by an aversive CS, but this suppression is lost after prolonged experience of cocaine.

Craving and habits both capture something of the casual definition of addiction as 'compulsive' behaviour. People who become drug addicts may be more vulnerable than other people to one or more of these neuropsychological effects, as well as being more predisposed to try drugs of abuse in the first place.

Substitutive Activities

From a behavioural-economic perspective, addicts weigh up the benefits and costs of drug-taking. They may do so rationally, or may exhibit decision-making flaws characteristic of humans, such as focusing inappropriately on short-term rather than long-term goals and being inconsistent in their choices. Drug addicts may be predisposed to act even more for short-term benefit than other people, or drugs may induce decision-making deficits. There is some evidence that self-control deficits may be a reversible consequence of cigarette dependence. None of these theories, or indeed levels of explanation, is adequate on its own. For example, although heroin may be taken to alleviate withdrawal, heroin self- administration can persist in the absence of withdrawal, and although heroin has euphoric effects, humans will work for doses that they cannot subjectively distinguish from a placebo. To seek a single theory of drug addiction is to miss the point that drugs of abuse have many effects, people take drugs for many reasons, and those reasons vary between people.

REINFORCEMENT LEARNING

Considerable progress has been made in establishing some of the mechanisms by which neural structures respond to appetitive and aversive events. To some extent these structures can be compared directly to the psychological processes known to influence animals' responding for rewards. A number of limbic cortical and subcortical structures in the brain play a role in assessing the value of reinforcers and of the stimuli that predict them, and in actions directed at obtaining those reinforcers or stimuli. Their relevance to addiction has been considered many times before, and influential theories of addiction have postulated that drugs of abuse 'short-circuit' or 'hijack' the neural mechanisms underlying reward or motivation

Nucleus Accumbens in Motivation and Learning

The discovery that rats would work hard to stimulate regions of their brain electrically (intracranial self-stimulation, or ICSS) was historically important. Many sites that support ICSS lie on the path of dopaminergic neurons from the substantia nigra pars compacta (SNc) and ventral tegmental area (VTA) to limbic sites including the ventral striatum (nucleus accumbens, Acb), and ICSS is substantially reduced after Acb DA depletion. The rate at which rats learn to respond for ICSS is correlated with the degree of potentiation of synapses made by cortical afferents onto striatal neurons, a potentiation that requires DA receptors.

The discovery that deep brain stimulation and transcranial magnetic stimulation can influence cognition, affect and motor performance in humans means that we cannot discount this means of altering brain function. Deep brain stimulation of the subthalamic nucleus has been successful with severe Parkinson's disease, and may have applications in obsessive compulsive disease and clinical depression but care is needed as the stimulation can be self-administered, and in the case of Parkinson's disease, dramatic emotional sequelae have been reported.

Pleasurable Aspects of Reward

An early suggestion was that Acb DA mediated the pleasurable aspects of reward. There is now strong evidence against this simple idea. Certainly, DA is released in response to appetitive reinforcers, intra-Acb DA agonists are reinforcing, animals may adjust their drug-taking to maintain high Acb DA levels, and some aspects of naturally- and drug-reinforced responding depend on Acb DA. However, Acb DA does not mediate 'pleasure', though its release may correlate with activity in other systems that do, and reinforcement operates in its absence. Measured by microdialysis techniques, DA is also released in response to aversive stimuli, CSs that predict them, and other salient stimuli, which would be consistent with a more general motivational role. CSs that have been paired with reward also elicit approach; this effect also depends on the Acb and its DA innervation. DA may also be involved in learning this approach response, again perhaps under the control of the central nucleus of the amygdala (CeA). Acb DA also contributes directly to subjects' motivation to work hard. Hedonic assessment of rewards themselves, or 'liking', does not depend on dopaminergic processes. Instead, it involves opioid mechanisms in the nucleus accumbens shell (AcbSh) and other systems in the pallidum and brainstem.

Intra-Acb μ opioid agonists also affect food preference, increasing the intake of highly palatable foodstuffs including fat, sweet foods, salt, and ethanol, while chronic ingestion of chocolate induces adaptations in endogenous Acb opioid systems. However, the notion that 'pleasure' can be mediated by receptors in a sub-cortical nucleus is perhaps too simple. Activity in this circuitry is probably subject to further processing in cortical (particularly prefrontal cortical) circuits, before attribution and accompanying subjective commentary.

DA and Instrumental Behaviour

The notion that DA 'stamps in' the learning of stimulus-response connections has considerable support. It has acute effects that modulate corticostriatal transmission, and also lasting effects. The combination of presynaptic and postsynaptic activity normally induces long-term depression of corticostriatal synapses, but if the same pattern of activity is paired with a pulse of DA, then the active synapses are strengthened. Natural reinforcers, drugs of abuse, and CSs that predict either, trigger increases in DA release in the Acb. DA neurons fire to unexpected rewards, or to unexpected stimuli that predict reward. DA neuron firing may be a teaching signal used for learning about actions that lead to reward. The Acb similarly responds to anticipated rewards. Other parameters of tonic DA neuronal firing may signal reward uncertainty, possibly relevant to the understanding of gambling behaviour.

Targets of DA neurons certainly influence instrumental behaviour. Structures that learn from the DA '*teaching signal*' probably include the dorsal striatum and prefrontal cortex (PFC), but much attention has focused on the Acb. Blockade of *N*-methyl-D-aspartate (NMDA) glutamate receptors in the nucleus accumbens core (AcbC) has been shown to retard instrumental learning for food, as has inhibition or over-stimulation of protein kinase A (PKA) within the Acb. Concurrent blockade of NMDA and DA D1 receptors in the AcbC synergistically prevents learning. Once the response has been learned, subsequent performance is not impaired by NMDA receptor blockade within the AcbC. Furthermore, infusion of a PKA inhibitor or a protein synthesis inhibitor into the AcbC after instrumental training sessions impairs subsequent performance, implying that PKA activity and protein synthesis in the AcbC contribute to the consolidation in memory of instrumental behaviour. However, it is clear that the Acb is not required for simple instrumental conditioning but rather is implicated in providing extra motivation for behaviour, especially when such motivation is triggered by pavlovian CSs, or when reinforcers are delayed or require substantial effort to obtain.

Rats with Acb or AcbC lesions acquire lever- press responses on sequences of random ratio schedules at normal or slightly reduced levels and are fully sensitive to changes in the action-outcome contingency. Thus, the Acb is not critical for goal-directed action. Rather, it appears to be critical for some aspects of motivation that promote responding for rewards in real-life situations. For example, the Acb plays a role in promoting responding for delayed rewards and is required for pavlovian CSs to provide a motivational boost to responding, i.e. for PIT. PIT has sometimes been termed 'wanting', although 'wanting' could equally refer to the instrumental incentive value underpinning true goal-directed action or pavlovian arousal itself. PIT can be further enhanced by injection of amphetamine into the Acb and depends on DA, possibly under the control of the CeA.

Other motivational effects of pavlovian CSs also depend on the Acb, for example, the capacity of CSs to act as conditioned reinforcers of instrumental behaviour. The neural basis of conditioned reinforcement has been investigated using the 'acquisition of a new response' procedure, in which subjects work only for a conditioned stimulus that has previously been associated with a natural reinforcer such as food or water. From these studies, it is clear that the basolateral amygdala (BLA) and AcbC are important in the ability to respond normally for conditioned reinforcement. In naturalistic situations, rewards are frequently available only after a delay, require considerable effort to achieve, and are signalled by environmental stimuli. Thus, the Acb is central to a number of processes that require motivation. Functional neuroimaging evidence supports this conclusion in humans.

Contingency Activities

The prefrontal cortex (specifically, prelimbic cortex) is required for rats to represent the contingencies between actions and their outcomes, and acquisition of instrumental responses on a simple schedule is disrupted by blocking NMDA and DA D1 receptors in the PFC. This is relevant to evidence from cognitive neuroscience that sectors of the human PFC are important for volitional processes. The PFC is also involved in extinction, the cessation of responding when a CS or response is no longer paired with reinforcement. Extinction is not 'unlearning' but involves the learning of new, inhibitory associations. Lesions of the ventral medial PFC interfere with the extinction of pavlovian conditioned freezing in the rat.

The PFC interacts with the amygdala, an important site of CS–US association in this task, and may suppress conditioned freezing when it is no longer appropriate. The orbitofrontal cortex (OFC) is part of the PFC with a particular role in the assessment of reinforcer value. It has bidirectional connections to the amygdala, and both are heavily implicated in the retrieval of the value of primary reinforcers based on information from CSs. In humans, the OFC and amygdala are also activated during extinction of pavlovian conditioning. The amygdala regulates the DA signal to the Acb. Goal-directed action requires that action-outcome contingencies interact with the incentive value of goals and the connection between the amygdala and the PFC may provide this functional link.

Psychomotor Stimulants

It has been suggested that these motivational and learning processes are particularly significant in some addictions, and their modification may have therapeutic potential. The existence of dissociable parallel brain systems mediating the associative control over addiction sits comfortably within the classic dichotomy of behavior into pavlovian or instrumental learning. A number of influential theories of addiction have postulated the existence of multiple parallel processes, each with its own independent, but interacting, neural system. DA systems are affected by virtually all of the major classes of drugs of abuse, ranging from the psychomotor stimulants to opioids, alcohol and nicotine, as well as by natural reinforcers such as food. Some abused drugs are particularly potent in this regard. Both food and drugs of abuse increase Acb DA, but the DA response to drugs of abuse may not habituate to the same extent as that to food. Sensitisation occurs following psychostimulant administration directly into the VTA, which induces hypersensitivity to DA in the Acb and enhances the response to pavlovian CSs associated with reward.

In animal models of drug-seeking behaviour controlled by drug-associated stimuli, lesions of the AcbC or disruption of its glutamatergic neurotransmission reduce drug-seeking, probably by reducing the motivational impact of the CSs. DA D3 receptors are particularly concentrated in the Acb and amygdala, and D3 receptor antagonists and partial agonists reduce cue-controlled cocaine-seeking or relapse to cocaine-taking in animal models. Some manipulations that reduce drug-seeking or reinstatement of drug-taking in animal models, such as DA D3 receptor antagonists, do not reduce food-seeking in a similar manner. It is not yet clear to what extent sensitisation contributes to human addiction, but it has been suggested that a sensitised response to drug-associated cues contributes to drug craving – that this 'incentive motivational' system becomes sensitized.

In current animal models, drug sensitization enhances responding for food, or responding to CSs for food, but in human addiction, responding for non-drug reinforcement declines relative to that for drug reinforcement. However, pre-treatment with psychomotor stimulants results in animals being willing to work harder for cocaine and this may reflect an effect of sensitization on the motivation to seek drugs. The well-known ability of psychomotor stimulants to potentiate conditioned reinforcement, depends upon the integrity of the dopaminergic innervation of the Acb, especially the AcbSh. This might be one possible basis for understanding why psychomotor stimulant drugs are themselves reinforcing; they enhance the reinforcing effects of environmental stimuli. The importance of conditioned reinforcers is that they allow the mediation and maintenance of long chains of behaviour, including drug- seeking behaviour, over delays to primary reinforcement.

Although potent as conditioned reinforcers when presented contingently, CSs paired with drug infusions do not increase drug-seeking when presented noncontingently to animals. Thus, conditioned reinforcement appears to be reliant on the contingency between the response and the CS, irrespective of the motivational value of the US. Indeed, the reliance of drug-seeking and taking on drug-associated conditioned reinforcers is underscored by further findings that cocaine self-administration is lower in the absence of any contingent CS. Indeed, nicotine self-administration in animals is difficult to acquire in the absence of conditioned reinforcers, suggesting that conditioned reinforcers may form part of a powerful stimulus complex, along with the drug, in maintaining drug use. Similarly, conditioned reinforcement maintained by CSs previously paired with oral alcohol can be persistent and the ability to maintain responding is independent of the drug, suggesting that this type of drug-seeking has a habitual quality.

The impact of conditioned reinforcement on drug-seeking is persistent and relatively impervious to extinction. It can maintain responding independently from the drug with which it was paired, suggesting that it may depend upon a separate neural system from that which mediates the effects of the drug itself. In experimental models of addictive behaviour in which drug-associated conditioned reinforcers support and maintain prolonged bouts of seeking behaviour, the functional integrity of a neural system involving the BLA and AcbC is of major importance. Thus, lesions of the BLA or AcbC, but not the AcbSh, greatly impair the acquisition of cocaine-seeking behaviour.

There is also neurochemical specificity in these BLA and Acb mechanisms. DA receptor blockade, but not AMPA receptor blockade in the BLA, reduces established cue-controlled cocaine seeking. The reverse is true in the AcbC, where AMPA, but not DA receptor blockade, has this effect.

It has additionally been established that disconnection of the BLA and AcbC by blocking DA receptors in the BLA on one side of the brain and AMPA receptors in the AcbC on the other has the same effect of dramatically reducing cocaine seeking. These data provide the strongest evidence that the BLA and AcbC function serially as components of a neural system that mediates these conditioned influences on drug-seeking. In functional imaging studies of human drug addicts, the amygdala is consistently activated by exposure to cocaine-, heroin-, food- and sex-associated stimuli in a way that is correlated with drug craving. Other regions commonly activated by drug-associated stimuli include the anterior cingulate cortex, OFC and occasionally the Acb. These data show that in both animals and humans, limbic cortical-ventral striatopallidal circuitry is associated with emotional learning and in processes related to drug craving, addiction and relapse.

Development of Stimulus-response Habits

The development of stimulus-response habits may depend on dorsal striatal plasticity, which may in turn depend on DA receptors. The balance between habits and goal-directed behaviour may also be regulated by the prelimbic and infralimbic cortex, subdivisions of the rat PFC. Recent functional neuroimaging data in humans supports the hypothesis that the ventral and dorsal striatum are also involved differentially in pavlovian and instrumental learning. Dorsal striatal DA release to CSs is a correlate of well-established cocaine-seeking. By contrast, such DA release is not seen in the AcbC region. Consistent with the electrophysiological data, DA release is only observed there when the CS is presented in a surprising context.

Comprehensive studies of chronic cocaine self-administration in monkeys indicate a progressive involvement of limbic, association and sensorimotor striatal domains, with autoradiographic changes evident first in the ventral, and then in the dorsal striatum. These data support the notion that there may be a stage of stimulus–reward or action–outcome learning that precedes stimulus–response habit learning. These phases may be mediated respectively by the ventral and dorsal striatum, either successively, or more probably in a temporally overlapping manner, possibly via the recently characterised 'cascading' neural connectivity that links these different corticostriatal loops. Thus, drug addiction is conceived in terms of a switch between these modes of learning, operating across the corticostriatal circuitry, from ventral striatal (i.e. Acb) to dorsal striatal domains.

HIGH PROPENSITY TO RELAPSE

A key feature of drug addiction is the high propensity to relapse, even after protracted periods of abstinence. The prevalent animal model of relapse utilises the so-called extinction-reinstatement procedure recently reviewed by Shaham and colleagues in a double issue of *Psychopharmacology* devoted to this subject. The usual form of this procedure is to train rats to press a lever to self-administer a drug and then to extinguish the instrumental act of lever pressing by omitting drug infusions. Following extinction, three manipulations generally accepted to be of importance in precipitating relapse in abstinent human drug addicts can 'reinstate' drug-seeking responses by increasing lever pressing although the drug remains unavailable. They are exposure to drug-associated stimuli, experimenter-administered drug, or 'stress,' usually an electric shock to the feet. However, extinction of the instrumental act of drug self-administration is not generally a means by which human addicts achieve abstinence.

Abstinence is more likely to arise through an active decision to abstain or through abstinence imposed by the law or by treatment. Moreover, since the extinguished response is so readily reinstated, it is unlikely that extinction training will provide an effective clinical approach to treatment. Extinction of the acquired properties of drug-associated stimuli through their non-reinforced exposure has been attempted as a therapeutic strategy, but with limited success, most likely because cue exposure in the clinic is unlikely to reduce the properties of drug cues in the original drug-associated environment. Neurobiologically, these ways of inducing relapse in the extinction-reinstatement model depend upon both common and discrete elements of limbic cortical-ventral striatopallidal circuitry. Most studies have involved the reinstatement of cocaine-seeking behaviour, but there are also studies with heroin and nicotine.

Mechanism of Reinstatement

The neural basis of cue-induced reinstatement has been reviewed extensively. It is prevented by reversible or permanent inactivation of the BLA and reversible inactivation of the dorsal mPFC. Inactivation of the OFC also attenuates cue-induced reinstatement of drug seeking. Systemically injected D1 and D3 DA receptor antagonists block cued reinstatement, as do intra-BLA, but not intra-Acb, infusions of D1 DA receptor antagonists - consistent with the effects of D1 and D3 DA receptor antagonism in the BLA on cocaine-seeking measured under a second-order schedule. Perhaps surprisingly, inactivation of the Acb does not attenuate cue-induced reinstatement, yet this structure is important for conditioned reinforcement and other pavlovian influences on instrumental behaviour, while AMPA receptor blockade attenuates cocaine-seeking under a second- order schedule. Although limbic cortical-ventral striatopallidal systems are implicated in the conditioned control of drug-seeking and reinstatement after extinction, much remains to be established in terms of the processes occurring in cortical and subcortical structures and the ways in which different subsystems interact.

While the BLA mediates reinstatement following exposure to discrete, cocaine-associated stimuli, the hippocampus may underlie the motivational impact of contextual stimuli. Theta- burst stimulation of the hippocampus, a form of experimental deep brain stimulation, has been shown to reinstate extinguished cocaine-seeking in a manner that depended on glutamate transmission in the VTA. It was suggested that this might mimic the process by which reinstatement occurs when animals are placed in a context associated with drug-taking, rather than in response to discrete cocaine cues. Indeed, dorsal hippocampal inactivation attenuates context-induced reinstatement of drug seeking, as does inactivation of the dorsal mPFC. These data are in accord with the suggestion of a differential involvement of the amygdala in conditioning to discrete, and the hippocampal formation in conditioning to contextual, stimuli.

Moreover, electro-physiological and *in vivo* neurochemical studies have demonstrated that hippocampal, amygdala and PFC projections interact in the Acb in a way that is modulated by mesolimbic DA and that, in turn, can modulate the release of DA. Thus, hippocampal, amygdala and PFC mechanisms may influence drug-seeking through their convergent projections to the Acb, perhaps competing for access to response strategies subserved by different cortical-striato-pallido-thalamo-cortical re-entrant loops. The mPFC is clearly important in reinstatement - whether induced by cues, contexts, drugs or stress - following extinction of the instrumental seeking response. Determining the psychological process that the mPFC subserves in these settings is an important goal. The vigour of conditioned reinstatement of cocaine seeking increases with the duration of withdrawal, suggesting that neuroadaptations to chronic cocaine self-administration and withdrawal interact with the motivation to seek cocaine when cocaine cues are present in the environment. These findings may provide insight into the possible mechanisms that underlie the persistence or 'incubation' of cocaine- seeking reported to occur over time in abstinent cocaine addicts.

The mechanisms underlying this incubation effect have been shown to depend upon the upregulation of the extracellular signal-regulated kinase (ERK) signalling pathway specifically within the CeA. Thus, exposure to cocaine-associated stimuli increased cocaine-seeking and also ERK phosphorylation in the CeA, but not BLA, substantially more after 30 days than after one day of cocaine withdrawal, so the incubation effect was correlated with ERK upregulation in the CeA. Inhibition of ERK phosphorylation in the CeA, but not BLA, after 30 days of withdrawal greatly decreased cocaine-seeking, whereas stimulation of ERK phosphorylation in the CeA, but not BLA, increased cocaine-seeking after one day of withdrawal. Thus the mechanisms mediating drug-cue-induced relapse and its enhancement during protracted withdrawal appear to depend upon two dissociable mechanisms in the BLA and CeA, respectively.

Observing Drug-induced Reinstatement

Drug-induced reinstatement by 'priming' (i.e. non-contingent or experimenter administered cocaine or heroin - often given intraperitoneally and not intravenously) can be attenuated by D1-like dopamine receptor antagonists. In neuroanatomical studies it has been shown that drug-induced reinstatement can be blocked by inactivation of the VTA, dorsal mPFC, AcbC and ventral pallidum, called the 'motor subcircuit' by Kalivas and colleagues. Moreover, DA receptor antagonists infused into the mPFC or AcbSh also attenuate drug-induced reinstatement. Antagonists at AMPA, but not NMDA receptors in the ACb, block reinstatement induced by systemic or intra-mPFC cocaine and, by contrast, AMPA receptor agonists infused into the Acb reinstate cocaine seeking. The effects of cocaine or heroin to reinstate extinguished responding are mimicked by systemic injections of D2, but not D1, receptor agonists and by infusions of cocaine, amphetamine or DA itself directly into either the Acb or mPFC. Antagonists at μ opiate receptors prevent the effects of heroin and alcohol, but not cocaine, on reinstatement and a CB1 receptor antagonist has also been shown to prevent the reinstatement effects of cocaine.

Observing Stress-induced Reinstatement

Reinstatement can be induced by several stressors, including footshock, food deprivation and also CNS administration of corticotrophin releasing factor CRF. Inactivation of the dorsal mPFC prevents footshock-induced reinstatement; this area of the PFC is commonly involved in cued-, drug- and stress-induced reinstatement. Additional and unique neural circuitry appears to be critical for the effects of stress, including the CeA, bed nucleus of the stria terminalis (BNST) and the noradrenergic medullary tegmentum which innervates these structures. Thus, the following manipulations all block stress-induced reinstatement: intra-BNST, but not intra-amygdala, infusions of a CRF antagonist; systemic and intracerebroventricular, but not intra-locus coeruleus, injections of an alpha-2 noradrenergic receptor agonist; intra-BNST and intra-amygdala alpha-2 noradrenergic receptor antagonists; and destruction of the ventral noradrenergic bundle originating in the medullary noradrenergic cell groups.

The CRF-containing projections between the CeA and BNST have also been shown to be a critical link between these structures in mediating stress-induced relapse. Thus, two neural systems implicated in stress responses in general – one utilising NA and the other CRF – are implicated along with the mPFC in mediating relapse induced by footshock stress in the extinction-reinstatement procedure. The generally accepted mechanism is that stress activates the medullary noradrenergic neurons and leads to activation of the CRF system within the BNST and possibly the CeA. How this subcortical, neuroendocrine mechanism interfaces with the mPFC is not altogether clear, nor is how this impinges on limbic cortical ventral striatopallidal circuitry. One of the reasons for developing and studying the neural basis of relapse in experimental animals is to develop treatments that will promote abstinence.

Intensive experimental investigation of this area has yielded detailed information on the neural systems and neurochemical mechanisms underlying cue-, stress- and drug-induced relapse. An important issue for resolution is the extent to which the effects on reinstatement of cues, drug or stress actually depend upon the prior process of instrumental extinction. If they do, their utility in human addiction, where this extinction process does not occur, may be slight. D3 DA receptor antagonists appear to have efficacy in both the cued-reinstatement procedure and ongoing, cue-controlled cocaine-seeking suggesting that this dopaminergic target might affect the conditioned process common to both. In addition, the GABA-B receptor agonist baclofen both attenuates drug seeking that depends upon drug-associated conditioned reinforcers in rats and also attenuates cue-induced activation of limbic cortical areas in cocaine-addicted humans. New pharmacological treatments to prevent relapse may emerge from this rich preclinical data on experimental models of reinstatement in animals.

CHRONIC ADMINISTRATION OF DRUGS

The chronic administration of drugs of abuse results in the induction of intracellular cascades within the limbic corticostriatal circuitry. Although different drugs act at different receptor targets on the cell surface, there is a degree of convergence in their downstream signalling pathways. Interaction between the drug and its target results in either the opening of a ligand-gated channel, or the activation of a receptor-linked G-protein, both of which induce intracellular cascades. One common action of drugs is the activation of the transcription factor CREB and components of the cAMP signalling pathway, such as adenylyl cyclase (AC) and protein kinase A (PKA). CREB regulates the transcription of genes whose promoters contain the CRE element, and is thought to be a site of convergence of intracellular cascades as it can be activated through phosphorylation at serine 133 by several different protein kinases. Therefore alterations in CREB and the cAMP signalling pathway may represent common neuroadaptations of different drugs of abuse. Opiates and cannabinoids acutely inhibit adenylyl cyclase and the cAMP signalling pathway, resulting in a decrease in phosphorylated CREB. In contrast, acute administration of ethanol and stimulants increases the activity of the cAMP signalling pathway. However, in all cases, there is a common chronic upregulation of the cAMP signalling pathway that is accompanied by tolerance to the acute intracellular response to drugs of abuse.

The switch from acute inhibition to chronic upregulation of the cAMP pathway with repeated opioid administration is poorly understood, though it is known to involve adaptations in G-protein properties resulting from their persistent stimulation, and neuroadaptive changes in protein kinase systems. Furthermore, few downstream targets have been identified that mediate the functional effects of cAMP and CREB upregulation. Among the proteins whose levels are increased by chronic drug administration in a CREB dependent manner are AC, tyrosine hydroxylase (TH), the rate-limiting enzyme in DA synthesis, and the opioid peptide dynorphin. Dynorphin stimulates κ-opioid receptors, resulting in an inhibition of DA release, and the effects of CREB upregulation on drug reward are blocked by κ-opioid antagonists. Dynorphin mRNA levels are increased in the striatum of cocaine abusers *post-mortem*.

Therefore neuroadaptations in cAMP signalling, resulting in dynorphin upregulation, may partially underlie tolerance to the effects of drugs of abuse. Repeated intermittent administration of addictive drugs results in sensitisation of, rather than tolerance to, some of the behavioural and rewarding effects of drugs. The VTA is required for the initiation of behavioural sensitisation, and long-lasting adaptations in the Acb are correlated with the expression of sensitisation. Transient increases of GluR1 subunits in the VTA are important for the triggering of sensitisation. The resultant persistent increase in calcium signalling and calcium/calmodulin-stimulated (CaM) kinase activation have also been implicated in behavioural sensitisation. CaM kinase II stimulates the mitogen-activated protein (MAP) kinase signalling pathway, which is known to be involved in sensitisation.

Sensitisation may also be mediated by a chronic drug-induced downregulation of the *Homer* gene family. Developmental genetic knockout of *Homer1* or *Homer2* in drug-naïve rats mimics the sensitised response to acute drug administration observed in rats experiencing withdrawal. Specifically, *Homer* downregulation is critically important in the Acb, as localised antisense-mediated knockdown of *Homer1* expression in the Acb similarly induces sensitisation, and virally-mediated rescue of *Homer2* in the Acb of *Homer2* knockout mice reverses the drug-sensitised phenotype. One neuroadaptation that has attracted particular interest is the upregulation of the chronic Fos-related antigen ΔFosB. Levels of ΔFosB are increased in the Acb for up to four weeks following drug administration, and ΔFosB is also progressively upregulated with repeated drug administration. This suggests that it is involved in behavioural sensitisation, a hypothesis that is strongly indicated by the demonstration that ΔFosB overexpression in the Acb sensitises behavioural and rewarding responses to cocaine and morphine, whereas a reduction inhibits responses to cocaine and *fosB* knockout mice do not develop behavioural sensitisation.

Therefore, ΔFosB may be a *'molecular switch'*, that enables the sensitisation of responses to drugs of abuse and long-term adaptations underlying addiction that persist through withdrawal. Again, the current challenge is to identify downstream targets of ΔFosB signalling, one of which may be cyclin-dependent kinase 5 (cdk5). Upon withdrawal from drugs, which may be precipitated experimentally by the administration of an antagonist, there is a further increase in the activity of the cAMP signalling pathway beyond the level observed during tolerance. This reflects a state of dependence upon drugs of abuse, whereby when in withdrawal, the molecular cascades underlying reward are altered, resulting in an amotivational state.

Mice deficient in CREB display reduced opiate dependence showing that cAMP signalling is important for both tolerance and dependence. A focus for current and future research is the characterisation of the downstream targets of CREB, such as dynorphin, that are required for the development of tolerance and dependence. Neuroadaptations implicated in drug- induced reinstatement include the expression of *AGS3* (*activator of G protein signalling 3*), the blockade of which prevents cocaine-induced relapse to cocaine seeking, and a lowering of extracellular glutamate through reduction of cystine-glutamate exchange, the restoration of which also prevents cocaine-primed relapse. Drug-cue-induced reinstatement exhibits a time-dependent increase through withdrawal, with cue-induced cocaine, methamphetamine, heroin and sucrose seeking behaviours incubating over time.

Molecular changes that correlate with this incubation effect may be important for cue-induced relapse to drug seeking. With short periods of withdrawal, transient increases are observed in tyrosine hydroxylase activity and cdk5 protein levels in the VTA, and more persistent increases in PKA activity occur in the Acb. However, the closest correlate of incubation appears to be the progressive upregulation of BDNF protein in the VTA. BDNF appears to be involved in the persistence of the incubation effect rather than being critical for incubation itself, evidenced by the demonstration that intra-VTA infusion of BDNF protein increases cocaine seeking over and above the incubation-related elevation. BDNF has a well-established role in hippocampal LTP and learning and memory. Therefore there is a similarity between the molecular mechanisms of drug addiction and learning and memory that also applies to the other intracellular cascades described, particularly the involvement of CREB.

Furthermore, drugs of abuse induce changes in the VTA and Acb that are reminiscent of the influential cellular models of learning and memory LTP and long-term depression (LTD). One important issue that concerns research into both drug addiction and learning and memory is the longevity of both processes. All the molecular neuroadaptations described thus far are impermanent, and though some are indeed long-lasting, none can account for the compulsion and relapse that are observed months or even years after withdrawal. It is increasingly thought that morphological changes in synaptic structure are the only process by which the plasticity underlying both drug addiction and learning and memory can become near- permanent. BDNF is necessary for the neuronal growth and synaptic remodelling associated with learning and memory, and its putative role in incubation, as well as sensitisation, suggests that morphological plasticity may be critically involved in drug addiction. Many genes have been implicated in synaptic plasticity, and recently the involvement of *cdk5* in addiction-related permanent plasticity has attracted great attention. *Cdk5* is regulated by ΔFosB, providing a link between the longest-lasting molecular adaptation and permanent plasticity, while cdk5 mediates the proliferation of striatal dendritic spines in response to chronic administration of cocaine. Such structural changes are likely to involve neurofilaments, which are elements of the cytoskeletal architecture of neurons.

There is evidence for hyperphosphorylation of neurofilament proteins both in rodents and in human opiate addicts *postmortem.* The mechanisms underlying neuroadaptations in synaptic morphology will be a focus of future research investigating the mechanisms of the long-lasting plasticity mediating drug addiction. Synaptic morphology can also be altered by the production of new neurons. This neurogenesis is increasingly believed to play a role in drug-induced neuroadaptation. The few studies that have been conducted suggest that chronic exposure to drugs of abuse decreases neurogenesis in the hippocampus. A parallel is found in studies of depression, in which decreased hippocampal neurogenesis is observed. In contrast, learning and memory are associated with an increase in hippocampal neurogenesis, and one action of antidepressant drugs is to increase hippocampal neurogenesis and neuronal growth. This may suggest a potential avenue for the treatment of addiction.

Some antidepressants may work partly by increasing neurogenesis. Antidepressants are sometimes, but not always, effective medications for drug dependence. Though further delineation of the molecular pathways involved in drug addiction will be a focus of future research, an important challenge will be to integrate the resulting information, as the same molecular candidates are implicated in several aspects of drug addiction. BDNF is associated with incubation, relapse, sensitisation and permanent plasticity. Furthermore, neuroadaptations occur throughout the limbic corticostriatal circuitry. Although molecular changes have been localised to particular brain areas, their relevance to behaviour is only beginning to be determined in a spatially localised manner. Array technology has recently been used both *in vitro* and *in vivo* to produce large sets of information on the upregulation of genes following the administration of drugs of abuse, but it will be several years before it can be established whether such neuroadaptations are merely correlative, or critical for the development of addiction. The similarity between the molecular processes implicated in drug addiction and those firmly established in learning and memory will guide future research.

One exciting prospect is the possible manipulation of drug-associated memories long after they have been acquired. Studies of fear conditioning have demonstrated that previously learned memories for stimulus-aversive outcome associations can be disrupted in a retrieval-dependent manner, so that they are not expressed subsequently in retrieval tests. This impairment of the reconsolidation of memories has also been observed in several other learning and memory paradigms, including a study of appetitive incentive learning, and also in humans, and may be extended to drug addiction. Drug-associated environmental stimuli elicit strong craving and increase the chance of relapse in abstinent individuals.

The potential to reduce the impact of these cues through disrupting the reconsolidation of their association with addictive drugs may be a future avenue of research. Stimulus-addictive drug associations are supported by the same neuroanatomical substrates as both appetitive and aversive associations, further underlining the similarity between addiction and learning, and the upregulation of *Zif268* in the amygdala is strongly correlated with the reconsolidation of both stimulus- drug and stimulus-footshock associations, providing a prospective target for functional studies. *Zif268* has recently been shown to be a specific marker of the reconsolidation of hippocampal-dependent contextual fear memories. Moreover, it appears that the molecular mechanisms of consolidation and reconsolidation are doubly dissociable, at least in the hippocampus. It may be possible to target the reconsolidation of previously-learned maladaptive memories that are important in drug addiction.

RECREATIONAL USE OF DRUGS

A significant proportion of the population take drugs of abuse at least once in their lifetime. Many individuals are capable of maintaining prolonged recreational use. Only a few develop a true addiction. In the last few decades, the identification of the factors that determine these individual differences in propensity for addiction has become one of the major targets of drug abuse research. Emerging data from both clinical and animal experiments suggest that there exist 'vulnerable' phenotypes and genotypes that are more predisposed to drug abuse. Elucidating the nature of these vulnerabilities could help prevent addiction in the predisposed population.

Subjective and Reinforcement

Enormous differences between the subjective and reinforcing effects of drugs in humans are well-documented. Individuals who prefer the effects of amphetamine to placebo show increased ratings of euphoria and positive mood, compared to anxiety and depression in subjects that choose placebo over amphetamine. Recent advances in imaging technology have yielded exciting information about the neural correlates of these subjective differences. In one recent report, the intensity of the high induced by methylphenidate was significantly correlated with levels of released DA. Subjects who had the greatest increase perceived the most intense high.

Further, the magnitude of decrease in D2 receptor availability is significantly associated with the positive reinforcing effects of the psychomotor stimulant methylphenidate, and release of DA in response to *d*-amphetamine correlates with self-reports of 'drug wanting' and the personality trait of novelty-seeking. In support of these findings, rhesus monkeys with extensive cocaine self-administration history show significant decreases in D2 receptor densities throughout the striatum compared to monkeys with a history of food reinforcement. These data suggest that pre-existing differences between subjects in the rate of DA release and/or D2 receptor distribution may play a role in the predisposition to drug abuse. The cause and exact nature of these functional differences is not known.

Examples of Conditioned and Unconditioned Responses

Individual differences in conditioned and unconditioned responses to drugs of abuse have been reliably demonstrated in animals. In particular, the propensity to acquire intravenous self-administration (IVSA) in rats can be predicted by the behavioural reactivity of an individual rat to a stressful situation, such as exposure to a novel environment. In this model, the propensity to develop drug SA can be represented by dividing animals into subgroups based on their locomotor response to a novel environment.

Animals with an activity score above the mean for the entire group, so-called '*high responders*' (HRs), show enhanced acquisition of psychostimulant IVSA compared to animals with an activity score below the median of the group, the '*low responders*' (LRs). Further studies show that individual differences in drug intake originate from vertical shifts in the dose–response curve for intravenous cocaine- self-administration, and these vertical shifts can be predicted by reactivity to novelty. HR/LR groups also show differences in drug- mediated behaviours, such as increased locomotor response to systemic administration of cocaine, amphetamine, and morphine, enhanced psychostimulant-induced behavioural sensitisation, and stronger contextual conditioning to drugs. Behavioural differences between HRs and LRs appear to be mediated by differences in dopaminergic neuronal structure and function. For example, HRs show increased cocaine-, amphetamine-, and stress-induced DA levels in the Acb, as well as a higher 3,4-dihydroxyphenyl acetic acid (DOPAC)/DA ratio in this region.

Data from electrophysiological studies demonstrate higher basal firing rates and bursting activity of DA neurons in the ventral tegmental area and, to a lesser extent, the SNc in HRs. Structurally, HRs have increased DAT numbers and greater Bmax for D1 binding sites in the Acb. Regulatory mechanisms of the mesolimbic DA system also differ between HRs and LRs. Recent data indicate that HRs express lower levels of tyrosine hydroxylase levels and CCK-mRNA, part of the intrinsic inhibitory input to dopaminergic VTA neurons, but higher levels of PPE-mRNA, part of the extrinsic facilitating input to these neurons. These behavioural and neurochemical differences are accompanied by differences in activity of the hypothalamic-pituitary-adrenal axis (HPA), the system primarily activated under stressful situations.

Animals designated as HR have higher novelty- induced corticosterone secretion compared to LR rats, and self-administration of amphetamine is positively correlated with corticosterone levels after two hours of exposure to stress. The work of Piazza and his colleagues suggests that individual differences in vulnerability to addiction can be modelled in animals, and that these differences are related to altered structure and function of the DAergic and HPA systems. Nonetheless, the developmental cause of these behavioural and neural differences is not known.

Vulnerability to Drug Addiction

Environmental experience may contribute to individual differences in vulnerability to drug addiction. Early adverse experience, such as childhood sexual or physical abuse, is one of the most important biological and environmental factors that are associated with vulnerability to substance abuse. The prevailing view is that these stressors influence the development of neural systems that underlie the expression of behavioural and endocrine responses to stress and reward. Although clinical data confirm a relationship between early adverse experience and substance abuse, it is not known whether this relationship is direct or indirect. Recent developments using animal models of early adverse environmental experience have been important in elucidating the causal nature of this relationship.

Prolonged Maternal Separation

One animal model of early adverse experience takes the form of disrupted maternal care, whereby infant rats experience repeated episodes of prolonged maternal separation (MS) during the first two weeks after birth. This consistently gives rise to profound behavioural, neural, and endocrine differences in adult animals. It leads to increased behavioural reactivity in response to stressors, and these behaviours are accompanied by altered structure and function of neural regions involved in HPA activation. Recent work using these models has attempted to form a causal relationship between disrupted maternal care, reactivity to stressors, and individual differences in susceptibility to drug self-administration.

The data suggest that such separation leads to alterations in reward-related behaviours, such as a blunted response to both negative and positive contrast effects, attenuated locomotor activity to both a novel environment and *d*-amphetamine, and, in maternally separated females, blunted acquisition of a conditioned anticipatory locomotor response to food. MS rats also show altered acquisition and maintenance of cocaine self-administration when tested as adults, with dose and gender-dependent effects, and other studies report enhanced acquisition of cocaine self-administration. These behaviours are accompanied by structural differences in the mesolimbic DA system. One recent finding demonstrates that maternal separation leads to an enhancement of both stress-induced sensitisation to amphetamine and acute Acb DA release following stress and cocaine. This study also reported increased levels of D3 receptor mRNA in the AcbSh, but not AcbC, of MS rats. These findings dovetail nicely with data from a non-human primate model of disrupted maternal care.

When infant rhesus macaques are peer-reared (PR), rather than mother-reared, they exhibit a constellation of neurobehavioural dysfunctions at adulthood, such as reduced exploration and increased fear-related behaviours, as well as marked increases in HPA activity following stressful situations. PR rhesus macaques consume more alcohol than mother-reared subjects, but interestingly, acute stress in the form of social separation increases alcohol consumption in mother-reared animals to the level of their PR counterparts. Excessive alcohol consumption in PR animals also correlates positively with plasma cortisol levels and negatively with CSF 5-HIAA concentrations in infancy and adulthood. An interesting recent finding is that a functional variant of the rhesus serotonin transporter-linked polymorphism (rh5-*HTT* LPR) interacts with rearing condition and gender to influence adrenocorticotropic hormone (ACTH) response to stress, suggesting differential sensitivity to stress dependent upon a gene-environment interaction.

Separation of One Parent

Converging evidence suggests that early adverse social experience, such as maternal separation, may influence susceptibility to the effects of drugs. Like maternal separation, isolation rearing of infant rats (housed singly in small, opaque cages) leads to disruptions in a variety of reward-related behaviours when these animals are tested as adults. Isolation-reared rats show enhanced stereotypy to *d*-amphetamine and apomorphine, are more sensitive to the effects of negative and positive contrast in sucrose intake tests, and show increased locomotor activity following administration of psychostimulants. It is also noteworthy that these animals show sustained hyperactivity in a novel environment, a finding that converges with data from the maternal separation and HR/LR models.

Isolation rearing also increases self-administration of a variety of abused drugs, depending on dose, being more sensitive to low doses. Isolation rearing also alters the structure and function of regions involved in drug reinforcement. It leads to a reduced DOPAC/DA turnover in the frontal cortex, but a larger turnover in the Acb and striatum. Isolation rearing also leads to decreased opiate receptor binding in the frontal cortex, hippocampus, periaqueductal grey, and striatum, down-regulation of 5-HT in the hippocampus, and upregulation in catecholamine systems in the Acb.

Social and Environmentally Complex Environment

There is some suggestion that the behavioural and neural deficits observed following social isolation rearing or maternal separation may be reversed or altered by some period of enrichment rearing in which the animal is housed in a large, socially and environmentally complex environment. In one recent study, C57BL/6 mice, an inbred strain considered '*addiction-prone*', were more resistant to both cocaine and MPTP following enrichment rearing, and showed different patterns of *c-fos* expression in the striatum compared with mice raised in standard conditions. Further, after MPTP treatment, enriched mice showed less DA loss, less DAT binding, and increased BDNF expression in the striatum. In another instance, environmental enrichment during the peripubertal period completely reversed the effects of maternal separation on both HPA and behavioural responses to stress, with no effect on CRF mRNA expression.

Early Exposure to Drugs

Another important question is whether early exposure to drugs of abuse can influence individual differences in propensity to drug addiction. It is well established that prenatal exposure to cocaine, heroin, marijuana, nicotine, and alcohol can have profound effects on cognitive and motor function in adolescence and adulthood, and there is some suggestion that this exposure influences propensity to addiction. However, longitudinal or retrospective clinical studies are often problematic due to the high incidence of confounding variables. Thus, animal models are important in elucidating the individual contribution of early drug exposure to adult vulnerability to addiction. Data using animal models suggests that prenatal drug exposure has a profound effect on behaviour and neurochemistry when these animals are tested as adults and that these changes may predispose to addiction. Converging evidence using a variety of techniques suggests that prenatal exposure to alcohol, nicotine, opioids, or cocaine produces persistent cognitive and neural deficits in adults.

In particular, these data suggest that prenatal exposure to drugs of abuse leads to downregulation and tolerance in neural systems involved in drug reward. Although these changes may indirectly influence propensity to addiction, only a few studies have addressed the question of whether prenatal exposure to drugs of abuse affects later drug self-administration. Prenatal cocaine exposure in rodents increases intravenous self-administration for moderate doses of cocaine, but the rate of cocaine intake does not differ between cocaine and saline-exposed animals. This result suggests that prenatal exposure to cocaine does not alter the reinforcing value of cocaine, a finding that is consistent with a study examining addiction-prone versus addiction-resistant rat strains.

Another important question is whether prenatal exposure to drugs of abuse facilitates the development of drug dependence in general. In one study, prenatal exposure to morphine enhanced rates of heroin and of cocaine self-administration. Adolescence is another period during which the brain undergoes many complex changes that can have a prolonged impact on decision-making and cognitive processes. In addition, adolescents are more likely to experiment with illicit drugs, which may be due in whole or in part to the increase in sensation- and novelty-seeking that is characteristic of adolescence.

Recent clinical data suggest that adolescent exposure to drugs of abuse is associated with increased risk of addiction in adulthood. Not unlike the data from prenatal exposure research, studies using animal models suggest that rodents with adolescent exposure to drugs such as methylphenidate, nicotine, cannabinoids, MDMA, and alcohol also show behavioural and neural changes indicative of tolerance to these and other drugs, which persist into adulthood. Adolescent-exposed animals tested as adults show a behavioural and neural profile that is different from adult animals administered a similar drug regime. This suggests differential long-term neuroadaptive responses to drugs, possibly related to immature or still-developing plasticity mechanisms in the PFC. One potential confound for these studies is that the majority of them look at effects of non-contingent drug administration during the periadolescent period on later adult behaviour and neurochemistry. Future studies should aim at exploring how self-administration drug experience during adolescence influences drug-mediated behaviours and neural changes when tested in adulthood.

Hereditary Susceptibility to Addiction

Individual differences in genetic make-up critically influence susceptibility to addiction. Although some aspects of vulnerability may be unique for specific substances, most known genetic influences are common to all drugs of abuse. Recent estimates suggest that genetic components explain 40-60 per cent of overall vulnerability to addiction. These data do not support single-gene models for the inheritance of addiction vulnerability. Contributions from allelic variations in several genes are likely to be involved. Genetic components do not necessarily impact upon the initiation of drug use, but instead influence progression from regular use to dependence. Recent data are yielding information on which chromosomal regions, genes, haplotypes, and allelic variants provide exactly what genetic influence on vulnerability to drug abuse. Although there are numerous candidate genes influencing addiction and addictive behaviours, human studies have generally focused on identifying genes associated with dopaminergic function. A number of studies report a significant association between substance dependence and polymorphisms of dopaminergic receptor genes (DRD1, DRD2, DRD3, DRD4). Subjects with a history of drug use show increased frequency of homozygosity for the restriction polymorphism Dde 1 of the DRD2 gene, and several studies have demonstrated a link between the presence of the A1 allele of the DRD2 Taq 1 polymorphism and drug dependence.

Polymorphisms of the DRD3 gene have recently been identified in 96 rat strains and substrains, although there is good evidence that this gene does not play a major role in the genetic vulnerability to alcoholism. Novelty-seeking, a personality trait often observed in addicts, is significantly associated with the 7-repeat allele of the DRD4 exonic polymorphism. Individual differences in genetic polymorphisms have functional outcomes. Individuals homozygous or heterozygous for the 7- (or longer) repeat allele (DRD4 L) report significantly higher craving after consumption of alcohol compared to individuals classified as DRD4 S. Further, although olanzapine reduces craving for alcohol in both DRD4 S and DRD4 L individuals, it only reduces cue- and alcohol priming-induced craving in DRD4 L individuals.

Hereditary Neurovulnerability to Addiction

The role of genetic factors in contributing to drug-related behaviours can be examined using inbred rodent strains, which, in contrast to outbred strains, provide a stable genotype. Two inbred rat strains that differ in responses to drugs of abuse are Lewis (LEW) and Fischer 344 (F344) rats.

In comparison to F344 rats, LEW rats show greater behavioural responses to several drugs, including oral self-administration, intravenous acquisition of self-administration of morphine and cocaine, place conditioning, and locomotor sensitisation. These strains also differ in properties of their mesolimbic DA systems. LEW rats have lower basal extracellular DA metabolite levels in the Acb and lower numbers of spontaneously active DA neurons in the VTA. They also show a more prolonged elevation of DA levels in the ventral striatum following acute cocaine administration. At a biochemical level, LEW rats express higher levels of tyrosine hydroxylase in the VTA, but lower levels in the Acb, than F344 rats.

Finally, strain differences are also observed in HPA function. Although F334 have higher basal and stress-induced levels of glucocorticoids, LEW rats show a more prolonged elevation of corticosterone following exposure to a stressor. Differences in susceptibility to the reinforcing properties of cocaine, amphetamine, morphine and ethanol have been described among inbred strains of mice. A number of studies have demonstrated that mice belonging to the inbred strains C57BL/6 (C57) and DBA/2 (DBA) differ in their behavioural and neural responses to drugs of abuse. The data suggest that the C57 genotype can be characterised as '*drug-preferring*' and the DBA genotype as '*drug-resistant*'. Sensitivity to the unconditioned locomotor effects of amphetamine and level of locomotor activity in a novel environment are both susceptible to the influence of environmental manipulations such as food restriction when measured in these strains. These data provide information on genetic-experience interactions, and suggest a possible homology between these phenotypes and psychostimulant-induced place preference.

Sex Differences in Vulnerability to Drugs

Epidemiological data suggest a greater prevalence of substance use disorders among men, but recent surveys show increased rates of substance dependence in women. These reports are supported by recent evidence that drug-dependent women are more vulnerable to the deleterious effects of drugs and show a faster progression to drug dependence. Women have a more pronounced subjective response to psychostimulants, become more rapidly addicted to cocaine, heroin, and alcohol, and experience greater perceived severity of withdrawal from addictive substances such as nicotine. Additionally, despite often having a shorter experience with drugs of abuse than men, substance-dependent women show either a comparable or increased severity of addiction, as well as a faster progression to treatment entry. Analogous differences are observed in experimental animals. Female rats show enhanced acquisition, but not necessarily maintenance, of drug-taking behaviour, and increased vulnerability to relapse.

Although there are consistent sex differences during the acquisition phase of stimulant self-administration, the data are more equivocal with other abused drugs. The discrepancies may be due in part to dose differences. Sex differences are more apparent when low doses are used. In maternal separation, peer-reared, and social isolation models of early adverse environmental experience, female rodents show different behavioural and neural effects. The nature of sex differences in vulnerability to drug abuse is not known. Although it is probable that psychosocial factors have an impact, converging evidence from animal studies suggest that fluctuating ovarian hormones have an important role in these differences. There is little experimental work done in this domain in humans, apart from the finding that the subjective response to amphetamines is increased during the follicular phase of the menstrual cycle. In rats, cocaine self-administration varies across the oestrous cycle, and high doses of oestradiol facilitate cocaine self-administration. Moreover, the enhanced acquisition of cocaine self-administration in female rats is abolished in ovariectomised animals, but can be reversed with administration of oestradiol. These data suggest that oestrogens facilitate dopaminergic function.

DELETERIOUS EFFECTS OF DRUGS OF ABUSE

It is almost axiomatic that a suitably high dose of a psychoactive substance, administered either acutely or chronically, is likely to have deleterious effects. These can be transient or long-lasting, and impact on one or more of the body systems, including the brain itself. When the effects involve brain regions implicated in volition and executive control, such as the PFC, such damage may exacerbate the drive to abuse and addiction and retard attempts at rehabilitation. Types of drugs vary in their toxicity. This issue is crucial to the regulation of drug use, but the definition of adverse drug effects is difficult and controversial. It is sometimes problematic to infer the causal effects of psychoactive substances in poly- drug abusers and to distinguish them from possible premorbid factors present in the user prior to drug use.

Studies in experimental animals that control the exposure to specific doses of particular drugs avoid many of these difficulties, but may be compromised by the difficulty of comparing drug doses between species and selecting for investigation the doses most relevant to human drug users. Moreover, when the studies are limited to the influence of sensitising regimens of drug administration on fine details of neuronal organisation, such as dendritic branching in different regions of the rat PFC, the effects are sometimes apparently inconsistent with toxic effects, reflecting instead possible effects on neuronal plasticity. Thus, repeated treatment with psychostimulant drugs produces long-lasting increases in dendritic branching and spine intensity in some brain regions. However, it has also been reported that such effects subsequently limit the promotion of synaptic plasticity bestowed by housing in enriched environments.

Functional imaging studies of human abusers may likewise reveal abnormal patterns of brain activation, but in the absence of evidence of cognitive impairment in neuropsychological tests, the significance of these may be unclear. Notwithstanding these difficulties, much is now known about the toxic sequelae of chronic drug use, whether on the brain itself or on behaviour and cognition, based on the evidence of post-mortem neuropathology, neuroimaging and neuropsychology. Thus, among the stimulant drug class, high doses of methamphetamine can produce longterm neurochemical and structural changes, including neurotoxic effects on DA- and 5-HT-containing neurons in several animal species. It also reduces markers of DA and 5-HT function in the human brain, postmortem studies indicating a reduction in the striatal DAT and reductions in 5-HT markers within the OFC.

This evidence is complemented by recent evidence of changes in striatal markers following chronic cocaine self-administration in monkeys. However, an abiding question is whether such changes are permanent and whether they generalise to other drugs of the same class, such as *d*-amphetamine and cocaine. It appears that chronic exposure to these agents can be associated with loss of grey matter in the PFC, impaired signs of brain activation in functional imaging, and impairment in certain aspects of cognition, including memory and executive function. Similarly, the balance of evidence indicates some toxic effects of MDMA (*'ecstasy'*) on 5-HT neurons based on studies using different techniques on rats, monkeys and humans.

It may be significant that, in monkeys self-administering (generally lower) doses of MDMA, such toxic effects are much weaker. However, assessing the possible functional significance of these effects is elusive because of considerable heterogeneity of the chronic drug-abusing population. Recent findings indicate that a genetic polymorphism of the 5-HT transporter is associated with pathological scores on a clinical rating scale of depression in a group of chronic MDMA users. Ecstasy abuse is also associated with a pattern of cognitive impairment, which however is subject to the caveats noted above. There is little doubt about the potential devastating effects of chronic alcohol abuse on brain function. Alcohol dementia is a clearly-defined syndrome associated with structural brain changes. Among the main actions of alcohol are the enhancement of GABA-ergic transmission in combination with a reduction of glutamatergic (including NMDA receptor) function, which are likely to promote sedation and impair learning and memory in regions such as the hippocampus.

There is also conclusive evidence of foetal alcohol syndrome (FAS) produced by heavy drinking in pregnancy that leads to a pattern of physical malformations and mental retardation. Indeed, with a prevalence rate of about 0.2 per cent in all live births (6 per cent of alcoholic mothers) reported in the US, FAS is one of the leading causes of mental retardation. Recent experimental evidence suggests that exposure to alcohol in developing rats led to severe reductions of glutamate receptors of the AMPA subtype in the neocortex. These adverse actions on the brain and intellect, as well as the social burden of domestic violence arising from the heightened aggression produced by abuse, and overall greater morbidity and mortality, place alcohol among the most behaviourally toxic of all psychoactive substances. However, not all alcohol consumption has adverse effects. There is consistent evidence of significant beneficial health effects of drinking alcohol (and associated substances such as the polyphenols of red wine) in small amounts, which reduces the incidence of strokes and dementia. By contrast with alcohol, the evidence for deleterious cognitive effects of cannabis intoxication is controversial. Any significant effects may depend on chronic use over many years in that small sub-population of users who become addicted. However, accumulating evidence suggests that cannabis can act as a triggering factor for schizophrenia.

ADDICTION OF DRUGS

Until recently, the gap between social cognition and molecular and cellular neuroscience has seemed unbridgeable, given the complexities of linking social constructs such as theory of mind with simple causal neural networks. However, the emergence of cognitive neuropsychology in the 1970s illustrated the potential of a productive synthesis of cognitive psychology and clinical neuroscience in addressing common questions of how the mind/brain works. Cognitive neuroscience will continue to prove important in the objective evaluation of cognitive effects of drugs and the intellectual sequelae of chronic drug use. A similar initiative in *'social cognitive neuroscience'*, embracing developments in *'affective neuroscience'* and *'neuroeconomics'* promises to be of considerable importance for understanding the nature of addiction in its social context. *Social cognitive neuroscience* (SCN) is a systematic and theoretically-driven approach designed to understand social and emotional phenomena in terms of the interaction between motivations and social factors that influence behaviour, information-processing mechanisms that underlie social-level phenomena, and the brain mechanisms that instantiate cognitive-level processes.

The concern with neural substrates underlying normal social cognitive mechanisms links social neuroscience to the basic neurosciences and has been facilitated by the increasing availability of methodologies for investigating neural function in non-brain-damaged adults. SCN bridges the gap between social cognition and neuroscience by exploring how the brain influences social process as well as how social processes influence the brain. Of particular interest is the issue of whether the processes that give rise to social cognition are a subset of more general cognitive operations, or whether instead there are unique processes governing social cognition. Although still in its infancy, the social cognitive neuroscience approach has already been applied successfully to a broad range of topics in the social sciences and neuropsychiatric conditions, which potentially include addiction. It is proving possible to elucidate the neural and cognitive mechanisms underlying more complex social constructs such as volition; attribution theory; self-regulation; cognitive reappraisal; attitudes; mental representation of self; reward; beliefs; emotions, deception; empathy; theory of mind; cognitive control; intuition; moral emotions and complex social and economic judgements such as decision-making.

Role of Compulsive Craving in Drug Use and Relapse

Viewing it as a complex brain disorder, it is possible to consider the main behavioural characteristics of drug addiction in terms of at least four impairments of social cognition: Many modern theories of drug use and dependence assign central prominence to the role of compulsive craving in drug use and relapse. Until recently it was believed that addiction was predominantly driven by reward processes mediated by limbic circuits. However, results from recent neuroimaging studies implicate a highly interconnected network of brain areas including orbital and medial PFC, amygdala, striatum and dopaminergic mid-brain in reward processing. Distinct reward-related functions can be attributed to different components of this network.

The OFC is involved in coding stimulus reward value and in concert with the amygdala and ventral striatum is implicated in representing predicted future reward. These frontal areas are frequently activated in addicted subjects during intoxication, craving, and bingeing, but deactivated during withdrawal. The same regions are also involved in higher-order cognitive and motivational functions, such as the ability to track, update, and modulate the salience of a reinforcer as a function of context and expectation and the ability to control and inhibit prepotent responses. Cognitive theories have been influential by embedding craving within a network based on social learning theory. According to Goldstein and Volkow (2002) these results imply that addiction involves brain areas involved in several cortically regulated cognitive and emotional processes including the overvaluing of drug reinforcers, the undervaluing of alternative reinforcers, and deficits in inhibitory control for drug responses.

Social cognition

The PFC has been implicated in guiding social cognition (decision making and inhibitory control) by eliciting emotional states that serve to bias cognition, a role that is further supported by investigations of normal decision making and social reasoning studies. The effects of damage to medial and orbital PFC are consistent with a role for these regions in guiding the strategic adoption of someone else's point of view and impaired performance in reasoning about social exchange. Compromised decision making could contribute to the development of addiction and undermine attempts at abstinence.

The behaviour of those addicted to drugs could be viewed as demonstrating faulty decision making given their inability to discontinue self-destructive drug-seeking behaviours. A go/no-go response inhibition task in which working memory demands were varied demonstrates that the compromised abilities of cocaine users to exert control over strong prepotent urges were associated with reduced activity in both anterior cingulate and right prefrontal cortices. The results suggest a neuroanatomical basis for this dys-executive component in addiction, supporting the importance of cognitive functions in prolonging abuse or predisposing users towards relapse.

Abnormalities in the PFC are found consistently in most drug-addicted subjects using imaging studies. Thus, one might expect that the disruptions of self-monitoring and decision-making processes observed in drug-addicted subjects might possibly arise from drug dependent disruption of these prefrontal functions. However, an alternative possibility is that the deficits are not a consequence of drug-taking, but that both arise from premorbid changes in the PFC. Furthermore, it is even possible that the drug-taking behaviour might arise from a propensity to self-medication.

Loss of control

The issue of volition is central to social cognition since most consider willed action as essential to social democracy and to social constructs such as guilt, responsibility, accountability, law and sanctioning deviant behaviour. Drug addiction is typically portrayed as a compulsive drive to take drugs despite awareness of serious adverse consequences. The self-perceived '*loss of control*' where the addict seems unable or unwilling to control their drug use is traditionally viewed as 'voluntary' despite studies showing long-lasting changes in the brain that could compromise crucial elements of the volitional system. Campbell (2003) has argued that addiction should be considered a disease of volition caused by a cognitive impairment involving an inability to recall the negative effects of the addictive behaviour. Historically however, the construct of 'will' has been generally defined as the capacity to choose what action to perform or withhold and in a recent review Zhu (2004) distinguishes three stages of volition: the mental act of decision making; the mental act of initiating voluntary action; and the mental activity of executive control.

According to Zhu (2004) the essential engagement of the ACC in all three types of volition suggests a pivotal role in sustaining the volitional function. Other imaging studies implicate PFC, SMA and lateral PFC. Dysfunction in these regions has been associated with neuropsychiatric disorders of action including hysterical weakness, alien hand and schizophrenia, and have also been found in relation to issues of deception and malingering. Spence and Frith (1999) suggest that 'the dorsolateral PFC and the brain regions with which it is connected are essential to performing willed action, and that diseases or dysfunction of these circuits may be associated with a variety of disorders of volition, such as Parkinson's disease, 'utilisation' behaviour, 'alien' and 'phantom' limbs, delusions of 'alien control', and the passivity phenomena of schizophrenia. The issue of impaired volition raises possible ethical issues about the capacity of addicted persons to give 'free and informed' consent to participate in studies that involve detecting neural abnormalities in addiction and treatments designed to reduce their addiction.

Research involving persons who are cognitively or physically impaired in their decision making or volitional control would require special ethical consideration because they may not be capable of providing informed consent. The view among addiction researchers has been that drug-dependent people are able to give free and informed consent to participate in research studies and clinical trials so long as they are not intoxicated or suffering acute withdrawal symptoms at the time that they give consent. However Cohen (2002) controversially argues that 'the nature and pathology of untreated substance dependence make the condition inherently incompatible with a rational, internally uncoerced and informed consent on the part of those volunteering to receive addictive drugs in a non-therapeutic research setting'.

Self awareness

Drug-addicted individuals use drugs despite apparently knowing the long-term physical and psychological consequences. Rinn *et al.* (2002) tested the hypothesis of this lack of apparent awareness by suggesting that it was a product of cognitive failure rather than an emotion-driven rejection of the truth. In their study they found persistent denial to be significantly correlated with greater impairment of executive function, verbal memory, visual inference, and mental speed. Self-awareness deficits are common after traumatic brain injury and reflect a person's 'inability to recognise deficits or problem circumstances caused by neurological injury'. Such awareness disorders are believed to reflect a complex interaction between neurological, psychological and social factors depending on lesion location and cognitive dysfunction.

NEUROBEHAVIOURAL ECONOMICS OF ADDICTION

Behavioural economics, a merging of traditional economic theory with psychological studies of choice, offers different perspectives on addiction. Much of economics is based on utility theory, which assumes that agents are rational and exhibit certain reasonable attributes of preference. For example, one assumption is transitivity of preference: if an agent prefers A to B and B to C, then it must prefer A to C or it would easily be exploited by more rational agents. Given these assumptions, there must exist a utility function that assigns unidimensional values to real-world multidimensional events or outcomes, such that the agent prefers outcomes with higher utility. Psychologically and neurally, a similar process must also happen if only to decide access to motor output. Agents can then use their knowledge about the world, and about the consequences of their actions (which may be uncertain), to act so as to maximise their expected utility.

Rational behaviour need not require complex, explicit thought. Conversely, if people are logical, then we can infer their value system by observing their behaviour. A direct application of traditional economics to addiction is the calculation of elasticity of demand for goods, such as drugs. In a barter economy, and therefore in animal experiments, the 'price' of a commodity has no absolute meaning. We can speak of price only in terms of what other commodities an animal will give up to obtain the good, and that may depend on the specific commodities being traded. In humans, elasticity has a more general meaning, since humans use a monetary economy. Money is a single commodity that is substitutable for almost all others (fungible), so we can calculate elasticity as the change in consumption as money price changes. Elasticity is defined as the proportional change in consumption divided by the proportional change in price. Elasticity is usually negative (we consume less as price goes up), so elasticities between -1 and 0 represent relatively 'inelastic' demand (consumption is not reduced much by price increases) and elasticities below -1 represent relatively 'elastic' demand (consumption is strongly affected by price).

Development of Inelastic Behaviour

An obvious way to think about addiction is that demand for drugs is inelastic compared to demand for other things. The more someone is addicted, the more inelastic their demand is - they will therefore sacrifice other commodities (work, money, social interaction) rather than sacrifice the drug. Alcohol demand in rats can be more inelastic than demand for food. Yet drug demand is not completely inelastic, and addiction is not an all-or-nothing phenomenon. Most users of heroin, cocaine, and alcohol do not use extremely large amounts, as the stereotype of an addict would suggest. Instead, most use infrequently, or 'chip'. Furthermore, over 75 per cent of those dependent on an illicit drug recover. In fact, the elasticity of demand for cigarettes is typically about -0.4; that is, if the price goes up by 10%, consumption goes down by 4%. When the price goes up, some people quit altogether and others smoke less.

As for most commodities, elasticity varies with price: smokers working for cigarette puffs in the laboratory are fairly inelastic when the price is low ($\varepsilon = -0.56$), but become more elastic when the price goes up ($\varepsilon = -1.58$). Probably for this reason, elasticity is greater for poorer smokers, for whom cigarettes are proportionally more expensive. In the UK, elasticity of demand for alcohol varies from -1.69 for wine through -0.86 for spirits to -0.76 for beer. Participation price elasticities (the effect of price on the number of people using a drug) are about -0.90 to -0.80 for heroin and -0.55 to -0.36 for cocaine.

Overall elasticities (the effect of price on the total amount consumed) are about -1.80 to -1.60 for heroin and -1.10 to -0.72 for cocaine. Elasticity also varies with motivational state and other factors. Animals' demand for food is more inelastic when they are hungry and if there are no alternative ways of obtaining food. Similarly, demand for cigarettes is more inelastic when smokers have been abstinent. From a policy perspective, it is also important to consider cross-price elasticity. If a policy reduces consumption of drug A, will the benefits be mitigated by increased consumption of drug B? In the case of alcohol and cigarettes, the two are either complements ($\varepsilon < 0$) or independent, so reducing consumption of one tends to reduce (or not affect) consumption of the other. Similar analyses have been conducted for other drugs and non-drug reinforcers.

Rational Addiction Therapy

Some economists have described addiction as rational, in that addicts take the future consequences of their behaviour into account and have stable preferences. In rational addiction theory, addiction arises because the quantities of the addictive good consumed at different time periods are complements, which can lead to unstable states. This accounts for binges of consumption. Assuming rationality allows us to predict behaviour much better than not assuming it unless we can predict the specific way in which people will be irrational. A contribution of rational addiction theory was therefore to consider price as a major influence on the consumption of addictive drugs. However, the premise that drug addicts choose rationally, maximising their total happiness, has been criticised.

Certainly, humans do not always choose according to rational norms. They deviate from the optimum when making decisions because human cognitive abilities are limited ('*bounded rationality*') and because people frequently make choices that are not in their long-term interest ('*bounded willpower*'). In particular, humans and animals do not discount the future in a consistent way. It is rational to value future rewards somewhat less than immediate rewards. Steep temporal discounting (temporal '*myopia*' or short-sightedness) leads to short-termism and impulsive choice. The shape of the temporal discounting function is also very important. Simple economic models assume exponential temporal discounting, which leads to preferences that are temporally consistent (what is preferred at time x is also preferred at time y). But animals and people actually exhibit hyperbolic

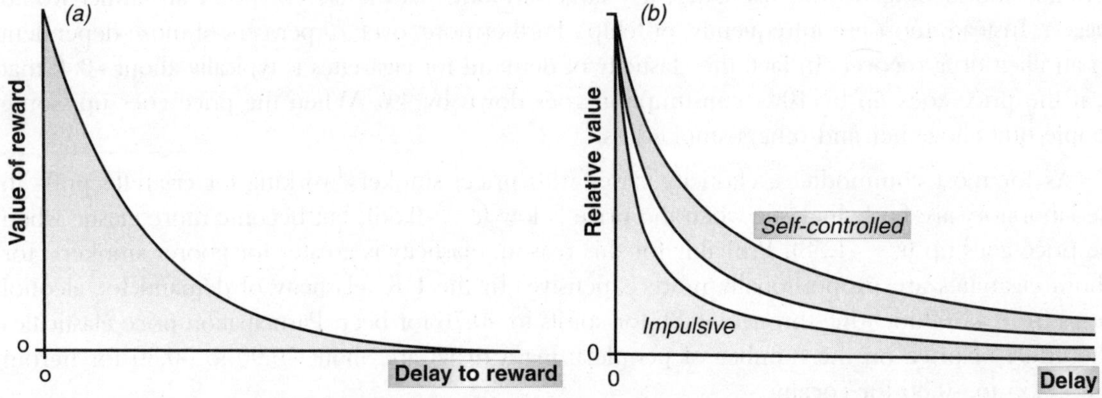

Fig. 1.1. Temporal discounting. (a) The value of a reward declines the more it is delayed. (b) Some individuals do not value future rewards very much and are impulsive. (c) Animals and people tend to discount the future in a hyperbolic, not exponential, way. (d) This leads to preference reversal.

temporal discounting. This leads to preferences that depend on when a choice is made. Therefore, many major behavioural economic theories of addiction emphasise that addiction results from the maximisation of short-term rather than long-term utility, with preferences that are inconsistent over time thanks to hyperbolic discounting, and that drug addictions are bad because short-term selection of drugs leads to lower long-term overall utility.

Consumption of drugs reduces the value of future activities – the '*primrose path*' to addiction. Knowing that one is predisposed to be temporally inconsistent allows the use of self-control strategies, such as precommitment to a particular course of action, which improve long-term utility. Economic theories of addiction are also relevant when considering the extent to which drug use is voluntary. The diagnostic criteria for drug dependence include a compulsion to take a drug, yet drug use can be voluntary. Drug use certainly has utility to the user; this may be in the form of euphoria, enhanced social experiences, or enhanced intellectual performance.

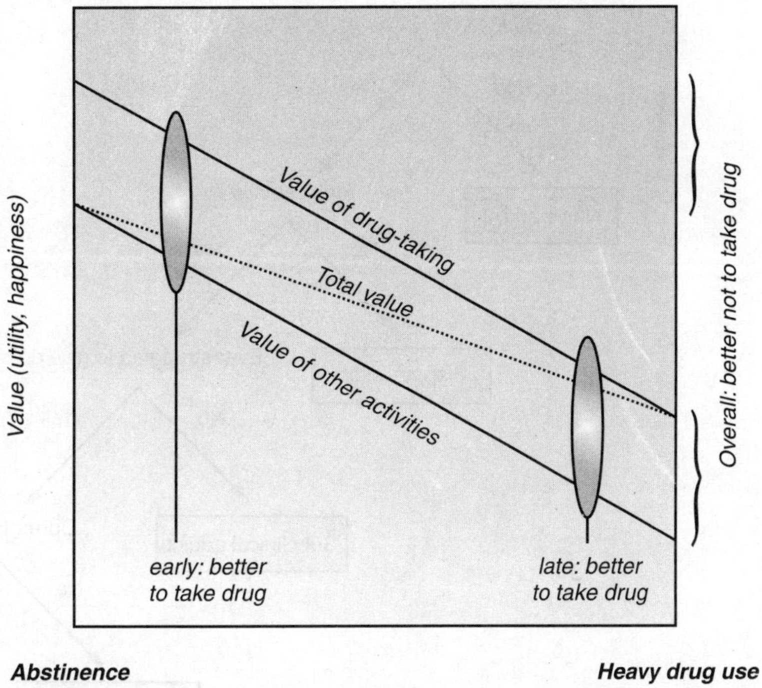

Fig. 1.2. Good now, bad in the long run – the 'primrose path' to addiction.

It is debatable whether even addicts take drugs involuntarily. Just because someone says they don't want to smoke and then later smokes doesn't mean they're smoking involuntarily – it might simply be that they're inconsistent. Furthermore, not everyone who smokes wants to give up. Appreciating these differences leads to a broader classification of addiction. The fact that people do not act to maximise their total long-term expected reward can explain a number of otherwise counterintuitive results. For example, cigarette taxes can make smokers happier because they serve as a valuable self-control device, helping them to avoid smoking. Such self-control strategies are not merely a human phenomenon. Such short-termism can explain relapse. Since one cigarette doesn't cause cancer and one shot of heroin doesn't condemn you to a junkie lifestyle, a person can correctly reason that since it's 'just for one last time', the drug is the better choice. A series of 'one-last-times' is a relapse.

Neuroeconomics of Addiction

Research into the neural basis of decision-making and the way the brain processes economic variables is a large field. Some studies have sought neural correlates of utility or hedonic evaluation

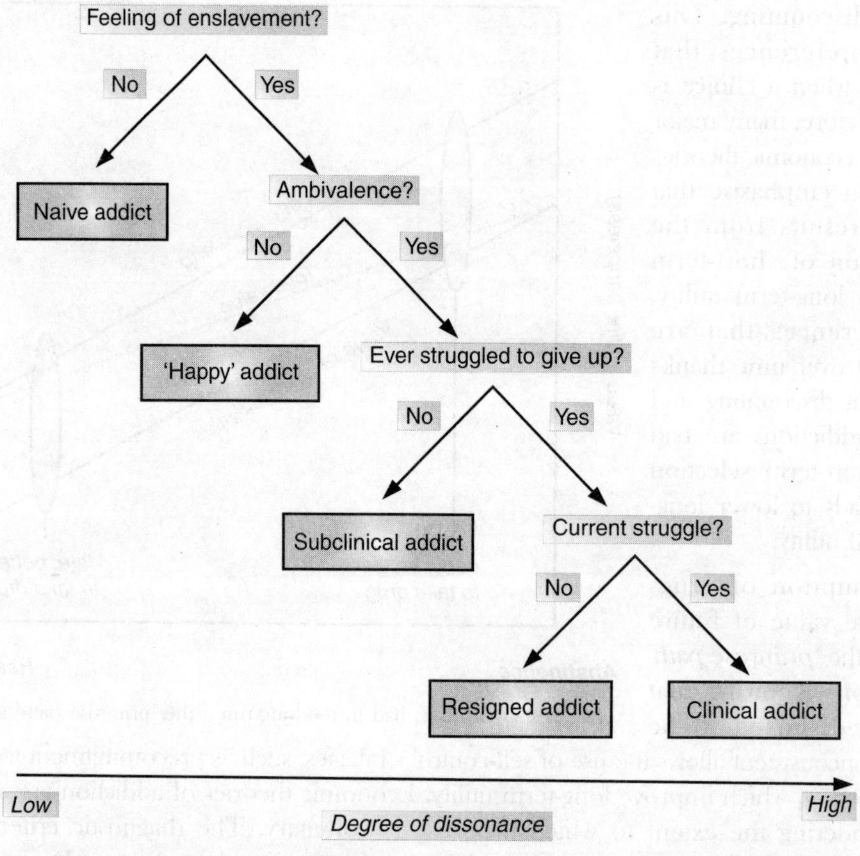

Fig. 1.3. Skog's view of addiction. A person may be unaware that it is difficult for him or her to live without a drug.

directly. Others have attempted to map the neural structures corresponding to psychological processes such as action-outcome contingency evaluation, instrumental incentive value, stimulus-response habits, and PIT. Lesions have also been used to establish the contribution of different brain structures and neuromodulators to choices between reinforcers differing in size, probability, delay, or the effort required to obtain them, while imaging studies have correlated human preferences with the activation of specific brain regions. However, some basic psychological principles remain unknown. We have not discovered how the hyperbolic temporal discounting process operates neurally, or whether hyperbolic discounting is explicable as the overall effect of two different systems – for example, a cognitive, declarative system that exhibits minimal or exponential discounting, plus phenomena such as PIT or '*visceral factors*' that make rewards more salient and promote their choice when they are immediately available.

Recently, such a two-factor model was used in the analysis of a functional magnetic resonance imaging (fMRI) study of choice involving rewards differing in magnitude and delays, with delays ranging from less than a day to 6 weeks: lateral prefrontal and intraparietal cortical regions were activated independently of the delay, and were suggested to be part of a system that evaluates both immediate and delayed rewards according to a rational temporal discounting system, while limbic regions including the ventral striatum and medial OFC were preferentially activated by the relatively immediate rewards, and were suggested to be part of a system that promotes the choice of imminent rewards without consideration of delayed alternatives. These limbic regions were more likely to be activated than the '*delay-independent*' areas by trials in which an earlier reward was chosen. A knowledge of the operation of these neural systems may offer opportunities for pharmacological treatment of addiction, but probably would not change the fact that the simplest and most powerful way to influence these neural systems is often through conventional economic tools.

FUTURE DRUGS OF ABUSE

Abused drugs of the future are likely to arise from:

1. Refinement of the properties of known drugs,
2. Synthesis of novel therapeutic compounds with abuse potential, such as euphorigenic or cognition-enhancing effects,
3. Synthesis of drugs acting on newly defined molecular targets, especially within relevant areas of the brain such as the reinforcement ('reward') system and cortical areas devoted to cognitive or affective processes.

Cognition-enhancing drugs

Powerful and effective addictive drugs are readily available. It is not understood whether the very non-selective actions of some drugs, such as cocaine and amphetamine, which affect the release or re-uptake of all the monoamines, enhance or limit their positive effects and therefore their abuse potential. It is frequently assumed that effects on the DA system are of paramount importance in their addictive potential and in mediating their positive effects. If this is the case, highly selective DA re-uptake inhibitors (a '*super-cocaine*') or releasers (a '*super-amphetamine*') may have special abuse potential. By contrast, the very nonselective actions of these drugs may underlie their potent effects, in which case novel drugs will have broader, not more selective, actions. Direct agonists at specific receptors mediating the reinforcing effects of drugs by acting within the 'reward system', may also have abuse potential.

Direct μ-opiate receptor agonists have exceptional efficacy, but DA receptor agonists are not generally abused, and this may represent a basic difference in the abuse potential of stimulants as opposed to CNS depressants such as opiates. However, DA receptor subtype-selective drugs, e.g. acting at the D3 receptor, may be abused because of their selective actions within key circuitries. Novel non-peptidergic agonists at opiate receptors may also have great abuse potential, particularly if they are devoid of effects at downstream systems mediating tolerance and physical withdrawal. Similarly, non-peptidergic drugs (or drugs from other classes able to penetrate the brain) acting on known peptide transmitter systems within key structures, such as the Acb, may also have abuse potential (e.g. drugs interacting with CCK, neurokinin, as well as opioid peptide systems). One of the key factors involved in addiction is the progressive development of neuroadaptations in the brain which may underlie withdrawal phenomena, impair functioning of specific areas of the brain (e.g. OFC and striatum) and induce persistent and maladaptive forms of learning. Limiting or preventing these neuroadaptations may enhance the abuse potential of known drugs. Thus the newly-introduced CB1 receptor antagonist rimonobant may ameliorate opiate withdrawal and also the cognitive impairments and intoxication caused by smoking cannabis. Thus, the acute, positive, effects of heroin may be achieved without a severe withdrawal syndrome and cannabis may be smoked with less deleterious effects through use of drugs such as rimonobant.

In general, drugs that might prevent long-term neuroadaptations to chronic drug intake could enhance the potential for acute use by reducing the chances of dependence. In particular, combining addictive drugs such as heroin or cocaine with others designed to prevent or reverse neuroadaptive changes in intracellular signalling cascades may be especially powerful – either as a way of minimising harm and managing abuse or of countering the adaptations, through tolerance, that minimise the acute hedonic effects. This notion can be seen as a form of '*combination drug use*' – rather as heroin and cocaine are already combined by some addicts to limit the adverse effects of one drug while enhancing those of the other (the aversive effects of cocaine and the sedating effects of heroin).

There is great potential for combining drugs from different classes to enhance the positive subjective effects and the advent of new agents will amplify this potential. GABA-A receptor subtype selective benzodiazepines (BZD) and related compounds are now being developed and have specific effects; e.g. alpha-1 subtype-selective compounds are hypnotic but not anxiolytic; alpha-5 subtype-selective compounds affect cognition, but are not hypnotic or anxiolytic. There is considerable interest in producing specific anxiolytic compounds that not only have no sedative or cognition-impairing effects, but also have no abuse potential; i.e. no euphorigenic effects and no dependence liability.

Understanding the mechanisms of BZD dependence may result in the development of more effective anxiolytics, and define the targets for novel BZD ligands with non-anxiolytic abuse potential. It seems likely that understanding genetic determinants of the responses to drugs will enhance therapeutics by allowing 'individualised' medicine. This may allow individuals to identify their own risks, possibly via counselling, for adverse effects of specific drugs and find drugs that are 'safer' to abuse. This may be achieved in conjunction with information derived from functional markers, such as cognitive test performance and brain neuroimaging. Genetic polymorphisms affecting the 5-HT transporter can already be used to predict responses to SSRIs and also the likely susceptibility to depression following chronic ecstasy abuse. Understanding the polymorphisms in genes encoding proteins that regulate the efficiency of chemical neurotransmitter systems is a major goal for future tailored therapeutics, but will also carry the risk of more optimal selection of drugs for abuse by individuals to minimise harm or optimise positive effects.

Development of antidepressant drugs

Intense research activity is directed towards the development of conceptually novel antidepressant drugs. The majority of today's antidepressants are designed to treat depressed mood and sadness, rather than the dysphoria or anhedonia which also characterises depression. Drugs that reverse these symptoms may therefore have hedonic effects and thus considerable abuse potential. Indeed, there is a growing interest in prodopaminergic drugs to target anhedonia, such as bupropion, which has weak antidepressant effects but is also used to diminish cravings in smoking cessation. But a DA- selective reuptake inhibitor may both alleviate the apathy and anhedonia of depression while carrying marked abuse potential. Whether such drugs have this potential will depend upon their pharmacokinetic properties, since rapid onset short half-life stimulant drugs are those most readily abused and have greater addictive potential.

Thus prodopaminergic drugs with more favourable (slower) pharmacokinetics may be effective therapeutically, but have limited abuse potential. CB receptor agonists developed for their analgesic and other properties may carry abuse potential. If they have appropriate pharmacodynamics, they may be safer than smoking cannabis but bring with them the risk of inducing psychosis in vulnerable individuals. Drugs primarily targeted at appetite reduction or weight loss (e.g. CB1 receptor antagonists, leptins) will increasingly become available and may be abused. It is interesting to speculate whether our knowledge of the modulation of central neurotransmitters (cholinergic, serotonergic and dopaminergic) by dietary manipulation could become sufficiently sophisticated to allow significant subjective or cognitive effects in normal subjects.

Similarly, novel drugs that enhance cognition are likely also to carry abuse potential, perhaps especially for short-term advantage under conditions of high demand such as examinations. Such drugs might affect glutamatergic transmission directly or indirectly via cholinergic or catecholaminergic neuromodulation, or by novel mechanisms of action (e.g. modafinil, CREB pathway memory consolidators). However, we do not anticipate major overlap between future recreational drugs and cognitive enhancers for the workplace, because of their generally distinct neural domains of action.

Synthesis of new drugs with abuse potential

As indicated in this project's pharmacology review new drugs with abuse potential will always be synthesised either by the pharmaceutical industry or by illicit laboratories. Many of these drugs will be directed at known targets but it is likely that new targets will also be discovered – either new chemical transmitters or new receptors ('orphan receptors') that have no known endogenous ligand. The presence of these transmitter systems and novel receptors within the mesolimbic DA system, Acb or limbic cortical areas, such as the orbital prefrontal cortex that define the neural systems underlying reward, hedonics and addiction, or those underlying cognitive processes, will be especially interesting targets for new psychoactive compounds for therapeutic or recreational use.

Neuroscientific advances

Addiction is not an all-or-nothing problem, and it will not be sensible to search for a single cure. In the face of predictions of elevated drug use, policymakers should seek methods to reduce consumption and minimise the harm from drug taking. Focusing only on prevalence (the number of people using a drug) may be inappropriate. A strategy of total harm reduction should also consider ways to reduce the average quantity used and the amount of harm per use. Many neuroscientific addiction theories focus on the way in which drugs change the brain. As Kelley and Berridge (2002) have noted, drugs may activate the same circuits as natural rewards, perhaps in a more potent manner; they may create new states, such as the motivational state of withdrawal; and they may differentially affect the balance of processes that normally contribute to responding for natural rewards, such as habits, goal-directed actions, and cue-induced motivation. There may be other effects too. Food makes you full and exercise makes you tired, but not all drugs will satiate you to the same extent.

Acute intoxication impairs decision-making, so the decision to have the sixth pint of beer may not be made in the same way as the decision to have the first. Chronic use of some drugs may alter the brain so as to impair the ability to make good choices. Some forms of brain damage may make you more likely to choose impulsively, maximising short-term rather than long-term gain. Future treatment strategies will focus on these effects, attempting to reduce drug consumption and reduce the frequency of relapse. Neuroscientific advances may contribute to the diagnostic process and the matching of treatments to addicts.

Techniques ranging from genetics to functional neuroimaging may become useful as a way of predicting which treatments will work best for an individual patient, and in assessing the likely efficacy of that treatment at preventing relapse before the patient is discharged. Both would be important advances. However, neuroscientific strategies are just one avenue of attack. Once addictive behaviours are recognised to be sensitive to drug price and to the relative value of drugs and other activities, it is clear that many options currently available may be further refined. When treating individual addicts, it is important to realise that neuroscientific strategies will potentially be increasingly interpreted in economic terms, allowing their comparison with other macroeconomic strategies.

Pharmacological techniques can already reduce the value of specific drugs. For example, methadone treats opiate withdrawal symptoms and reduces the 'high' produced by concurrently-administered heroin, thus reducing the value of heroin. Heroin prescriptions reduce the value of contaminated, street heroin. Nicotine patches treat nicotine withdrawal, reducing the value of nicotine. Disulfiram alters ethanol metabolism temporarily so that ethanol consumption induces illness and so reduces the value of alcohol. Vaccination against cocaine is being tried at the moment. This reduces the 'high' and therefore the value of cocaine. All of these can be seen as self-control tactics, and depend on the choices made by the addict. Because the addict would prefer a drug-free lifestyle in the long term, he deliberately adopts a strategy (e.g. taking disulfiram) that reduces the future value of the drug.

It is also possible to target the brain's motivational systems directly. Thus, chemicals that reduce drug-seeking in animals (e.g. D3 receptor agents) may be another line of therapy. Better knowledge of the risks of drug-taking could also help reduce the perceived value of drugs, and effective advertising of risk should take advantage of human reasoning biases, perhaps by using vivid images of the potential unpleasant outcomes of drug use. Taken to the opposite extreme, overestimation of the risks of drug-taking may also help some people avoid addiction. A personal theory that cocaine use inevitably leads to full-blown destructive addiction might not be true, but this belief is a self-control device that may prevent some people taking any cocaine.

Misinformation is clearly not a useful public health strategy, since the credibility of advisers depends upon providing accurate information, but clear and vivid statements of genuine risks are of value. Finally, the addict pays for drugs with money and therefore forfeits other alternative commodities, and may also forfeit commodities that cannot be bought with money, such as social support. Therefore, other strategies can be used to treat addiction. For example, making it easier for an addict to obtain substitutes for drugs may in the future be as effective as making it harder for the addict to obtain drugs. Rewarding abstinence directly with money or other tangible rewards also promotes abstinence.

Addicts may also learn to use self-control techniques such as pre-commitment to improve their sensitivity to the long term. Neuroscientific research aims to understand the neural mechanisms behind addiction. In the long run, this research is likely to identify a series of molecular mechanisms that operate to promote drug-taking in the addicted brain. Some will prove to be therapeutic targets, for example to reduce drug craving, and may be useful in the treatment of established addiction. Some potential therapies may be specific to the effects of drugs of abuse, but others will not be – for example, reducing strong cravings for all reinforcers, not just abused drugs.

The potential to erase drug-related memories selectively might be of substantial benefit if it can be translated to clinical practice. Other molecular markers may indicate individual vulnerability to addiction, indicating to potential users which drugs might be relatively safe to use and which would be likely to lead to strong addictions. Furthermore, techniques may become available to predict which treatments will be best suited to an individual addict by analysing the patient's genetic makeup or neural responses. Thus the most important policy decision to be made regarding the neuroscience of addiction will be how much to spend on research that may lead to treatments, and how much to spend on the treatment of addicts who seek help. However, we predict that the overall level of consumption of addictive drugs, and therefore the harm to society that such drugs cause, will be determined instead by macroeconomic decisions about drug regulation.

2

Psychostimulants

The acute administration of psychomotor stimulants, such as cocaine and amphetamine, produce behavioral activation in humans and increased locomotor activity in laboratory animals. These agents are also self-administered by various species by virtue of their reinforcing effects. It is generally accepted that these actions results, at least in part, from an increase in dopaminergic (DAergic) neurotransmission in the *nucleus accumbens* (NAc), a terminal projection region of dopamine (DA) neurons comprising the mesocorticolimbic system. Cocaine increases extracellular DA concentrations by binding to the DA transporter and inhibiting the uptake of DA from the synaptic cleft, whereas amphetamine causes a reversal of the DA transporter and increases DA release. Repeated exposure to psychostimulants and other drugs of abuse results in a progressive and enduring enhancement of their locomotor stimulatory effects, a phenomenon referred to as sensitization. A prior history of psychostimulant exposure also results in an augmentation of the reinforcing effects of these agents, indicating that sensitization also develops to those behavioral effects of drugs more directly related to their abuse liability.

Behavioral sensitization can persist for weeks or months after the last drug exposure and is postulated to contribute, in part, to drug craving and relapse to addiction. Evidence, although more limited, indicating the development of sensitization to cocaine and amphetamine in human subjects has been obtained. Behavioral sensitization is associated with marked and long-lasting alterations in the functional activity of the mesocorticolimbic DA system. These include: a transient subsensitivity of A10 somatodendritic D2 autoreceptors as well as presynaptic D2 autoreceptors in striatum and nucleus accumbens; an enhancement of the ability of psychostimulants to increase extracellular DA levels in the Nac; and a persistent enhancement of D1 receptor sensitivity within the NAc. No consistent changes in D1 receptor number or affinity have been observed after chronic cocaine administration, suggesting that supersensitive D1 responses reflect changes in postreceptor signal transduction.

Since D1 receptors are positively linked to adenylate cyclase via G_s and G_{olf}, it is noteworthy that chronic cocaine administration increases levels of adenylate cyclase and cAMP-dependent protein kinase in the NAc, while decreasing levels of G_i. More recently, in vivo microdialysis studies have shown that chronic pretreatment with cocaine potentiates the ability of a subsequent amphetamine or cocaine challenge to release DA in the NAc. The repeated administration of psychostimulants and other drugs that increase DA neurotransmission in the mesocorticolimbic system is also associated with marked alterations in the activity of endogenous opioid peptide systems.

These opioid systems modulate DA neurotransmission, and increasing evidence suggests that alterations in the activity of these opioid systems can profoundly affect the response to psychostimulants. This chapter will review data regarding the interaction of opioid peptide systems with mesocorticolimbic neurons and the relevance of this interaction to adaptations in behavior and neurochemistry that occur as a consequence of repeated psychostimulant administration.

HISTOLOGICAL DETAILS

DA neurons project to various brain regions, including the dorsal and ventral striatum, medial prefrontal cortex, amygdala, and hippocampus. The mesocorticolimbic DA system, originating in the ventral tegmental area, consists of numerous topographically organized positive and negative feedback circuits. As our understanding of the organization of these projections has advanced,

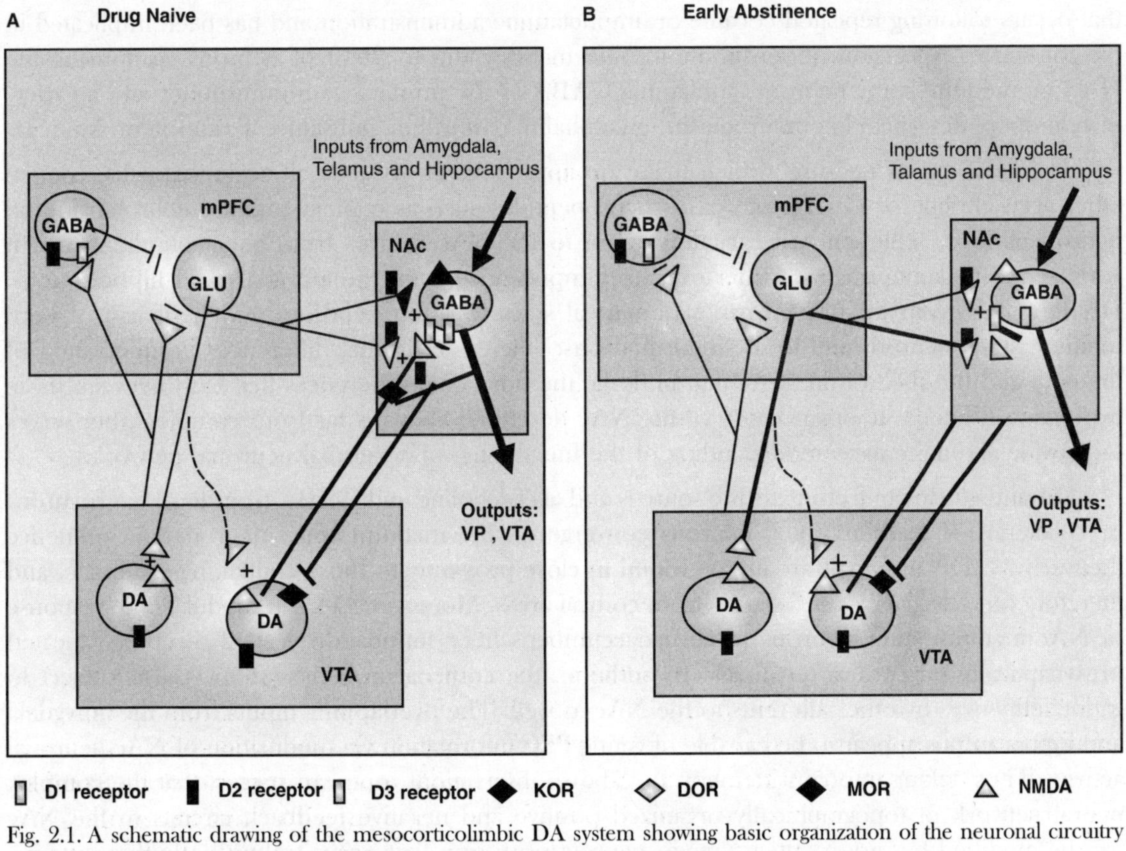

Fig. 2.1. A schematic drawing of the mesocorticolimbic DA system showing basic organization of the neuronal circuitry and receptor localization in naive and cocaine pretreated animals.

new insights have emerged as to the neuronal circuitry mediating the acute actions of psychostimulants as well as the sensitized behavioral responses that occur following their repeated administration. The vast majority of the *ventral tegmental area* (VTA) projections to the NAc are DAergic, whereas about 60% of mesoprefrontal projections contain GABA. However, prefrontal cortex (PFC) glutamatergic terminals selectively synapse onto GABA cells that project to the NAc and DA cells that project to the PFC. These PFC projections exert an excitatory influence on target cells. There are data indicating the existence of a direct excitatory PFC drive on mesoaccumbens DA neurons, but recent data indicate that this drive is mediated by indirect glutamatergic projections to the VTA.

Alterations in the activity of this PFC-VTA projection may be one mechanism leading to the induction of behavioral sensitization. White et al. have hypothesized that, during repeated treatment with cocaine, DA-induced inhibition of mPFC excitatory amino acid neurons projecting to the VTA diminishes, leading to greater activation of glutamatergic receptors on VTA DA neurons and the induction of cocaine-induced sensitization. Moreover, it has been shown that abstinence from repeated cocaine administration is associated with a decrease in the mPFC DA response to a subsequent cocaine challenge. This decrease, by increasing the glutamatergic drive on mesoaccumbens DA neurons, may underlie the increase in cocaine-evoked DA release in the NAc that occurs following repeated cocaine or amphetamine administration and has been implicated in the long-term expression of sensitization. The majority (up to 90%) of neurons comprising the NAc are medium spiny neurons, containing GABA as the primary neurotransmitter and a variety of neuropeptides, including neurotensin, enkephalin, dynorphin, substance P, and neurokinin B.

The local circuit neurons, which make up approximately 10% of all accumbal cells, contain ether acetylcholine or GABA as well as neuropeptides such as cholecystokinin, neurotensin and neuropeptide Y. The primary excitatory input to the NAc derives from glutamatergic axons of cortical and thalamic origins. Other excitatory inputs arise from the amygdala and hippocampus. DA terminals from the VTA form symmetrical synapses onto dendrites and perikarya of both medium spiny neurons and local circuit neurons. There are distinct differences in innervation of the core and the shell of the NAc that underlie the functional differences that exist between these two regions. The basic organization of the NAc described above is far from complete, but serves to provide a rudimentary understanding of the functioning of accumbal neuronal network.

DA and glutamate from extrinsic sources and acetylcholine and GABA from local interneurons or collaterals of medium spiny neurons converge on the medium spiny neurons and influence their activity. DA terminals are always found in close proximity to those containing glutamate, and therefore can effectively "*gate*" signals from cortical areas. Moreover, DA can modulate the response of NAc medium spiny neurons to corticoaccumbens fiber stimulation via D2 receptors, located presynaptically on cortical terminals. In addition, the corticoacumbens system is also subject to gating influences by other afferents to the NAc as well. The overlapping inputs from the amygdala and hippocampus appear to be capable of gating PFC information via modulation of NAc neuronal activity. Thus, taking into consideration the above observations, one can suggest that the complex neural network of topographically organized positive and negative feedback circuits in the NAc can be modified by various alterations in neurotransmission that occur within the limbic system.

CLINICAL STUDY OF RECEPTORS

The complexity of the NAc neural network is further complicated by the presence of an array of neuropeptides that are co-localized in medium spiny neurons and also found in various afferent projections. This review will focus on the endogenous opioid peptides: β-endorphin, methionine and leucine enkephalin, the dynorphins, the endomorphins, and orphanin FQ. Genes encoding these opioid peptide families and the G-protein-coupled receptors to which they bind (μ-opioid receptor, MOR; δ-opioid receptor, DOR; and κ-opioid receptor, KOR; opioid receptor like-1, ORL 1) have been cloned and characterized. β-Endorphin binds with high affinity to both the MOR and DOR receptors.

The enkephalins and the dynorphins bind with highest affinity to the DOR and KOR receptors, respectively, and are considered to be endogenous ligands for these receptor types. However, these peptides bind with varying affinities to other opioid receptor types, and the correspondence between peptide and receptor is not invariant. The recently discovered opioid peptides, endorphin-1 and -2, although structurally distinct from other opioid peptides in which the first four amino acids are Tyr Gly Gly Phe followed by either methionine or leucine, bind with high affinity and selectivity to MOR, suggesting that they may be the endogenous ligands for the MOR. In addition, the superfamily of opioid receptors and peptides was recently extended by the discovery of a novel G-protein-coupled receptor, termed opioid receptor like-1 (ORL1), which is widely distributed throughout the CNS, and its endogenous peptide orphaninFQ or nociceptin (OFQ/N).

Immunoreactivity

A dense plexus of β-endorphin processes, arising from the arcuate nucleus of the hypothalamus, is observed in the NAc. Fibers rich in endomorphin-1 and -2 nerve fibers are also apparent, as are enkephalin-containing cell bodies that exhibit a patchlike distribution. Cell bodies and terminals immunoreactive for dynorphin A and B are also observed. OFQ/N and its mRNA are sparsely localized in the cells of the core and shell of the NAc, and more densely localized fibers and terminals are found in these same brain regions, where they are especially dense in the NAc core. An extensive collocalization of opioid peptide precursors and DA receptors is observed in the various subdivisions of the NAc (rostral pole, shell, and core). However, the expression pattern of these proteins is region-dependent.

In the NAc core, proenkephalin (PENK) and prodynorphin (PDYN) mRNA, the precursors of enkephalin and dynorphin, respectively, are localized in separate cells. PDYN cells in the core express D1 and D3 receptor mRNAs, while PENK cells primarily express D2 mRNA. In contrast, the rostral pole and the shell of the NAc contain specific clusters of cells that express PENK, PDYN, D1, and D3 mRNA. A very small portion of NAc PENK cells that also express PDYN/D1 mRNA has also been observed. Acb PDYN medium spiny neurons primarily send projections to the VTA and substantia nigra (compacta and reticulata). PENK mRNA neurons innervate the ventral pallidum, which can influence mesoaccumbens neurotransmission via projections to the mediodorsal thalamus as well as by shorter feedback loops to the VTA and NAc. MOR immunoreactivity is observed throughout the NAc. The shell compartment contains vast clusters of cells and fibers, with the highest density seen in fiber clusters along the ventral edge, the lateral border of the NAc, and the dorsal septal pole region. Scattered among the clusters are numerous medium spiny neurons that are MOR-immunoreactive.

Ultrastructural studies have shown that in the shell, MORs are localized prominently on extrasynaptic somatodendritic plasma membranes (65%) and axon terminals (28%) of GABA-containing neurons. By contrast, DORs are localized mainly to non-DAergic terminals and dendritic spines that contain GABA and possibly acetylcholine, as well as one or more neuropeptides. The proportion of KOR within dendritic spines is similar to that of MOR and DOR, but the primary sites for KOR are small axons and terminals having the morphological features of DA- or acetylcholine-containing neurons or larger axon terminals that contain either GABA and dynorphin or excitatory amino acids. Interestingly, KOR mRNA is expressed in PDYN-, but not in PENK-positive cells, and significantly overlaps with MOR mRNA.

A low to moderate density of ORL1 is observed in the core and the shell of the NAc, with slightly greater receptor binding in the shell compared with the core. The findings that distinctive cell clusters in the shell of the NAc contain opioid peptide precursors, KOR, MOR, and D1 and D3 receptors, is consistent with functional data showing extensive interactions between opioid and DA systems in this brain region. Although the distribution of opioid peptide and receptors has been examined in other regions that modulate mesoaccumbal DA neurotransmission, an in-depth knowledge of their ultrastructural localization is lacking. In comparison to the NAc, opioid input to the VTA in the rat is relatively modest. β-Endorphin fibers are present in limited quantities throughout this region. Fibers containing enkephalin are somewhat denser than those for the other opioid peptides, and monosynaptic contacts between axon terminal-containing enkephalin and DAergic neurons have been demonstrated. Although largely devoid of dynorphin- and endorphin-containing cell bodies, a modest number of enkephalin-positive perikarya are seen.

The VTA contains also numerous neurons moderately stained for orphaninFQ/nociceptin as well as lightly stained fibers. In contrast to other areas in the rat, opioid receptor density is low in the VTA. MOR and MOR RNA are present in both the parabrachial and paranigral regions of the VTA. Rostral portions have light fiber and cell staining, whereas a high frequency of densely labeled fibers and a few darkly stained perikarya are observed more caudally. MOR is preferentially targeted to the plasmalemma of non-DAergic dendrites and axon terminals. Although several lines of evidence indicated that these dendrites and axons are GABAergic, MOR immunoreactivity is also observed in some unmyleinated axons and terminals that form excitatory synapses onto neurons expressing tyrosine hydroxylase or MOR. Scattered DAergic cells expressing MOR are also seen. These data indicate that although non-DA cells are principal sites for MOR activation in the VTA, MOR ligands can influence DA output, indirectly, through modulation of presynaptic axon terminals apposing DA terminals and directly in a subpopulation of MOR expressing DA neurons. In contrast to MOR, the density of DOR and ORL1 receptors is low.

Anatomical Details of Mesocortical Zone

PDYN mRNA-expres sing neuronal populations are present throughout the neocortex, with a notably higher expression in the medial prefrontal and anterior cingulate cortices, regions that have been implicated in the modulation of motivation and goal-directed behavior. These regions have a strong reciprocal connectivity to the amygdala, NAc, and VTA. PDYN mRNA-expressing cortical neurons are localized primarily in layer V and are presumed to be corticofugal fibers that innervate subscortical regions such as the NAc.

The anatomical localization of the KOR mRNA and binding sites is, however, consistent with a greater local intrinsic role of KOR in the deep cortical layers, though cortically derived KOR terminals are also evident in the striatum. In the rat, the mPFC contains a limited number of fibers and cell bodies positive for dynorphin and enkephalin. Dynorphin staining is found in nonpyramidal cells of layers 2-4 as well as in layer 5, which contains scattered cells and fibers. OFQ/N is found throughout the frontal cortex, with densest staining in area 4 and moderate staining in layers 2 and 6. Although MOR, DOR, and KOR are present in the prefrontal cortex, MOR binding is particularly dense relative to other receptor subtypes.

Pharmacological Study of Amygdaloid Complex

The *amygdaloid complex* provides substantial inputs to the VTA and NAc and can thus profoundly affect mesocorticolimbic neurotransmission and the response of DA neurons to psychostimulants. This complex is rich in opioid peptides and receptors. In the human brain, expression of PDYN is particularly high in the accessory basal nucleus, amygdalohippocampal area, cortical nucleus, and periamygdaloid cortex. The accessory basal and basal nuclei target the NAc, giving rise to some of the KOR binding sites located on the presynaptic terminals of the NAc. These nuclei also have a strong interconnectivity with the medial prefrontal and orbitofrontal cortices and provide the main intra-amygdala input to the central nucleus, which, in turn, receives the prominent amygdala DAergic innervation from the VTA.

In the rat, staining of dynorphin-and enkephalin-containing fibers and cells is also high in the central nucleus, as is that for OFQ/N. Interestingly, a recent study has shown that OFQ/N decreases neuronal excitability in a large number of spiny projection neurons in the central and lateral amygdala, resulting in a net dampening effect of amygdaloid output. MOR immunoreactivity is high in the central nucleus, intercalated nuclei of amygdala, and the posteriormedial cortical amygdala with only a few immunoreactive fibers seen in the basolateral nuclei. In contrast, DOR staining is light in the central nucleus, dense in the basolateral nucleus, and intermediate in the cortical amygdaloid nucleus. Like the MOR, KOR labeling is dense in the central nucleus. However, profuse labeling is also observed in the medial, central, and cortical nuclei. Low to moderate ORL1 binding observed in the anterior part of the medial and basolateral nuclei, and the central nucleus, is devoid of binding at all levels. Taken together, the distribution of opioid systems in the mesocorticolimbic sytem as well as in afferent and efferent projection areas clearly suggests that endogenous opioid peptides systems are strategically situated to modulate DA, glutamate, and GABA neurotransmission in the NAc.

OPIOID RECEPTOR ACTIVATION

Acute systemic administration of KOR agonists decreases DA levels in the NAc and dorsal striatum. In vitro studies assessing the modulation of electrically evoked [^{3}H] DA by opioid receptor activation revealed that in the NAc, olfactory tubercle, and PFC, DA release can be inhibited by activation of KOR. Evidence that the acute systemic administration of selective KOR agonists decreases dialysis levels of DA in the NAc and dorsal striatum has also been obtained. In vivo microdialysis studies in the rat have also shown that the intra-NAc infusion of the selective KOR agonist U-69593 decreases basal DA overflow in the NAc, whereas the selective blockade of KOR in this region significantly increases basal DA overflow. Infusion of KOR ligands into the VTA fails to modify basal DA overflow in the NAc, indicating the existence of a tonically active KOR system in the NAc that regulates basal DA tone in the NAc. In view of the localization of KOR on DA terminals, these effects have been attributed to a direct effect on DA neurons. However, the NAc shell receives significant glutamatergic input, and KORs are present on the presynaptic terminals of presumed excitatory synapses as well as on the dendrites of medium spiny neurons.

This anatomical arrangement raises the possibility that the behavioral effects of KOR activation in the NAc are due, in part, to the regulation of glutamatergic excitatory transmission. Indeed, NAc KOR activation produces a dose-dependent inhibition of glutamatergic excitatory postsynaptic currents (EPSCs). KOR activation also causes an increase in the paired-pulse ratio as well as a decrease in the frequency of spontaneous miniature events, consistent with a decrease in presynaptic glutamate release. A KOR-mediated inhibition of calcium-dependent glutamate release has also been observed in the dorsal and ventral striatum. A recent study has shown that KOR also modulates DA uptake in the NAc. Using the technique of quantitative microdialysis, which permits simultaneous assessment of drug-induced alterations in DA uptake and extracellular DA levels, Thompson et al. have shown that acute KOR activation increases DA uptake. This effect is delayed, occurring 1-2 h after drug administration, is dose-dependent, and is reversed by a selective KOR antagonist.

These data are particularly interesting in view of ultrastructural studies showing collocalization of KOR and DA transporter in axon terminals, small axons of NAc neurons, and suggest that KOR agonists regulate mesoaccumbens DA neurotransmission by two distinct mechanisms, inhibition of release and stimulation of uptake. In contrast to acute KOR activation, a decrease in DA uptake is observed 24-72 h following repeated KOR antagonist treatment. This effect is due to a decrease in the maximum capacity of uptake rather than a change in the affinity. As discussed below, these effects on DA uptake and release are functionally opposite to those produced by the acute and repeated administration of psychostimulants and provide an anatomical basis for the modulation of behavioral sensitization by KOR ligands. In contrast to KOR agonists, systemically administered MOR agonists increase DA overflow in the NAc.

Infusion of MOR agonists into the VTA produces similar effects indicating an involvement of VTA opioid receptors in producing this effect. Since MOR induces membrane hyperpolarization, this increase is consistent with the location of MOR on VTA GABA interneurons and disinhibition of DA neurotransmission. Anatomical and neurophysiological studies indicate that MOR activation may also stimulate the activity of mesoaccumbens DA neurons, indirectly, by enhancing the response of NAc medium spiny neurons to the excitatory effects of NMDA. Medium spiny neurons that contain enkephalin and GABA project to the ventral pallidum and onto GABA interneurons in the VTA. An increase in their output would enhance DA release in the NAc.

Consistent with this hypothesis, a recent dialysis study showed that the intra-NAc infusion of the MOR agonists DAMGO and fentanyl increases DA overflow in the NAc. In contrast to MOR activation, the blockade of MOR receptors in the VTA has been reported to decrease DA overflow in the NAc. This finding is consistent with the existence of a tonically active VTA MOR system that functionally opposes the actions of the NAc KOR system. Like MOR agonists, the intracerebroventricular infusion of the DOR agonist, DPDPE stimulates DA overflow in the NAc. A similar effect is also observed in response to intra-VTA infusions, suggesting that the activation of either MOR or DOR receptors in the VTA increases DA release in the NAc.

Infusion of the enkephalinase inhibitor thiorphan also increases DA overflow, suggesting that the VTA is one site mediating DOR agonist-induced increases in DA neurotransmission. Interestingly, however, other studies have shown that the DOR agonists deltorphin and DPDPE are also effective in increasing DA levels when dialyzed into the NAc, suggesting that DOR activation in the VTA or NAc can modulate mesoaccumbens DA neurotransmission. Recent studies have shown that OFQ/N can also modulate DA levels in the NAc. Like KOR agonists, this peptide inhibits locomotor activity. In accordance with this inhibitory effect, the intracerebroventricular or intra-VTA infusion of OFQ/N reduces DA overflow in the NAc. This effect is associated with an increase in GABA and glutamate overflow in the VTA.

Administration of the GABAA receptor antagonist bicuculline into the VTA prevents the effect of OFQ/N on NAc DA levels, suggesting that GABAergic interneurons located in the VTA may mediate this effect. Since, however, ORL1 activation, like MOR activation, induces membrane hyperpolarization, it appears that OFQ/N reduces the release of another neurotransmitter that inhibits GABA release in the VTA. Likely candidates in this regard are the enkephalin fibers that project to the VTA and synapse with GABA neurons.

MODULATORY EFFECT OF OPIOID

In addition to the modulatory effects of opioids on DA neurotransmission, it is also apparent that psychostimulants and other drugs that increase DA neurotransmission, affect the activity of endogenous opioid systems. The acute administration of cocaine, amphetamine, methamphetamine, or the DA uptake inhibitor GBR 12909 increase PDYN expression in the rodent NAc and dorsal striatum. This increase is delayed relative to the increase in DA levels, produced by these drugs and, in view, of the inhibitory effects of KOR agonists on extracellular DA levels, is consistent with it representing a compensatory mechanism that opposes stimulant-induced increase in DA neurotransmission. Increases in PDYN gene expression are also observed in the NAc and dorsal striatum immediately following the cessation of repeated cocaine self-administration or a binge pattern of cocaine or amphetamine administration.

Although specific subterritories of the NAc have not been examined, the increase in gene expression in striatum is generally greater than that in the NAc. The increase persists for some days following the cessation of cocaine treatment, and although it is one of the more persistent changes in gene expression that occur in response to repeated psychostimulant administration, it is no longer apparent at later phases of abstinence, and, at later time points, a decrease in PDYN gene expression may be seen. Interestingly, tissue levels of dynorphin are also increased during the early phase of cocaine abstinence. Importantly, manipulations that reduced DA but not serotonin neurotransmission prevented this effect of cocaine, suggesting that the effects on PDYN result from DA uptake inhibition. KOR density is increased in the NAc, caudate putamen, and ventral tegmental area of rats immediately following repeated cocaine administration. These effects persist for at least 48 h following the cessation of cocaine treatment. Importantly, postmortem studies have also revealed an upregulation of KOR in the NAc and its afferent projections (e.g., cingulate cortex, basolateral amygdala) in humans following an overdose of cocaine.

At present, methodological limitations have precluded the assessment of dynorphin release. Therefore, it is unclear as to whether the observed increases in receptor and endogenous ligand reflect an increase or decrease in the activity of the KOR system. In view, however, of the dysphoric and aversive effects of KOR agonists in humans and experimental animals, it has been hypothesized that the unopposed actions of an upregulated KOR system may be one mechanism underlying the "crash" that occurs in human subjects following binge cocaine use. To date, the effects of psychostimulants upon changes in POMC or endomorphin-1 and -2 have not been examined. Studies in rats have shown that PENK expression is increased or unchanged immediately following the cessation of repeated cocaine administration.

A very recent study has shown that PENK expression is increased in the NAc for at least 10 d in animals that have been allowed to self- administer cocaine for some weeks, as well as in animals that received passive injections of cocaine. Interestingly, and in contrast to the NAc, PENK expression was decreased in the central nucleus of the amygdala. Several studies have shown that repeated psychostimulant administration affects the density of MOR opioid receptors in the NAc as well as in its afferent and efferent projections. In rats sacrificed immediately after a 2 wk binge pattern of repeated cocaine administration, the density of MOR-binding sites is increased in the rostral nucleus accumbens, basolateral amygdala, cingulate cortex, and caudate putamen. In contrast, DOR binding was unchanged.

An increase in MOR density in the NAc can also be detected when the duration of cocaine exposure is reduced. Alterations in MOR systems have also been observed in human subjects. Positron emission tomography revealed increased MOR binding in several brain regions, including the frontal cingulate cortex, 1-4 d after the cessation of cocaine use. A positive correlation between MOR binding and self-reports of the severity of drug craving was also observed. The upregulation of MOR observed in this study may result from an increase in binding sites or a reduction in endogenous opioid peptide release.

In view, however, of the positive reinforcing effects of MOR agonists and the dysphoria that cocaine addicts experience during the early phase of drug abstinence, a reduction in peptide release or in receptor/effector coupling appears likely. Data regarding the effects of other manipulations that alter opioid gene expression shed some light on the possible mechanisms mediating the above changes. D1 DA receptors are located on medium spiny neurons that contain dynorphin, and their activation increases PDYN gene expression in the caudate putamen. In contrast D2 receptors are located predominantly on PENK neurons and their activation inhibits PENK gene expression.

The repeated administration of cocaine is associated with D1 receptor supersensitivity and a transient downregulation of D2 DA receptors that regulate DA release. If, as has been suggested, the early phase of abstinence from cocaine and other psychostimulants is associated with an increase in DA release, then increased release in the face of supersensitive D1 DA receptors would be expected to increase PDYN gene expression. Although DA release is increased, there appears to be a functional downregulation of D2 receptors. Thus, PENK expression would be unchanged or even increased, depending on the extent of downregulation.

SYSTEMIC ADMINISTRATION OF KOR

The systemic administration of KOR agonists with cocaine prevents the sensitization that develops to the locomotor activating and conditioned reinforcing effects of this psychostimulant. KOR agonists also prevent changes in basal and cocaine-evoked DA levels in the NAc that are associated with the induction and longterm expression of sensitization. As discussed above, quantitative dialysis studies have shown that repeated administration of KOR agonists produces changes in DA neurotransmission that are opposite to that produced by cocaine. Abstinence from repeated cocaine use is associated with a transient increase in DA uptake and release in the NAc core, whereas abstinence from repeated KOR agonist treatment produces decreases in these parameters. In animals that received a 5-d cocaine treatment regimen with the selective KOR agonist, U-69593, DA uptake was not different from controls.

Thus, repeated activation of KORs during cocaine administration may prevent cocaine-induced alterations in NAc DA neurotransmission by producing changes in basal DA uptake and release that are opposite to those produced by cocaine. In the mPFC, the early phase of abstinence from cocaine is associated with an elevation of basal DA uptake and a blunted response of mesocortical DA neurons to a subsequent cocaine challenge. This reduction in cocaine-evoked levels of DA in mPFC can be an important element in the mechanism of behavioral sensitization, since disinhibition of EAA neurons in mPFC leads to increased excitatory drive on VTA neurons. The co-administration of a selective KOR agonist with cocaine prevents these changes in mPFC DA neurotransmission as well as the sensitized behavioral response, possibly by decreasing basal DA uptake and increasing drug-evoked DA release in the mPFC. This, in turn, would oppose and normalize changes in mPFC DA neuro-transmission induced by repeated cocaine administration. Interestingly, although KOR agonist treatment, by itself, does not modify basal DA uptake in the mPFC, the co-administration of a KOR agonist with cocaine is effective in preventing increases in DA uptake, which occur in this brain region during early abstinence. Taken together, these data indicate that repeated activation of KOR profoundly affects the function of the DA transporter in terminal areas of the mesolimbic DA system.

At present the mechanism by which KOR activation regulates DA transporter function is unknown. However, KORs are collocalized with DA transporters in the NAc, and their activation modulates protein kinase C and other intracellular cascades that has been implicated in both phosphorylation and internalization of the transporter. Therefore, activation of these cascades and resulting changes in transporter trafficking may underlie the KOR agonist-induced changes in transporter function that occur in the absence of changes in transporter protein expression. Anatomical mapping studies provide some insights regarding the localization of KOR that mediate the prevention of sensitization produced by systemic agonists. Infusions of a KOR agonist into the NAc suppressed behavioral sensitization to cocaine and prevented cocaine-induced increases in basal and drug-evoked levels in the NAc, while activation of KOR in the mPFC potentiated the expression of behavioral sensitization and increased cocaine-evoked DA levels.

The exact mechanisms mediating these opposing effects of KOR activation in the two projections fields of DA neurons are unclear. However, KOR agonists decrease DA release in vitro and in vivo, and downregulate D2 autoreceptors. Inhibition of DA release and D2 receptor downregulation in the mPFC would, through D2-mediated changes in GABA release or disinhibition pyramidal cell firing, increase glutamatergic drive on VTA DA neurons, and enhance the behavioral and neurochemical effects produced by the repeated administration of cocaine. In contrast, a decrease in DA release and downregulation of D2 autoreceptors within the NAc would oppose the effects of cocaine and attenuate behavioral sensitization. Recent studies in our laboratory have shown that the selective blockade of KORs in the NAc with the selective KOR antagonist norbinaltorphinine (nor-BNI), does not affect the development of behavioral sensitization to cocaine, but significantly exacerbates the locomotor response to a subsequent cocaine challenge in cocaine naive animals.

This fact is consistent with the hypothesis that NAc KOR plays an important role in inhibiting those processes leading to behavioral sensitization. It is also consistent with the notion that changes

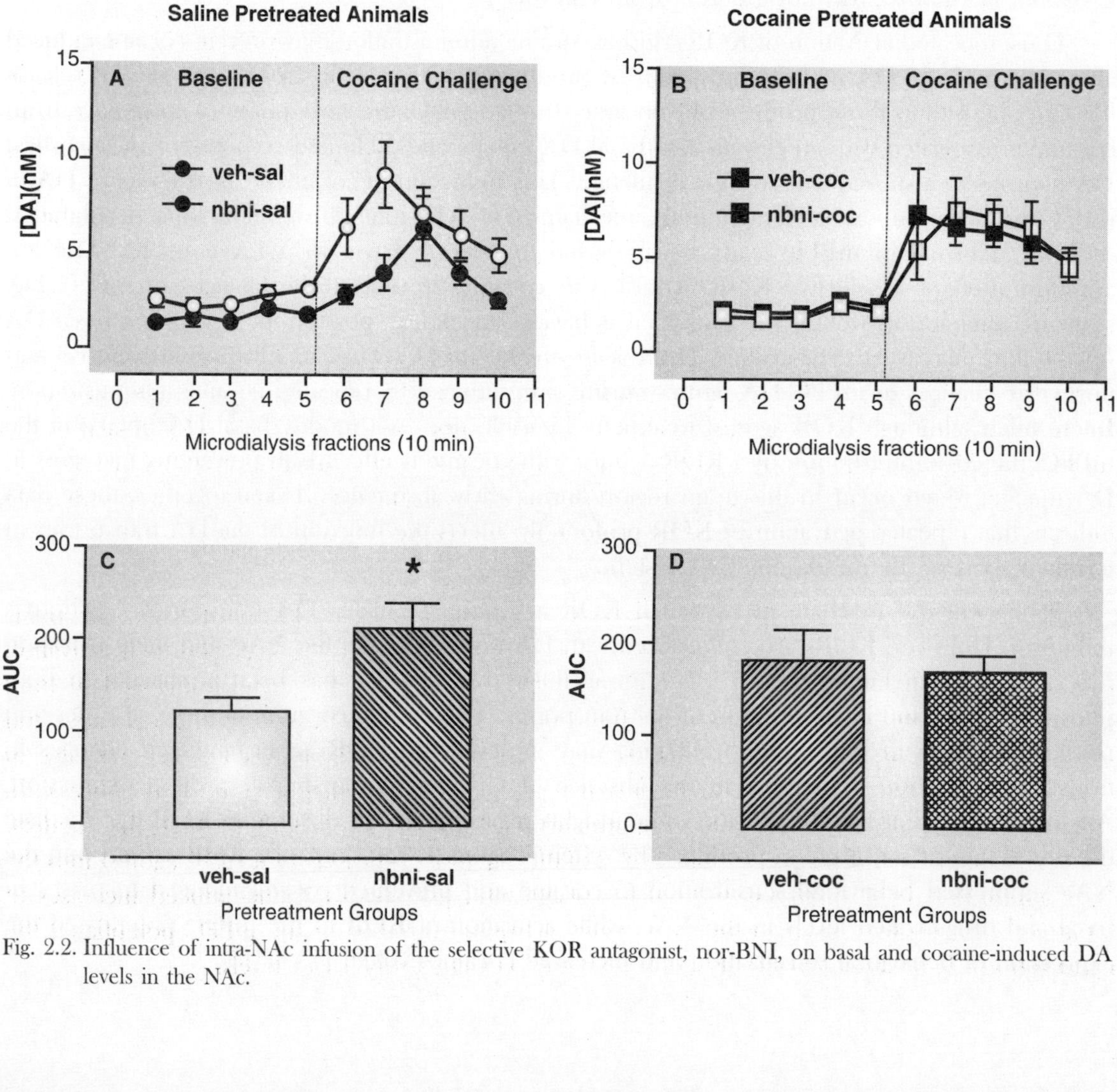

Fig. 2.2. Influence of intra-NAc infusion of the selective KOR antagonist, nor-BNI, on basal and cocaine-induced DA levels in the NAc.

in DA neurotransmission in axon terminal fields such as the NAc are critical for the expression of sensitization, while the somatodendritic region of VTA is critical for its initiation. These findings are particular noteworthy, because selective and irreversible blockade of KORs by nor-BNI can be seen as a type of functional deletion of KORs and can be compared with emerging data from mice with a constitutive deletion of the KOR1 receptor. Recent advances in molecular biology have enabled the construction of mice lacking different types of opioid peptides or receptors.

Since this technique, in contrast to pharmacological approaches, is not limited by the lack of in vivo selectivity and CNS penetration of ligands, it may be particularly useful in dissecting the roles of individual components of the opioid system. On the other hand, one cannot exclude possible compensatory adaptations that occur in response to the absence of particular proteins throughout development in mediating an expressed phenotype. Knockout mice for several genes of the opioid system have been generated: β-endorphin, enkephalin, and dynorphin, as well as the MOR, the DOR, and the KOR. To date, only one study has examined the influence of opioid-receptor deletion on the effects of psycho stimulants. Conventional and no-net-flux microdialysis was used to assess basal and cocaine-evoked levels of extracellular DA as well as the rate of DA uptake in the NAc of KOR knockout, heterozygous, and wild-type mice.

Conventional microdialysis revealed that basal as well as cocaine-evoked dialysate levels of DA in the NAc were significantly higher in KOR knockout mice compared with wild-type controls. No-net-flux microdialysis revealed no difference in basal extracellular DA between genotypes. However, the extraction fraction for DA (an indirect measure of the rate of DA uptake) was greater in KOR knockout animals as compared to wild-type controls. Because the extracellular levels of DA are determined by its release and uptake, the fact that the extracellular DA levels were not different, whereas the rate of DA uptake was higher in KOR knockout mice, attest that these animals have greater DA release then wild-type controls. Therefore, one can assume that DA release is increased in the NAc of knockout animals due to the absence of KOR inhibitory control and the apparent increase in DA uptake is a compensatory adaptation to this change in DA neurotransmission.

INTRACEREBROVENTRICULAR APPLICATION

The intracerebroventricular application of OFQ/N dose-dependently attenuates both basal and cocaine-induced locomotor activity as well as basal and cocaine-induced DA levels in the NAc. In accordance with the notion that changes in mesolimbic DA neurotransmission are involved in mechanisms of behavioral sensitization and the finding that OFQ/N can modulate this neurotransmission, OFQ/N does not prevent the development of behavioral sensitization when administered into the VTA prior to cocaine injection, but induces behavioral sensitization to a cocaine challenge when administered alone.

As one of the possible explanations for this effect, the authors hypothesize that OFQ/N, directly or indirectly, increases somatodendritic DA release, thereby reducing DA levels in the NAc. This increase would, presumably, result in subsensitivity of somatodendritic D2 autoreceptors in the VTA and a sensitized response to a subsequent cocaine challenge, when the peptide is administered alone, but would not be additive to the effect of repeated cocaine. Interestingly, however, preliminary studies in OFQ 1 knockout mice revealed no alteration in basal DA overflow in the NAc. Whether or not, however, a compensatory increase (e.g., increased KOR activation) occurs in other systems and functionally opposes the primary effect of the gene deletion awaits clarification.

INFLUENCE OF REPEATED ADMINISTRATION OF MOR

Studies examining the influence of repeated administration of MOR and DOR agonists on the locomotor activating effects of amphetamine demonstrate an important role of these opioid receptor types in the modulation of sensitization. The repeated intra-VTA infusion of enkephalin, which results in sensitization to its locomotor-activating effects, also results an enhancement of amphetamine- and cocaine-evoked activity. Exposure of animals to an environment previously paired with repeated systemic injections of low, but not high, doses of morphine also results in sensitization to the activating effects of amphetamine. Using the conditioned place preference and conditioned reinforcement paradigms, morphine-induced cross-sensitization to the rewarding effects of cocaine and amphetamine has also been demonstrated.

Although anatomical mapping studies are limited, the findings that intra-VTA or intra-NAc infusions of MOR and DOR agonists facilitate the development of sensitization and increase mesoaccumbens DA neurotransmission are consistent with a DA-dependent mechanism of opiate-induced cross-sensitization. The co-administration of the opioid receptor antagonist, naltrexone, with cocaine attenuates sensitization to cocaine-induced increases in locomotor activity and stereotypy. Similarly, in animals that received once-daily injections of cocaine in combination with naltrexone for 14 d, the conditioned reinforcing effects of cocaine were prevented. Although naltrexone produces conditioned aversive effects in rodents, this action did not underlie the prevention of the conditioned reinforcing effects of cocaine, since lithium chloride, another drug that produces aversive effects, did not modify the conditioned response to cocaine. These findings indicate that naltrexone can prevent locomotor sensitization as well as context-dependent sensitization to the conditioned reinforcing effects of cocaine.

Parallel biochemical studies revealed that the ability of DA to stimulate adenylate cylcase is increased in cocaine-treated animals, a finding consistent with previous reports of D1 receptor supersensitivity following repeated psychostimulant administration. This increase was not observed in animals that had received naltrexone in combination with cocaine, indicating that naltrexone also can inhibit a biochemical correlate of sensitization. In view of the relatively large doses of naltrexone employed, it is unclear whether the observed effects reflect an interaction with the MOR and/or the DOR. However, these findings and those described above clearly indicate an involvement of one or both receptor types in the modulation of cocaine-induced behavioral sensitization. Studies employing selective DOR antagonists suggest the specific involvement of DOR receptors in the development of sensitization to cocaine.

Co-administration of low doses of naltrindole with cocaine prevented the development of sensitization to the locomotor-activating and conditioned reinforcing effects of cocaine. When, however, naltrindole was administered only once, immediately prior to a challenge dose of cocaine, no alteration in the expression of sensitization was seen, suggesting an involvement of endogenous DOR systems in the development, but not the expression of cocaine-induced sensitization. Evidence that MOR and DOR systems contribute to the sensitization that develops to cocaine-induced seizures and lethality has also been presented. Daily injections of a high dose of cocaine (e.g., 50 mg/kg) are associated with a progressive increase in the percentage of rats exhibiting convulsions, indicating the development of sensitization.

Intracerebroventricular administration of naltrexone, the MOR antagonist CTOP, or the DOR antagonist naltrindole significantly reduced the incidence of seizures across days as well as the mean seizure score. Both naltrexone and naltrindole, but not CTOP, decreased cocaine-induced lethality. To date, only one study has examined the interaction of endomorphin with psychostimulants. The repeated intra-VTA infusion of either endomorphin-1 or -2 resulted in a sensitized locomotor activity response to an acute challenge dose of amphetamine administered 1-3 d following the cessation of peptide treatments. Thus, in rats that had received repeated infusions of endomorphin alone, amphetamine-evoked activity was significantly greater that that of control (saline-treated) animal and did not differ from that of animals that had received repeated injections of amphetamine. In contrast to the ability of endomorphins to induce sensitization, the intracerebro-ventricular infusion of these peptides failed to modify its expression when administered following the cessation of repeated amphetamine administration.

MOLECULAR MECHANISMS

At present, the *molecular mechanisms* by which opioid peptide systems modulate the development of psychostimulant sensitization are unknown. It is apparent, however, that psychostimulants alter protein phosphorylation and gene expression in medium spiny neurons of the NAc. Their acute administration is associated with phosphorylation of cAMP response element-binding protein (CREB) and increased expression of both AP1 and non-AP1 family immediate early genes. As discussed by Berke and Hyman, it is likely that these alterations induce a series of down-stream molecular events that contribute to the process of sensitization. D1 DA receptor signaling has been implicated in the initiation of psychostimulant sensitization. Medium spiny neurons containing PDYN express D1 DA receptors and psychostimulants induce the expression of PDYN in these cells. PDYN gene expression is also induced in response to other manipulations that elevate extracellular DA.

In contrast, D1 receptor deletion decreases PDYN gene expression. As is observed in the NAc, induction of the PDYN gene by DA in primary cultures is dependent on D1 DA receptor stimulation. It is also correlated with serine-133 phosphorylation of CREB. This same serine residue is phosphorylated in response to amphetamine. After chronic amphetamine administration, the time course of CREB phosphorylation is prolonged; consistent with the documented upregulation of other components of the cAMP system after chronic psychostimulant treatment. An upregulated D1 receptor-camp-CREB pathway may serve as an important mechanism of adaptation, contributing not only to PDYN gene induction and an attenuation of sensitization, but also to the regulation of other target genes that contain cAMP response elements and whose induction facilitates or opposes the behavioral and neurochemical consequences of repeated psychostimulant use.

The postulated role for CREB and PDYN in mediating neuroadaptations produced by drugs of abuse has a parallel in recent work implicating these proteins in the maintenance (CREB) and attenuation (PDYN) of some forms of long-term synaptic plasticity, including long-term potentiation and long-term depression. Long-term synaptic plasticity has been demonstrated in the NAc and, more recently, in the ventral tegmental area. Although the role of endogenous opioid peptides in modulating plasticity in the mesocorticolimbic system has not yet been examined, the influence of MOR, DOR, and KOR agonists on long-term potentiation and long-term depression in the hippocampus is not inconsistent with the hypothesis that modulation of synaptic plasticity by opioid peptide systems may occur in regions comprising this system. Consistent with the opposing effects of MOR–as compared to KOR–agonists on the development of sensitization, MOR agonists facilitate mossy fiber long-term potentiation and KOR activation blocks its induction.

HORMONAL AXIS

Glucocorticoid hormones (cortisol in humans and corticosterone in rodents) are the final step of the activation of the hypothalamic-pituitary-adrenal (HPA) axis, one of the major systems implicated in responding to environmental modifications. Activation of the HPA axis is determined by brain inputs to the hypothalamus, which releases corticotropin-releasing hormone (CRH); CRH reaches the hypophysis via the hyphophyseal portal system and activates the release of ACTH in the bloodstream, which, in turn, triggers the secretion of glucocorticoids by the cortical part of the adrenal gland. The secretion of glucocorticoids is characterized by a circadian cycle. Concentrations of these hormones are low during the inactive phase (dark phase in humans and light phase in rodents) and rise during the first hours that precede the active phase.

The secretion of glucocorticoids is also activated by practically all forms of stress, which induces a rapid and large increase in hormone levels. Stress-induced glucocorticoid secretion has been extensively studied and is considered one of the principal adaptive responses to environmental challenges. Glucocorticoids have many peripheral effects that principally involve modulation of energy metabolism and of the immune system. These hormones also readily reach the brain, where they exercise a negative feedback on their own secretion and where they regulate many behavioral and neurobiological activities. These effects are mediated by binding of glucocorticoids to two receptors: the mineralocorticoid receptor (MR) and the glucocorticoid receptor (GR). These receptors are hormone-activated transcription factors belonging to the family of the nuclear receptors.

In the inactive state, GRs and MRs are located in the cytoplasm. Upon activation by glucocorticoids, these receptors migrate to the nucleus, where they bind to specific responsive elements located on the promoters of many genes and activate or repress transcription. In the brain, MRs are principally located in the septo-hippocampal system, whereas GRs have a more widespread distribution. Because of their different affinity for glucocorticoids, MRs are practically saturated by low basal levels of the hormone; in contrast, GRs are only activated during the circadian elevation of the hormones and after stress. Consequently, it is generally believed that the effects of stress-induced corticosterone secretion are mediated by GRs.

STEROIDAL HORMONES AS PSYCHOSTIMULANT DRUGS

Studies of the role of glucocorticoids on drug responding have been performed by manipulating glucocorticoid levels in animals and by studying the behavioral consequences of these manipulations.

In this section we will review the role of glucocorticoids on behavioral responses to drugs such as locomotor activity, self-administration, and relapse.

Psychostimulants in Movements

Locomotor response to drugs is an interesting parameter, as it constitutes an unconditioned response to drugs that correlates well with drug self-administration. Studies on the role of corticosterone in drug-induced locomotor activity are consistent, showing a facilitatory effect of this hormone. This has been shown over a wide range of drug doses and with different methods of manipulating HPA activity. Suppression of glucocorticoids by adrenalectomy reduces the psychomotor stimulant effects of cocaine and amphetamine. These effects are corticosterone-dependent, as they are reversed by administration of corticosterone. Detailed dose-response studies have shown that adrenalectomy does not modify the locomotor response to low doses of cocaine, but decreases the response to higher doses; that is, adrenalectomy produces a vertical downward shift in the effects of psychostimulants, and decreases the maximal locomotor response to these drugs by approximately 50%.

The decrease in drug effects caused by adrenalectomy is reversed by corticosterone replacement (a subcutaneous corticosterone pellet delivering constant basal levels of the hormone) in a dose-dependent manner, and the response to cocaine is fully restored when basal concentrations of corticosterone are reached. The decrease in drug effects following adrenalectomy has been confirmed using pharmacological manipulations that reduce corticosterone levels. Thus, acute or repeated treatment with metyrapone, a corticosterone synthesis inhibitor, has been shown to decrease the locomotor response to cocaine. Further confirmation that decreased drug responding after adrenalectomy or pharmacological blockade of corticosterone synthesis is truly due to suppression of circulating glucocorticoids comes from studies using corticosteroid receptor antagonists. These

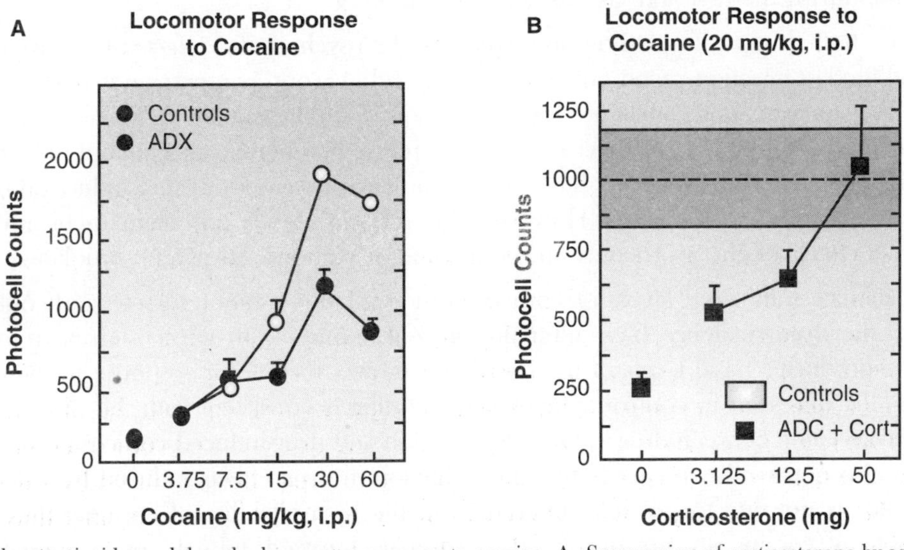

Fig. 2.3. Glucocorticoids modulate the locomotor response to cocaine. A—Suppression of corticosterone by adrenalectomy (ADX) decreases the maximal locomotor response to cocaine over a wide range of cocaine doses. B—The effects of ADX are reversed dose-dependently by corticosterone replacement.

studies show that blockade of central corticosteroid receptors (both MRs and GRs), by central administration of the MR antagonist spironolactone and of the GR antagonist RU 38486, also decreases the locomotor response to an injection of cocaine. This decrease is similar to that observed after adrenalectomy.

Effects of repeated doses of psychostimulants

Repeated doses of psychostimulants produce an increase in their psychomotor effects, which is referred to as behavioral sensitization. Although suppression of glucocorticoids reduces the acute locomotor response to psychostimulants, its effect on repeated drug treatment is more controversial. For example, regarding amphetamine sensitization, Badiani and colleagues have shown that although adrenalectomy attenuates the psychomotor effects of an acute injection of amphetamine, it does not prevent the development of sensitization to this drug. Rivet et al., however, concluded that adrenalectomy does inhibit development of amphetamine sensitization. Regarding cocaine sensitization, Prasad et al. reported that, similarly to amphetamine sensitization, adrenalectomy does not prevent behavioral sensitization tested at late withdrawal times. At early withdrawal times, however, different groups have shown that adrenalectomy does prevent the development of cocaine-induced sensitization.

In these studies, adrenalectomy decreased sensitization only if it was performed before, but not after, the sensitizing paradigm. However, blockade of GRs by selective antagonists after sensitization has been shown to prevent the expression of amphetamine sensitization. These results illustrate the controversial role of glucocorticoids on drug-induced sensitization. One factor that could explain some of the discrepancies among different studies is the parameter used to measure sensitization. For example, the studies by both Badiani et al. and Rivet et al. show a similar phenomenon: adrenalectomized rats, compared to controls, have a lower locomotor response to amphetamine during the first and last injections of the drug.

However, both groups of rats show an increase in the psychomotor effects of the drug between the first and the last injections. Consequently, when a within-group comparison is performed, as in the case of Badiani et al., one concludes that sensitization is still present after adrenalectomy. Instead, if one compares, as in the case of Rivet et al., the response between groups, it could be concluded that sensitization is reduced. Drugs of abuse, and in particular psychostimulants, induce an important increase in corticosterone secretion. However, this response does not seem to be involved in mediating the effects of glucocorticoids on the locomotor response to psychostimulants.

Thus, animals with basal levels of corticosterone and that cannot secrete corticosterone in response to the drug challenge (i.e., adrenalectomized animals with corticosterone replacement treatments reproducing basal levels of the hormone) show a locomotor response to the drug that is similar to the one seen in control animals. This finding is consistent with the observation that there is no correlation between drug-induced locomotion and drug-induced corticosterone increase. Finally, it is also noteworthy to specify that the reduction in drug effects induced by a decrease in glucocorticoids is not due to possible differences in the bioavailability of cocaine; thus cerebral concentrations of cocaine are not decreased in adrenalectomized animals, or in animals treated with metyrapone.

Individual Differences in Drug Responding

Drug self-administration in animals undoubtedly is one of the best models of drug- taking in humans. Animals self-administer most drugs that are addictive in humans and, like humans, show large individual differences in drug responding. Several studies in rodents have shown that suppression or reduction of circulating corticosterone decreases the reinforcing effects of psychostimulants as measured by self- administration. For example, suppression of glucocorticoids by adrenalectomy abolishes acquisition of cocaine self-administration over a wide range of cocaine doses, an effect that is reversed by corticosterone replacement. Other dose-response studies have shown that adrenalectomy induces a vertical downward shift of the dose-response curve to cocaine during the maintenance phase. This reduction in drug effects is reversed dose-dependently by exogenous administration of corticosterone to adrenalectomized rats and the response to cocaine is fully restored when stress levels of corticosterone are reached.

Blockade of corticosterone secretion by metyrapone also reduces self-administration of cocaine, in both the acquisition and retention phases. Another corticosterone synthesis inhibitor, ketoconazole, has similar effects and reduces acquisition of self-administration of low but not high doses of cocaine. These effects are not due to nonspecific decreases in motor behavior or motivation, as these treatments do not modify seeking behavior in food-related tasks. These findings that suppression of corticosterone decreases self-administration behavior are in contrast with studies by Suzuki and colleagues showing that adrenalectomy does not modify cocaine-induced place preference, a behavior that is often considered to be an index of the rewarding effects of drugs of abuse. However, this discrepancy can be explained by the fact that place preference induced by intraperitoneal injections of cocaine does not depend on the same neurobiological systems as those induced by intravenous injections of cocaine (as used in self-administration experiments).

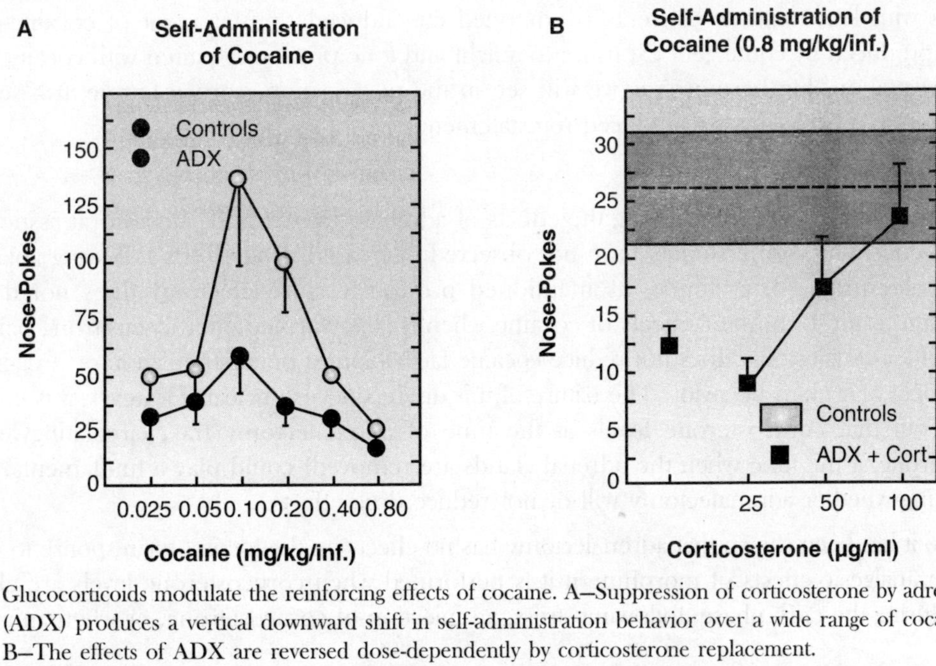

Fig. 2.4. Glucocorticoids modulate the reinforcing effects of cocaine. A–Suppression of corticosterone by adrenalectomy (ADX) produces a vertical downward shift in self-administration behavior over a wide range of cocaine doses. B–The effects of ADX are reversed dose-dependently by corticosterone replacement.

In particular, the dopamine system does not seem to mediate the place preference induced by intraperitoneal injections of cocaine, although it mediates psychostimulant self-administration. Furthermore, place conditioning is a test that is not very well adapted to measure changes in the intensity of reinforcing or rewarding effect of drugs. Thus, with this test, changes in the threshold dose of psychostimulant required to produce conditioning can be measured, but once the response is induced, the intensity of its effects does not change as a function of drug dose. In other words, place conditioning is adapted to measure horizontal shifts in dose-response functions, but not vertical ones. In contrast, in drug self-administration, the intensity of responding is highly dependent on changes in drug dose, and this test can clearly evaluate both horizontal and vertical shifts in dose-response functions. Because suppression of glucocorticoids induces a vertical shift in the dose-response to cocaine self-administration, it is not surprising that this manipulation has no effects of cocaine-induced place conditioning.

Relapse Behaviour

Relapse behavior is generally studied following an extinction period from drug self-administration. During this period, the drug is not available, so the animals extinguish responding to obtain the drug that is no longer available. Different priming factors such as exposure to the drug or the drug cue can induce the reinstatement of drug-seeking behavior. Although corticosterone plays a facilitatory role in many behavioral responses to drugs of abuse, its role in relapse is a little more controversial. Suppression of glucocorticoids does not seem to have important effects on relapse induced by drug priming. Thus, cocaine-induced reinstatement of cocaine self-administration is only minimally decreased by adrenalectomy, and is not modified by ketoconazole, a corticosterone synthesis inhibitor that reduces circulating levels of corticosterone.

Instead, corticosterone plays a significant role in cue-induced reinstatement of seeking. Thus, treatment with ketoconazole prevents conditioned cue-induced reinstatement of cocaine-seeking behavior produced by contingent exposure to a light and tone previously paired with cocaine during self-administration. Furthermore, as we will see in the next section, corticosterone also seems to play an important role in stress-induced reinstatement.

Observing Behavioural Responses

Although most studies examining the effects of adrenalectomy on the behavioral responses to drugs are consistent, some studies have not observed decreased drug effects following ablation of the adrenal glands. For example, as mentioned previously, adrenalectomy does not decrease sensitization to the locomotor effects of cocaine when it is performed after a sensitizing paradigm. In addition, adrenalectomy does not reduce cocaine facilitation of brain stimulation or drug-induced reinstatement of seeking behavior. The nature of this discrepancy is unclear. However, it is important to point out that corticosterone levels at the time of adrenalectomy (i.e., circulating levels of corticosterone at the time when the adrenal glands are removed) could play a fundamental role in determining whether adrenalectomy will or not reduce drug effects.

Thus, it has been shown that adrenalectomy has no effects on the locomotor response to cocaine or on the analgesic effects of morphine if it is performed when corticosterone levels are elevated, such as during the dark phase, following stress, an injection of corticosterone, or when animals are anesthetized with pentobarbital (because of the longer induction of anesthesia with barbiturates, adrenals are removed several minutes after the animals have been removed from the colony room—a time long enough for corticosterone levels to rise). Adrenalectomy seems most efficient in reducing drug effects when levels of the hormone are low, that is, when it is performed rapidly, under inhalant anesthetics. Although the mechanisms underlying this state-dependent effect of adrenalectomy are not known, it is likely that this effect could explain, at least in part, the discrepancies in the literature. For example, it is possible that high levels of corticosterone after withdrawal from drug self-administration could prevent the effects of adrenalectomy on drug-induced reinstatement of seeking behavior, or that removing the adrenal glands during a drug-sensitizing paradigm (when levels of the hormone are elevated) could also prevent the locomotor-suppressing effects of adrenalectomy.

EFFECTS OF STRESS TO PSYCHOSTIMULANTS

Stressful conditions have been shown to increase the psychomotor stimulant effects of drugs of abuse, a phenomenon often referred to as stress-induced sensitization. Mild food restriction (10-20% of body-weight loss) increases locomotor activation induced by cocaine and amphetamine, and other stressors such as mild tail pressure, foot shock, restraint/immobilization, handling, social isolation, or social defeat also have similar effects. The enhancing effects of stress on drug self-administration have been described for decades. Again, different stressors such as tail pinch, foot shock, social isolation, social stress, and food restriction all increase intravenous self-administration of amphetamine and cocaine. These effects have been observed for different doses of the drugs, during the acquisition phase, the retention one, as well as in progressive ratio schedules.

Stress-induced Sensitization

Several studies have shown that *stress-induced sensitization* to the locomotor effects of psychostimulants depends on the increase in corticosterone levels induced by the stressor. Thus, treatments that block stress-induced corticosterone secretion, but maintain basal levels of the hormone, have been shown to inhibit stress-induced sensitization. For example, adrenalectomy associated with replacement of basal levels of glucocorticoids (via subcutaneous corticosterone pellets) prevents the increase in locomotor response to amphetamine observed after food-restriction stress. A similar effect is seen on the locomotor response to amphetamine following social isolation, restraint stress, or different repeated stressors. These findings are confirmed by the observation that pharmacological blockade of stress levels of corticosterone (with acute or repeated metyrapone treatment) also decreases sensitization to the psychomotor effects of cocaine or amphetamine.

Overall, using different stressors and different methods to block stress-induced corticosterone secretion, these works show that stress levels of corticosterone secretion are essential to produce stress-induced sensitization. In addition to these studies, the role of stress levels of corticosterone on drug responding has been confirmed by research examining the effects of repeated exposure to corticosterone on drug effects. Thus, repeated administration of corticosterone at doses producing blood levels of the hormone that are comparable to those observed in stressful conditions also increases (similarly to stress) the psychomotor effects of amphetamine. It is noteworthy to point out, however, that these glucocorticoid treatments were performed repeatedly, and not acutely. It

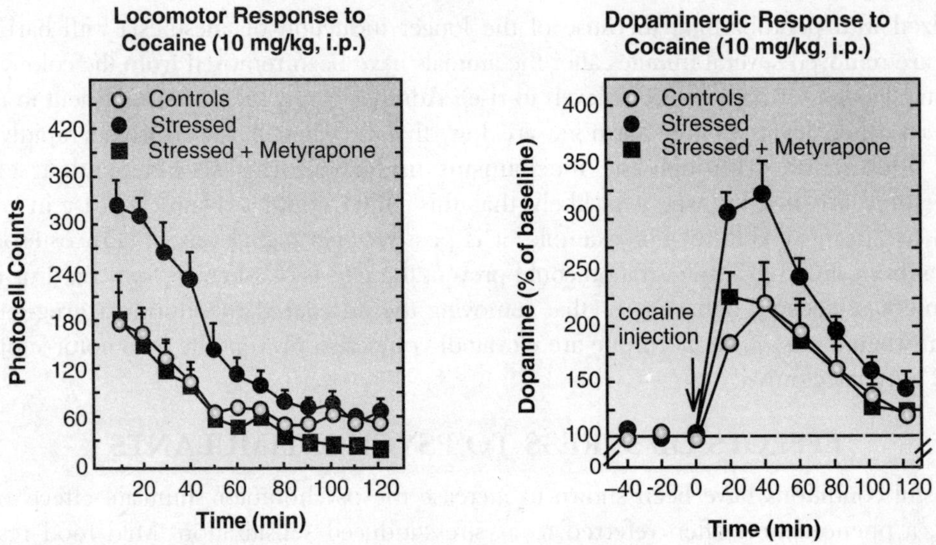

Fig. 2.5. Stress-induced corticosterone secretion mediates stress-induced sensitization to cocaine. A–Stress increases the locomotor response to an injection of cocaine (10 mg/kg). B–Food restriction increases dopamine levels in response to cocaine (10 mg/kg).

is therefore possible that a long-term exposure to high levels of corticosterone is necessary for stress-induced sensitization to develop.

Stress-induced Corticosterone Secretion

Similar to locomotor activity, the increase in self-administration induced by stress also depends on stress-induced corticosterone secretion. Thus, blocking the rise in corticosterone during stress prevents the increase in drug responding induced by the stressors. To our knowledge, these effects have only been examined following treatment with ketoconazole, the corticosterone synthesis inhibitor. Repeated ketoconazole treatment reduces both the rate of acquisition of cocaine self-administration and the percentage of rats meeting acquisition criterion following food-restriction stress. The effects of corticosterone reduction are not due to nonspecific changes in motivation or motor behavior, because operant responding for food is not modified in groups whose corticosterone levels have been modified. A larger number of studies examined the consequences of administering stress levels of corticosterone. These studies show that repeated administration of high (stress-level) doses of glucocorticoids increases drug self-administration, and reproduces the increase in drug responding seen during stressful situations. For example, rats receiving repeated injections of corticosterone intraperitoneally acquire cocaine self-administration at a lower dose compared with vehicle-treated controls.

An intravenous injection of corticosterone before a self-administration session can also increase drug responding in animals that would not readily acquire cocaine self-administration. The importance of stress levels of corticosterone in drug self-administration is confirmed by work showing that the decrease in cocaine intake induced by adrenalectomy is reversed by administration of high (stresslike) doses of corticosterone, via corticosterone in the drinking solution of the animals. It is still uncertain whether these effects of stress levels of corticosterone are mediated by MRs or

GRs. Mantsch et al. suggested a role for MRs. In their study, treatment with GR agonist dexamethasone prevented acquisition of cocaine self-administration. This effect, attributed to the decrease in corticosterone following dexamethasone, was reversed by the MR agonist aldosterone. It is also possible, however, that GRs are involved in these effects.

Adrenalectomized rats receiving replacement of low levels of corticosterone (probably only saturating MRs) do not maintain self-administration. Instead, high levels of the hormone (probably activating GRs) are necessary for cocaine self-administration. In addition, blockade of GRs by RU 38486 dose-dependently decreases cocaine self-administration, further suggesting a role of GRs in drug self-administration.

Reinstatement of Behavior

Following extinction training, reinstatement of drug-seeking behavior can be elicited by exposure to different stressors such as foot shock or food restriction. Regarding foot shock stress, it was shown that basal levels of corticosterone are necessary for foot shock to induce cocaine seeking, but that stress-induced increase in corticosterone does not play an important role on this type of reinstatement. Thus, adrenalectomy decreases foot shock-induced reinstatement, but basal levels of corticosterone are sufficient to reverse this effect. Similarly, in a different study, treatment with ketoconazole has been shown to decrease foot shock-induced reinstatement while only partially decreasing stress-induced corticosterone secretion.

Studies on foot shock-induced reinstatement show that this behavior is blocked by administration of CRH antagonists or α_2-adrenergic receptor agonists, suggesting that extra-hypothalamic CRH and the central noradrenergic system are the important players in this type of relapse. Using a different stressor to induce reinstatement of drug-seeking behavior (food restriction), the role of corticosterone was reconsidered. Preliminary data show that adrenalectomy prevents the effects of food restriction-induced reinstatement to cocaine-seeking behavior. These effects are not reversed by restoring basal levels of the hormone, but only by restoring higher levels, suggesting that more elevated levels of corticosterone might be necessary for food restriction to produce reinstatement of drug-seeking behavior. Overall, these results suggest that corticosterone plays an important role in relapse to drug-seeking behavior induced by stress. However, it is possible that different stressors require different threshold doses of corticosterone to produce reinstatement. The implication of stress levels of corticosterone in drug-seeking behavior has also been assessed in studies using other relapse paradigms.

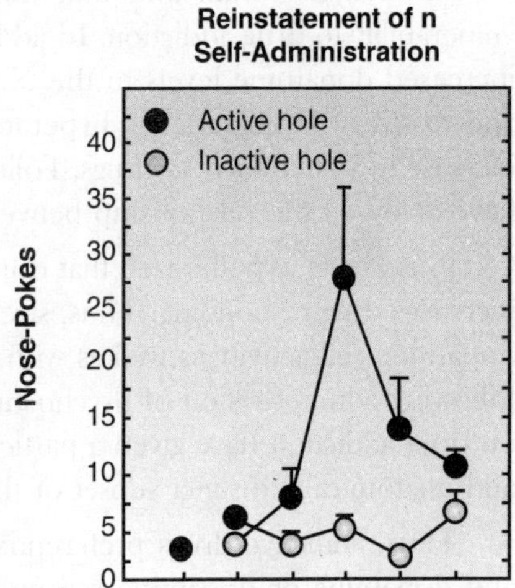

Fig. 2.6. Injection of corticosterone induces reinstatement of drug-seeking behavior.

This work shows that intravenous injections of corticosterone can precipitate reinstatement of drug-seeking behavior following extinction training. This effect is dose-dependent, and shows an inverted U-shaped curve. Thus, lower doses of corticosterone produce mild reinstatement, higher doses produce greater effects, and even higher doses will decrease reinstatement. Peak effects are obtained for doses of corticosterone that are similar to those observed during stress, whereas the descending limb of the curve is obtained for supra-physiological doses of corticosterone, suggesting that physiological stress levels of corticosterone facilitate drug-seeking behavior. In another study, it has been shown that after acquisition and stabilization of cocaine self-administration, food-restricted animals treated with metyrapone during withdrawal and reexpo sure to drug self-administration show decreased drug-taking during the reexposure phase, suggesting that corticosterone facilitates relapse to drug-taking in animals undergoing food-restriction stress.

MESOLIMBIC DOPAMINE SYSTEM

Research on the substrates of the addictive effects of drugs of abuse has underlined the role of dopamine cells, and in particular of the neurons originating in the *ventral tegmental area* (VTA) and projecting to the nucleus accumbens (NAc). This dopamine pathway plays an important role in natural reward-related behaviors, such as seeking for food, sexual partners, or novel stimuli. Numerous studies have shown that different addictive drugs, such as psychostimulants, opioids, nicotine, and alcohol, all share the common property of increasing NAc dopamine. In addition, lesions of the NAc, or of neurons projecting to the NAc, decrease self-administration of these drugs, and dopamine receptor agonists or antagonists, can respectively increase or decrease the rewarding properties of psycho stimulants.

The role of the meso-accumbens dopamine projection in drug addiction is further underlined by findings showing that the activity of dopamine neurons codes for the rewarding aspects of environmental stimuli and that increased activity of these cells is associated with enhanced vulnerability to drug addiction. In addition, animals with increased vulnerability to drugs also display increased dopamine levels in the NAc in basal conditions, in response to psychostimulant drugs and to stress. Dopaminergic hyperactivity also seems to play an important role in stress-mediated increase in vulnerability to drugs. Following the work by Thierry and collaborators, numerous studies have analyzed the relationship between stress and dopaminergic activity.

It has been hypothesized that dopaminergic hyperactivity could be a mechanism by which stress increases drug responding. Thus, stress-induced increase in drug effects is associated with increased dopamine cell activity as well as with increased levels of dopamine in the NAc in basal conditions, following administration of psychostimulants, or a subsequent stressor. In the past decade, studies on drug addiction have given a particularly important role to the "shell" of the NAc, a functionally and anatomically distinct subset of this nucleus.

Thus, addictive drugs preferentially increase dopamine transmission in the shell of the NAc, and dopamine or dopamine receptor antagonists administered in the shell can modify drug self-administration behavior. In addition, animals will self-administer drugs with addictive potential preferentially in the shell of the NAc. It should be noted that although the shell seems to play a key role in brain reward mechanisms, the "core" of the NAc also has an important role; this substructure seems particularly involved in the processing of reward-related cues.

Glucocorticoids and the Dopamine Dependent Responses

Here we will review several observations suggesting that glucocorticoid hormones could facilitate drug-related behaviors by acting on the dopamine system. The first evidence of dopamine involvement in the interaction between glucocorticoids and psychostimulants probably comes from studies investigating changes in behavioral effects of centrally injected psychostimulants. Thus, the locomotor response induced by the injection of psychostimulants in the NAc is dopamine-dependent. It was shown that the locomotor response to intra-NAc cocaine is decreased by suppressing glucocorticoid hormones and reestablished by restoring basal levels of the hormone. In addition, this response is also modulated by glucocorticoids in stress conditions. In food-restricted animals, blockade of stress-induced increase in corticosterone (by adrenalectomy and replacement of basal levels of corticosterone) prevents the stress-induced increase in locomotor activity following intra-NAc amphetamine. These effects are reversed by reproducing stress levels of the hormone.

Modulating effects

As noted above, glucocorticoids facilitate dopamine-dependent behaviors. Several studies tried to determine whether these hormones act by modulating dopamine transmission in the NAc. The effects of both suppression and enhancement of corticosterone have been examined using in vivo microdialysis and expression of Fos-related proteins. Suppression of glucocorticoids by adrenalectomy reduces extracellular concentrations of dopamine in the NAc, both in basal conditions and in response to psychostimulants. These effects are corticosterone-dependent, as they are reversed by corticosterone replacement. Interestingly, glucocorticoids have a specificity of action in the NAc. Thus, adrenalectomy selectively and dramatically (over 50%) decreases dopamine in the shell of the NAc, without modifying dopamine concentrations in the core.

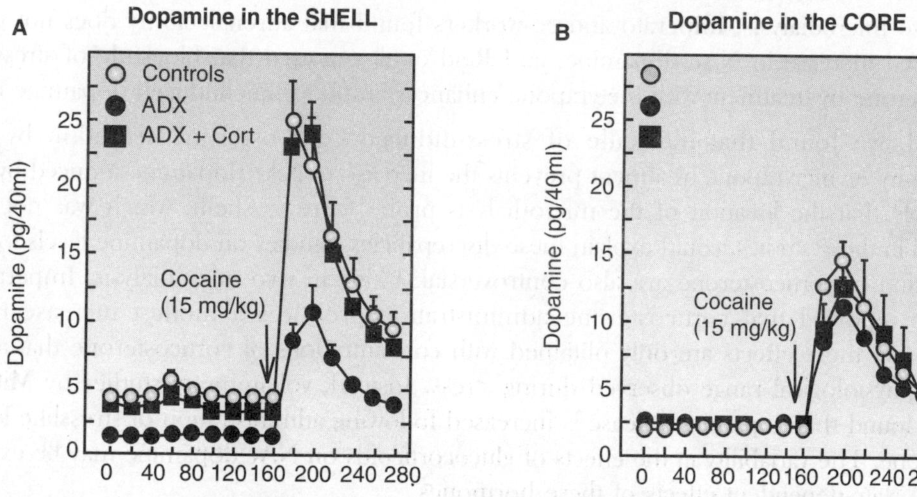

Fig. 2.7. Glucocorticoids selectively modulate dopamine concentrations in the shell of the NAc. A–Suppression of corticosterone by adrenalectomy (ADX) decreases dopamine concentrations in the shell of the NAc, both in basal conditions and in response to cocaine (15 mg/kg). B–Adrenalectomy does not modify dopamine concentrations in the core of the NAc.

The reduction in NAc shell dopamine following adrenalectomy has been observed for basal levels of dopamine, as well as for dopamine increases following cocaine or stress. Detailed studies of the role of MRs and GRs in these effects have proposed a role for GRs. Thus, the administration of the MR antagonist spironolactone does not modify extracellular levels of dopamine, whereas the administration of GR receptor antagonist RU 38486 or RU 39305 dose-dependently decreases basal levels of dopamine in the shell of the NAc. These effects are very similar to those produced by adrenalectomy, as these antagonists reduce the basal levels of dopamine by more than 50%. The decrease in dopamine levels in the NAc shell following glucocorticoid suppression is also translated postsynaptically. Thus, in adrenalectomized animals, Fos expression, an index of cellular activation that depends mainly on dopamine D1 receptor activation, is also decreased in the shell of the NAc after administration of cocaine, whereas no changes are observed in the core.

In other words, decreased levels of dopamine in the NAc shell following adrenalectomy produce decreased postsynaptic activation. On the other hand, adrenalectomized and sham controls show similar Fos activation in the shell in response to the D1 receptor agonist SKF 82958, suggesting that postsynaptic D1 receptors are functionally unaltered in this structure. Changes in NAc D1 receptors have, however, been reported following corticosterone manipulations. Thus, adrenalectomy decreases D1 receptors in the NAc, and this effect is reversed by administration of the GR agonist dexamethasone, suggesting, once again, a role for GRs. Furthermore, decrease in circulating corticosterone by administration of metyrapone, the corticosterone synthesis inhibitor, also decreases D1 receptor binding and mRNA in the NAc. Overall, these results indicate that glucocorticoids, through GRs, have a facilitatory action on dopamine transmission presynaptically, probably by modulating dopamine release, and that this effect results in reduced postsynaptic activity. Studies on the effects of stress levels of corticosterone on NAc dopamine are more controversial. Thus, using in vivo microdialysis, Imperato and co-workers found that adrenalectomy does not prevent stress-induced increase in NAc dopamine, and Reid et al. reported that blockade of stress levels of corticosterone by treatment with metyrapone enhances amphetamine-induced dopamine release.

Instead, we found that blockade of stress-induced corticosterone secretion by either adrenalectomy or metyrapone treatment prevents the increase in NAc dopamine induced by stress. It is possible that the location of the microdialysis probe (core vs shell), which was not clearly determined in these studies, could explain these discrepancies. Studies on dopamine levels following administration of corticosterone are also controversial. Using in vivo microdialysis, Imperato and co-workers reported that corticosterone administration produces a modest increase in NAc dopamine, but these effects are only obtained with concentrations of corticosterone that are well above the physiological range observed during stress. Instead, voltammetry studies by Mittleman et al. have found that dopamine release is increased following administration of stresslike levels of corticosterone. The variability in the effects of glucocorticoids on NAc dopamine may be explained by possible state-dependent effects of these hormones.

Thus, it has been shown that corticosterone administration increases NAc dopamine when it is administered during the dark phase, but not during the light phase, and these effects are greater in the dark phase if the hormone is administered just before eating. In addition, after administration of corticosterone, there is a greater increase in NAc dopamine in rats that spontaneously show higher dopamine release than in those with lower dopaminergic activity. In other words, it appears that corticosterone can only increase NAc dopamine if the hormone is administered in conditions in which the dopamine system is activated, such as during the dark phase, during food intake, or in animals with a spontaneously increased dopaminergic tone. In conclusion, glucocorticoids, via GRs, modulate extracellular concentrations of dopamine in the NAc. These hormones specifically modulate dopamine transmission in the shell of the NAc, without influencing dopamine transmission in the core. In addition, the effects of glucocorticoids are state-dependent, and are greater when the dopamine system is activated. These observations suggest that glucocorticoid hormones could enhance drug responding by selectively facilitating dopamine transmission in the shell of the NAc.

Dopamine Transmission

The mechanisms by which glucocorticoids facilitate dopamine transmission and increase NAc dopamine are unknown, but various hypotheses can be made on possible effects of glucocorticoids on dopamine metabolism, dopamine reuptake, and dopamine cell activity. It is still unclear whether these effects of glucocorticoids are direct or indirect, but the presence of glucocorticoid receptors in a subset of dopamine neurons in the VTA suggests a possible direct action of glucocorticoid hormones on these cells. A strict relationship between glucocorticoids and dopamine neurons is also comforted by the observation that mesencephalic dopamine cell cultures respond to blockade of corticosteroid receptors by decreasing dopamine release in basal and stimulated conditions. One possible action of glucocorticoids is modulation of dopamine synthesis and degradation. Thus, it has been shown that suppression of corticosterone by adrenalectomy decreases the activity of tyrosine hydroxylase (TH), the rate-limiting enzyme in dopamine synthesis. Conversely, administration of glucocorticoids increases TH activity in mouse brain, in the locus coeruleus, and in the VTA. However, total TH protein content in the midbrain was not modified by adrenalectomy in other studies.

Regarding dopamine catabolism, it has been shown that hydrocortisone or the GR receptor agonist dexamethasone decreases the activity of monoamine oxidases (MAO) in vivo and in vitro, but have no effects on catechol-O-methyltransferase (COMT). This is in agreement with the effects of glucocorticoids on the degradation products of MAO and COMT. Thus, treatment with dexamethasone decreases DOPAC (the deamination product of dopamine by MAO), and increases 3MT (the O-methylated product of dopamine by COMT). It is also possible that glucocorticoids increase dopamine levels by blocking dopamine transporter sites, or that these hormones modulate the sensitivity or the number of dopamine transporter sites in dopamine terminal regions. For example, Gilad et al. have shown that dopamine reuptake is decreased in striatal synaptosomes incubated with methylprednisolone, a glucocorticoid analog. More recently, Sarnyai et al. reported that removal of corticosterone by adrenalectomy decreases the number of dopamine-binding sites in the shell of the NAc, without modulating those in the core.

These effects depend on corticosterone, as they are reversed by replacing basal levels of corticosterone. In addition, these findings are consistent with the selective role of glucocorticoids in the shell of the NAc. Finally, glucocorticoids could modulate dopamine transmission by acting on the impulse activity of dopamine cells. An increase in the impulse activity of dopamine cells, and in particular the bursting mode, is associated with increased dopamine release in dopamine terminal regions. Although there are only a few electrophysiological studies about the influence of glucocorticoids on the activity of midbrain dopamine cells, in vivo extracellular recordings by Overton and co-workers have shown that glucocorticoids facilitate glutamate-induced bursting of mid-brain dopamine cells. Thus, adrenalectomy reduces the activity of VTA dopamine cells when these cells are stimulated by iontophoretically administered glutamate, and corticosterone administration reverses this effect. In addition, in vitro extracellular recordings by Cho and Little also showed that corticosterone potentiates the response of dopamine cells to excitatory amino acid activation. Thus, corticosterone enhances NMDA-induced increase in dopamine cell impulse activity, but has no effects on spontaneous firing.

Once again, these studies confirm the state-dependent effect of glucocorticoids on dopamine transmission. It is probable that the effects of glucocorticoids on the activity of dopamine cells are mediated by GRs, as in the work by Cho and Little the effects of corticosterone were reversed by the GR antagonist RU 38486. Overton and co-workers, however, suggested an involvement of MRs, but in this study plasma levels of corticosterone were not measured, so it is possible that the dose of corticosterone used to reverse the effects of adrenalectomy was high enough to induce GR activation. Corticosterone's facilitation of cell activity could also depend on intrinsic components of dopamine cells that modulate bursting activity, such as impulse-regulating somatodendritic receptors. These autoreceptors, primarily of the D2 subtype, are activated by somatodendritically released dopamine and reduce neuronal activity by hyperpolarizing the cell by activating K^+ channels, possibly G-protein inwardly rectifying potassium channels (GIRKs).

Other ion-gated channels, such as L-type Ca^{2+} channels and Ca^{2+}-activated K^+ channels, are also important modulators of dopamine cell impulse activity and could also be involved. Although no studies have yet examined the mechanisms by which glucocorticoids could modulate the impulse activity of midbrain dopamine cells, we could speculate that glucocorticoids alter cell excitability by modifying these ion-gated channels, or even the intracellular signaling cascades, such as pertussis toxin-sensitive G proteins coupled to potassium channels. For example, in the hippocampus, it has been shown that low levels of glucocorticoids increase cell excitability, possibly by modifying Ca^{2+} conductances, Ca^{2}-activated K^+ channels, and the binding of the Ca^{2+}-calmodulin complex to cell membranes. In addition, in the hippocampus, these hormones are able to modulate GIRK protein levels and could thus participate in regulating cell excitability.

It is also possible that glucocorticoids modulate dopamine activity by acting on the density or function of impulse-regulating dopamine autoreceptors in the VTA; however, to our knowledge, no data are available on this possible action of glucocorticoids in the midbrain. Whatever the mechanism by which glucocorticoids influence dopamine transmission and facilitate dopamine release in the NAc, their effect does not seem to be derived from an action on the integrity of the dopamine system. In fact, although suppression of glucocorticoids reduces dopamine transmission, this manipulation does not induce changes in dopamine cell death or gliosis. Thus, the number of cells immunostained with TH or with glial fibrillary acidic protein in the midbrain is similar between adrenalectomized and control rats. The above studies suggest that glucocorticoids could modulate dopamine activity by acting directly on dopamine neurons, but we cannot exclude that these hormones could modify dopamine transmission via indirect mechanisms.

Glucocorticoids could influence, for example, glutamatergic, opioid, and GABAergic systems, which, in turn, are susceptible of modulating the activity of midbrain dopamine cells. Although we cannot exclude this effect, the hypothesis is rather improbable, as most of theses excitatory and inhibitory afferences to dopamine cells seem to implicate MRs, whereas many effects on dopamine activity have been shown to depend on GRs. In summary, the mechanisms by which glucocorticoid hormones facilitate dopamine transmission could be diverse and could vary from direct actions on dopamine neurons to indirect ones involving afferent regulation of dopamine transmission. Although a direct modulation of dopamine cell activity seems plausible, further electrophysiological, molecular, and biochemical investigations are definitely necessary to dissect the mechanisms underlying the interaction between glucocorticoids and dopamine that mediates the addictive properties of drugs of abuse.

Interactive Responses

The data reviewed above indicate that glucocorticoid hormones have a profound influence on the behavioral effects of psychostimulant drugs. In particular, glucocorticoids seem to modify the motivation of the subject to self-administer drugs. Thus, after suppression or inhibition of endogenous glucocorticoids, animals still self-administer psychostimulants, but the amount of work the subjects are ready to provide is reduced. An action of glucocorticoids on the motivational effects of drugs of abuse is also suggested by the ability of these hormones to induce reinstatement of drug-seeking after a period of extinction. Finally, this idea is also supported by the profound control that glucocorticoid hormones exert on drug-induced dopamine release in the shell of the NAc, a brain region that seems to be involved in the extremely powerful incentive value of these drugs and in their capacity to induce addiction. This striking relationship between glucocorticoids and drugs of abuse prompts one question: why it is so? In other words, which physiological function of these hormones allows them to interact so profoundly with drug-taking?

Glucocorticoids hormone are generally considered as “the” stress hormones, and it is certainly true that their secretion constitutes one of the major responses to external challenges. However, this does not necessarily imply, as is generally believed, that glucocorticoids mediate the primary effects of stress, that is, its aversive and avoidance-inducing properties. In fact, in our opinion, glucocorticoids have the very opposite effects. These hormones are more likely a component of the endogenous reward system whose primary role would be to “energize” goal-directed behaviors. Activation of glucocorticoids during stress would thus be a secondary compensatory response aimed at reducing the aversive effects of stress and increasing, in this way, the copying capabilities of the subject. In fact, glucocorticoids are not only activated by stress, their secretion also precedes many goal-directed behaviors and in particular food seeking. Furthermore, glucocorticoids, in the range of the concentrations induced by stress, have positive reinforcing effects and stimulate dopamine release. This theory of the role of the central effects of glucocorticoids allows for the unification of the central and the peripheral function of these hormones.

Indeed, it has already been hypothesized that the action of glucocorticoid hormones in the periphery is not part of the primary response to stress but a way of the organisms to protect itself against the primary responses to stress. For example, glucocorticoids increase lipolysis and glyconeogenesis, which increases blood glucose levels and help the organism under stressful conditions by increasing the availability of energy substrates. Similarly, the immunosuppressive effects of glucocorticoids allow avoiding potential adverse effects of overactivating the immune system and the inflammatory responses to external aggressions. In this context, the increase in vulnerability to drugs of abuse observed after repeated stress should be reinterpreted as the activation of a protective compensatory response during environmental challenges.

During chronic stress the repeated activation of glucocorticoid hormones and dopamine, aimed to compensate the aversive effects of acute stress, would result in sensitization of the reward system. This sensitized state, which can persist after the end of the stress, would render the subject more responsive to drugs of abuse and consequently more vulnerable to develop addiction. In conclusion, glucocorticoid hormones are an essential component of our capacity to endure stressful situations, probably by attenuating their aversive impact. The role of these hormones in drug abuse is likely related to the long-lasting sensitization they induce in the reward system when repeatedly activated during stress. Consequently, understanding the molecular mechanisms that mediate such long-lasting effects of glucocorticoids could help us to better understand addiction and to develop new treatments of this condition.

3

Problematic Recreational Substances

Science has provided a close-up view of the way in which psychoactive substances, whether illicit, pharmaceutical, or legal and used for recreational purposes, alter brain functioning. In the past four decades, advances in our understanding of the molecular basis of brain functioning have enabled the investigation and the development of psychoactive substances, and neurotransmitter theory and the production of novel psychoactive drugs have expanded almost in parallel. The rate of these advances is now accelerating. This progress has been accompanied by an increase in problematic psychoactive substance use, mainly substance use disorders that until recently were not usually amenable to drug treatment. Now we are entering an era of multiple drug treatments for substance use disorders - safer alternatives to drugs, agents that block problem drug actions, substances that do both, and therapies to treat the consequences of substance use disorders - which can be administered in conjunction with greater options for psychosocial support and behavioural intervention. The next two decades could witness a dramatic increase in the number, type, and range of effects of known psychoactive substances.

Individuals may employ tailored drug regimens for therapy, prevention, pleasure and enhancement, and combine various types of agents to novel effect. New substances will be developed from various known and possibly novel sources such as the pharmaceutical, chemical and other industries, academic institutions, ethnobotanical sources and illicit and 'underground' sources. These are likely to deliver new products that will increasingly vary in chemical make-up, and potentially in quality and availability. Increased use of psychoactive drugs could be accompanied by increased sophistication in individual and societal attitudes towards alterations in consciousness, such as aspiring to optimise brain functioning. New treatments and other psychoactive agents promise expanded benefits and harms to society. The advent of effective treatments for substance use disorders will expand available interventions, and may facilitate earlier intervention in problematic drug use.

Depending on social perceptions and legal policy, new treatments for substance use disorders (e.g., vaccines) could also raise the question of whether society is justified in imposing them in a compulsory fashion on affected individuals. The balance of risk in the use of psychoactive substances will depend on external factors (e.g., controls, availability) and on the properties of individual agents and users. Increased investment of resources and changes in regulatory mechanisms are likely to be needed to maximise the benefits and minimise the harms of current and future psychoactive substances for individuals and society. Despite this, new knowledge will undoubtedly bring unanticipated developments, problems and solutions. The key challenge will continue to be how to use this expansion of psychopharmacology for the most creative and least destructive effects.

HISTORICAL BACKGROUND

Human psychoactive substance use dates from prehistoric times, and a desire for intoxication may be an innate drive. Although other animals take intoxicants, widespread psychoactive substance use may be a defining feature of humankind. This overview focuses on the potential for psychoactive drugs to lead to substance use disorders, and includes illicit, pharmaceutical, legal recreational and other substances. Thus, antidepressants and other psychoactive agents whose adverse effects do not include substance use disorders are not included. The remainder of the review covers biological mechanisms of substance use disorders, and the ways this knowledge informs known and potential treatments for such disorders.

We include a summary of the state-of-the-art treatment recommendations from the British Association for Psycho- pharmacology (BAP). Finally, we propose an educated view of the future of problematic drug use, plus possible technologies and other strategies for management. Terminology in this area is confused, since terms like '*addiction*' and '*addicts*' are poorly defined and in some quarters stigmatising. Other terms, like '*substance dependence*', are better defined and are often used interchangeably. In reality, a spectrum of drug-taking behaviours exists including: appropriate therapeutic use; diversion (non- prescribed use) or other inappropriate use of prescribed agents; self-medication (informal use of drugs to mitigate symptoms); non-problematic recreational use (which may encompass regular use); hazardous drug use and substance misuse; harmful drug use and substance abuse; dependency and addiction.

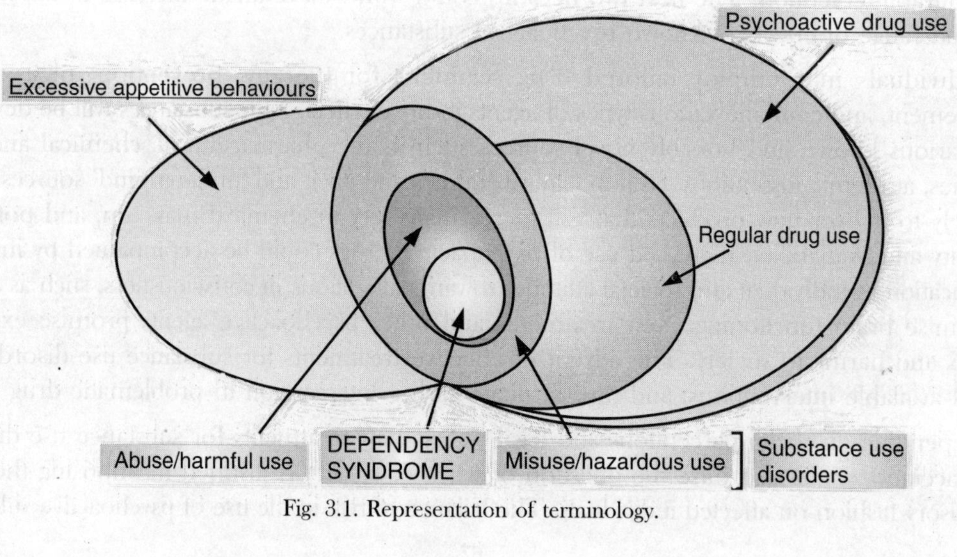

Fig. 3.1. Representation of terminology.

Reward and reinforcement: Drugs taken for pleasure are considered 'rewards' that activate brain reward systems similarly to pleasurable food or sex. Rewards have a stimulating aspect and also prompt positive reinforcement.

Addiction/substance dependence: Although distinctions are made between these two terms, we regard them as descriptors of the same syndrome, a complex phenomenon that is considered to involve biological (including molecular) processes that interact with social, cultural and political pressures. The term '*addiction*' is commonly used to represent an extreme pathological state involving loss of control over drug taking and its consequences. '*Dependency syndrome*' may be seen as a behavioural process, with individual components that include drug (or other pleasure) taking and seeking, tolerance, sensitisation, withdrawal, craving, dependence, rebound and relapse. Not all of these components may occur, although characteristically dependency syndrome develops by progression through at least some of these stages. Two international standards recognise the term '*substance dependence*', which is variably defined according to several criteria that can include substantial harm, impairment or distress from substance use, difficulty in controlling or compulsive substance use, withdrawal symptoms, and tolerance, though not all of these are necessary criteria. Overlooked in these definitions is the crucial point that dependency syndrome is invariably a chronic, relapsing condition.

Substance abuse/harmful use: International definitions recognise, with somewhat different criteria, a harmful but not dependent pattern of substance use that leads to substantial difficulties – eg legal or interpersonal problems, failure to fulfil obligations, or use in hazardous situations. This is known as substance abuse or harmful use depending on the criteria used.

Tolerance: Reduced effects of the same dose of drug on subsequent use or a need to increase dose to get the same effect are known as tolerance. Mechanisms of tolerance vary between drug classes, and may be neuronal or molecular, or psychological and behavioural.

Withdrawal: Withdrawal encompasses behavioural, psychological and physical symptoms and signs on stopping a drug or with pharmacological blocking of drug action. Usually withdrawal is an expression of processes that underlie tolerance, but the two states may be differentiated. Physical withdrawal is not a necessary criterion of dependency, since some agents, such as antidepressants, are linked with withdrawal symptoms but not dependence. However, the negative reinforcing aspects of withdrawal often perpetuate or worsen drug-seeking behaviours.

Sensitisation: Several agents engender a progressive increase in effects with repeated administration (also called reverse tolerance) or persistent hypersensitivity due to previous exposure.

Craving and relapse: These two processes are interrelated, although craving is not essential for relapse, the resumption of drug seeking and use after abstinence. The term 'craving' encompasses subjective desire for the drug, motivation towards drug seeking and avoidance of withdrawal, and behavioural changes. The neurochemical basis of craving is poorly understood, but craving can become conditioned to and elicited by environmental cues related to drug use.

Abuse potential: This is a qualitative measure of the likelihood of a specific agent to lead to dependency and cause harm. Thus, caffeine use is often dependent but usually causes little harm and has low abuse potential. Abuse potential is related to drug factors – agents with faster entry to and exit from the brain plus high efficacy (strong action) have higher abuse potential. The optimum method for determining abuse potential is not known, but includes the ability of a drug to maintain self-administration by animals.

Neurotransmitter: any chemical that is released from a neuron into a gap between neurons (synapse) or between a neuron and other cell type to communicate a change in state (excitation or inhibition) to the adjacent cell.

Psychoactive drugs: substances that affect brain function through neurotransmitter systems. Actions are described in terms of effects on the synthesis, storage, release, reuptake and metabolism of neurotransmitters and/or by their action at neurotransmitter receptors as follows:

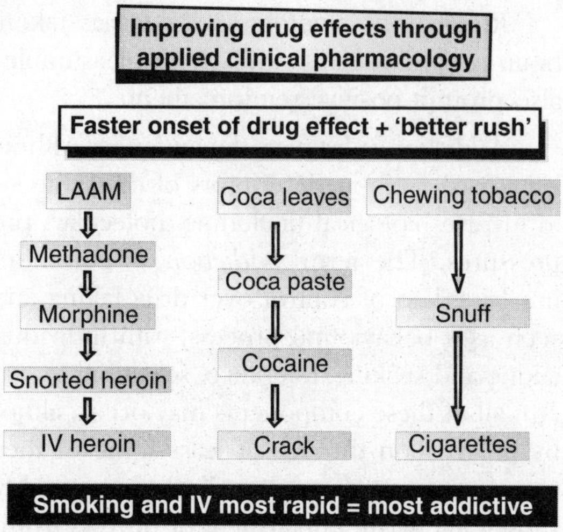

Fig. 3.2. Abuse potential is related to the form of the drug. The chemical make-up and formulation dictate the speed of entry into the brain.

1. *Agonist*: any agent that mimics the action of an endogenous neurotransmitter, including the neurotransmitter itself, is an agonist.
2. *Antagonist*: an antagonist blocks agonist action but has no intrinsic action itself.
3. *Partial agonist*: such substances partly mimic the effects of the neurotransmitter, but in the presence of agonists with greater efficacy, they act as antagonists.
4. *Inverse agonist*: for receptor systems that have ongoing (agonist-independent) activity, an inverse agonist will reduce such activity.
5. *Ligand*: this is the term for any substance that binds to a receptor.
6. *Analogues*: this pharmacological term describes structural relatives of a parent compound.
7. *Set and setting*: these lay terms respectively describe aspects of an individual and their environment that impact on psychoactive drug effects.

UNDERSTANDING NEUROTRANSMITTERS

Over the past four decades, major advances have occurred in our understanding of brain mechanisms such as the role of neurotransmitters, including those associated with substance use disorders. This knowledge has paralleled and influenced psychoactive substance use of all types, with an unprecedented increase in availability of all types of substances. Meanwhile, online technology has enhanced access to both substances and related knowledge of their use, effects and manufacture. Thus, overall drug use is increasing worldwide, with few exceptions. Use of multiple drugs (polypharmacy) is increasing and novel agents are appearing. These changes accompany increases in other excess appetitive behaviours such as overeating and '*Internet addiction*'. Already, the concurrent use of several psychoactive substances seems the norm in many populations, and may involve legal recreational substances, illicit drugs, prescription drugs, and other agents (e.g., herbs and nutrients). Problematic use of multiple substances ('*polydrug abuse*') seems common in individuals who both do and do not access treatment services.

Polydrug abuse may reflect a range of processes, from an individual tendency towards drug taking to the combining of certain substances for specific effects, including alleviation of withdrawal. Common examples include: alcohol with cocaine (synergistic effects); illicit heroin use with prescription methadone (additive); '*speedballs*' of heroin plus cocaine (offsetting sedative and stimulant effects); alcohol, cannabis or prescription sedatives taken with or after stimulants and ecstasy (alleviation of overstimulation or 'come-down' effects). Multiple substance use may also be sequential, with some individuals showing a progression in use through different classes of substances, which also may overlap. For example, alcohol use in young people may predict illicit drug use in later life, while many individuals who abuse illicit drugs in younger years switch to alcohol later, especially if illicit drug use stops. Such patterns could perhaps reflect an underlying neurochemical imbalance (e.g., untreated anxiety or ADHD).

Multiple and sequential drug use are likely to increase harms, complicate treatments and worsen the user's long-term prognosis. Internationally, the response to these trends has been diverse and variable according to the substance. Some increased controls have been developed related to tobacco use but not for alcohol. Instead, the availability and relative cheapness of alcoholic drinks has increased, combined with youth marketing. For illicit drugs, the UK has until the recent formation of the National Treatment Agency and the Drugs Intervention Programme, focused on reduction of supply as the main means of reducing prevalence, with changes in criminal law enforcement - this is now accompanied by an increasing focus on problematic illicit drug use as a health issue. Meanwhile, advances in pharmacology have expanded the drug treatments available for substance use disorders, including development of partial agonists (eg buprenorphine) and long-acting formulations of known drugs (e.g. naltrexone). Now, a realisation that treatments developed to target one problem drug may 'cross over' to prove beneficial for other drugs offers new possibilities for treatments. Already, a recognition exists of the potential for tailoring treatments to subgroups and individuals (e.g. for levels of dependency).

Clinical Aspects of Neurotransmitter

Psychoactive drugs are known to act through different brain neurotransmitter systems. Currently, at least eight neurotransmitter systems are definitely implicated in drug actions, but the importance of monoamines, particularly dopamine, as a final common pathway underlying drug seeking and craving has become prominent through research. Many other systems will probably be found to be involved in drug actions. In addition, some 200 genes in the human genome seem to code for receptors, with functions and ligands that remain to be discovered, known as orphan receptors. Already, some of the roles of different neurotransmitters in substance use disorders have been disentangled, although this view is simplistic compared with that likely to be available in 2025. Currently, dependency is viewed as a form of aberrant learning that hijacks brain reward systems formed of two major dopamine pathways.

Release of mesolimbic dopamine seems to be primarily involved in the anticipation and salience of, and appetite for, rewards such as food or sex and many abused drugs. This system also seems involved in the opposite state, aversion - e.g., to stressful stimuli. Low function (tone) in this system

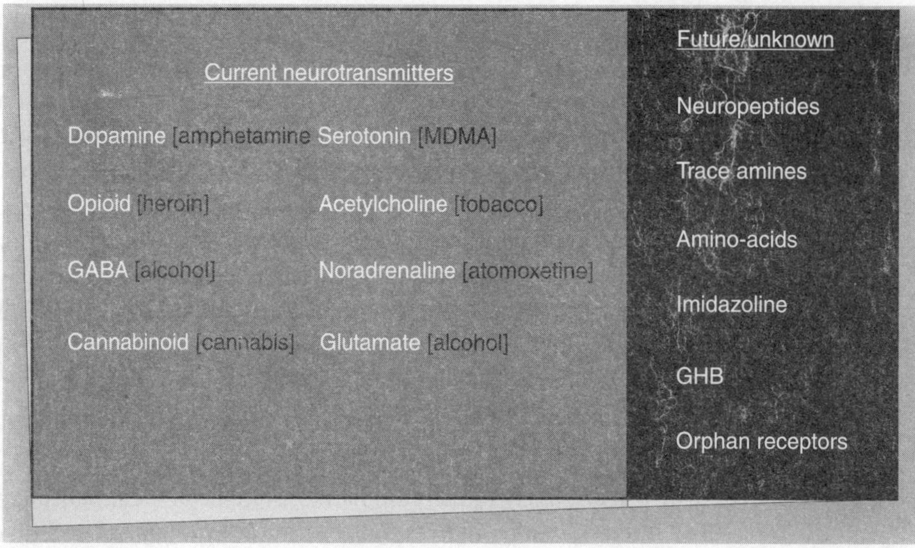

Fig. 3.3. State of scientific understanding on neurotransmitter targets for psychoactive substances.

is a feature of withdrawal, and might underpin craving. Mesolimbic nigrostriatal dopamine could be specifically important in generating drug-seeking stereotyped behaviours. However, various lines of evidence indicate that the involvement of dopaminergic neurons, receptors and the transporter in drug actions is not as straightforward as previously thought. The latest human research suggests that this system could be involved in the ability to generate cycles of learning and consequent behaviour according to beneficial or adverse outcomes. Other systems suspected to have key roles in dependency include neural pathways that release opioid neuropeptides, which modulate other neurotransmitter systems, and which may form part of brain pleasure circuits.

Three brain receptors recognise opioids - mu, delta and kappa - and recent evidence suggests that these receptors might have distinguishable properties, including enhancement or lack of pleasure and reward-related learning (reinforcement). The balance between diffuse excitatory neurotransmitters like glutamate, and inhibitory ones like GABA and cannabinoids, seems likely to impact on whether an individual learns

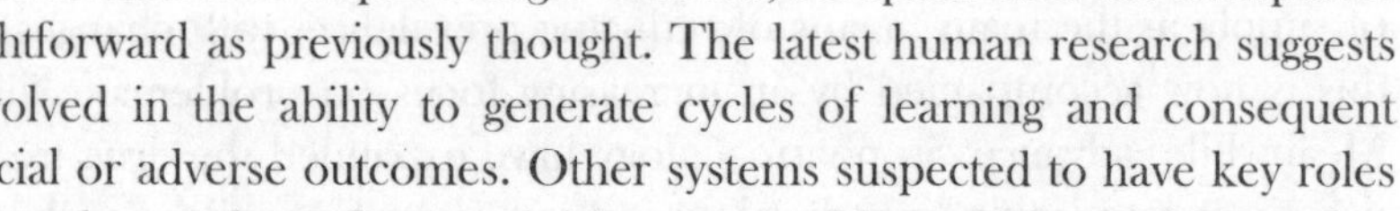

Fig. 3.4. Pharmacology of dependence-putative neurotransmitter roles. The key neurotransmitter systems are represented stylistically on this brain section, with labels that describe putative roles in symptoms of substance use disorders.

from an experience or is inhibited from doing so. These systems interact with reward and pleasure systems. So forms of aberrant reward-based learning may underlie symptoms such as craving, tolerance, sensitisation and reinstatement of drug-seeking behaviour.

Various neuropeptides that act in emotion-related areas of the brain might have longer-lasting roles in substance use disorders, e.g., substance P, which seems to mediate some of the pleasurable aspects of opiates and some effects of psychostimulants. The causes and consequences of substance use disorders are becoming clearer. For example, the stress hormone and neurotransmitter noradrenaline and the neuropeptide CRF have been linked with symptoms of withdrawal and the impact that withdrawal and other stressors have on the likelihood of relapse. Serotonin is involved in mood and anxiety, and has also been linked with impulsivity, including sensation seeking, as has dopamine.

ESTIMATING THE PREVALENCE OF DRUG

An important step in drafting any response to current and future problematic drug use is to estimate the prevalence of drug use and its associated harms. However, estimation of drug use is difficult, in part because many measures exist, including various crime statistics, seizures of illicit drugs, alcohol and tobacco sales, self-report from several surveys, plus estimates from health services data and death certification. In addition, many data estimates are not sufficiently accurate or responsive, while prevalence is not directly related to the extent of harms and costs. From available data, cigarette smoking remains stable at 26 per cent of adults in Britain, with the gap between men and women narrowing. At least 100 000 UK deaths per year are attributed to smoking, while lung cancer represents one in five cancer deaths in Europe. Overall, alcohol consumption is rising somewhat, especially in young people and women. Roughly 25 per cent of the English population drink over weekly limits (40 per cent of men and 23 per cent of women in Britain), at least 5 per cent are heavy drinkers, and 7 per cent or more are dependent, with dependent men outnumbering women by more than four to one.

Estimates of directly alcohol-related deaths in England and Wales vary from 5000 to 40 000 per year, while rates have risen dramatically. Between 1980 and 2000, the UK death rate from alcohol-related diseases more than doubled among men and almost doubled among women. The British Crime Survey 2002/2003 estimates around four million illicit drug users and around one million Class A drug users, with the majority of people using one type of illicit drug. In 2000, average estimates suggested at least 300 000 problem Class A drug users and under 300 000 injecting opiate users. In 2003 in England and Wales, 2445 deaths related to drugs other than alcohol and tobacco were reported, including 1338 deaths related to illicit drugs, 424 deaths involving antidepressants, 466 involving paracetamol-containing medication, 211 involving benzodiazepine sedatives, and 66 involving other sedatives. Some deaths involved multiple substances.

Pharmaceutical Importance of Psychoactive Substances

Use of tobacco

Cigarette smoking is probably the most harmful use of all psychoactive drugs. Harms extend to those passively exposed to smoke, including the developing foetus. Most smokers are dependent and have a long history of use. Harms include chronic lung disease, increased risk of several cancers, cardiovascular disease, and stroke. Pipe and cigar smoking and oral use of tobacco are also linked with oropharyngeal cancers. Smoking rates and dependency are greater in people with alcohol problems, and such use negates any health benefits of alcohol and increases many health risks. Smoking is increased in other substance use disorders, and in several psychiatric disorders - e.g., schizophrenia, depression - which might represent self-medication with nicotine.

Maintenance of high levels of smoking rates in young women has been linked to the appetite suppressant action of tobacco. Any cognitive benefits of nicotine in non-smokers are debatable. Smokeless tobacco products are less harmful than smoked products and induce less dependency - e.g., snus (snuff placed under the upper lip) has been linked with smoking reduction in Sweden. However, these products are not risk-free and may facilitate use of more harmful products, especially when started in childhood. Medicinal nicotine products are less harmful than smokeless tobacco and have low abuse potential, although dependency has been reported. Nicotine is widely considered the key addictive substance of some 4000 compounds in cigarette smoke. However, other substances in smoke, such as monoamine oxidase inhibitors, are also implicated. Recent research in rodents suggests that acetaldehyde, present in high amounts in tobacco smoke, increases the risk of dependency in adolescent animals.

Menthol, the only publicised additive in cigarettes, increases breath-holding and thus exposure to nicotine, tar, carbon monoxide, and other compounds. This action increases the likelihood of dependency and the risk of disease. Nicotine acts on nicotinic acetylcholine receptors, which in turn modulate other neurotransmitter systems, such as dopamine, noradrenaline, and glutamate. Changes in mesolimbic dopamine and possibly the release of endogenous opioids are thought to be responsible for the rewarding and pleasurable effects of smoking. Modulation of dopamine circuitry in the prefrontal cortex might be important in learning and in psychiatric diseases such as schizophrenia. A role for cannabinoid mechanisms in nicotine craving is suggested by the reported success of a cannabinoid antagonist for smoking cessation.

Use of alcohol

The brain actions of alcohol are numerous and complex. Binding to certain GABA-A receptors reduces anxiety and motor co-ordination, while effects at NMDA glutamate receptors lead to amnesia and sedation. The rewarding, stimulating and pleasurable effects of alcohol are thought to be due respectively to indirect stimulation of dopamine and noradrenaline circuits plus release of endogenous opioids. CRF is involved in stress-related alcohol consumption, while evidence suggests roles for other neuropeptides, such as urocortin, stresscopin and neuropeptide Y. In overdose, alcohol causes prolonged opening of the GABA-A receptor, which can paralyse respiratory drive neurons, causing respiratory arrest and death. Long-term alcohol use leads to an imbalance in excitatory and inhibitory systems. In withdrawal, the GABA system is left underfunctioning and the glutamate system is over-excited.

GABA hypofunction causes increased anxiety and restlessness, while glutamate over-excitation causes neurons to lose magnesium and gain excess calcium, which can prompt neuronal death, irritability and seizures. This process is exacerbated on subsequent withdrawals, and might if repeated contribute to alcohol-induced brain damage. Alcohol is considered a '*social lubricant*', so it is often (over)used by people with anxiety disorders, such as social phobia, for its anxiolytic effects. However, the overall effect of ongoing use can be an increase in anxiety due to withdrawal effects. Small amounts of alcohol are linked by some studies to reduced cardiovascular disease and overall mortality, especially from middle age onwards. Any excess alcohol is associated with increased cardiovascular disease, liver disease and neurological problems, including cognitive decline. Pregnant women who drink more than modest amounts of alcohol increase the risk of miscarriage, while alcohol use can lead to foetal effects, including the most severe form, foetal alcohol syndrome. Intoxication is associated with large social costs, including accidents, antisocial behaviour, sexual disinhibition or loss of functioning, and relationship breakdown. Alcohol is the drug most associated with aggressive behaviour and is a key cofactor in deaths from illicit drug use.

Use of opioids and opiates

Opiates from *Papaver somniferum* have been found to act on the brain's opioid system. Several synthetic and semi-synthetic agents (e.g., heroin, fentanyl, pethidine) plus naturally occurring substances (e.g., codeine, morphine) are available as medicines, mainly for their effective painkilling properties. Analgesia results from rapid, strong binding to μ-opioid brain receptors, which also leads to narcosis (sleepiness) and euphoria, while stimulation of κ receptors produces other adverse effects, such as hallucinations. With repeated use, peaks and troughs of opiate action lead to withdrawal, craving, and drug-seeking behaviour, depending on the rapidity of onset and offset of drug action. Withdrawal can be severe and can involve marked stress responses, which lead to further drug seeking.

Fast-acting opiates or routes of administration (e.g., intravenous, smoking) reliably lead to dependency when used recreationally, whereas orally ingested opium products and prescription opiate painkillers have less potential for dependency and abuse. When opiates are used in therapeutic contexts, dependency is uncommon. Exogenous opiates mimic the actions of brain endorphins, which act on opioid receptors, but due to a process known as agonist-directed signalling, regulatory mechanisms to terminate endorphin signalling are less effective with exogenous opiate binding. The opioid system interacts with many other neurotransmitters, including noradrenaline, acetylcholine, glutamate, GABA and dopamine. Opiates seem to increase the sensitivity of noradrenaline systems to stress responses, psychostimulants can potentiate the effects of opiates, while endogenous cannabinoid (CB1) receptors seem important to mediate the effects of opiates.

The neuropeptide nociceptin/orphanin FQ seems to be a functional inhibitor of opiate actions, while galanin reverses signs of morphine withdrawal. Potent full opioid agonists such as heroin and methadone can kill in overdose due to respiratory depression, and intravenous heroin has the greatest acute toxicity of common psychoactive drugs. Slower-onset agonists (e.g., codeine) and partial agonists (e.g., buprenorphine) have lower potential for overdose. Methadone is used as a long-acting heroin substitute in treatment but is implicated in similar numbers of deaths in overdose to heroin – coadministration of both substances is often a key factor in overdose. Other harms from illicit opiates are mainly related to the use of non-sterile injection syringes and needles, which carry risks including septicaemia, HIV and hepatitis C. Use in pregnancy can lead to withdrawal in the neonate. The use of prescription opiates for nonmedical purposes is thought to be possibly as common as illicit heroin use in the US. Rates are thought to be far lower in the UK but are probably rising, in part due to online availability. Risk of dependency is related to action and formulation – patches (e.g., fentanyl), longer-acting or partial agonists (e.g., buprenorphine) and slower-delivery formulations (e.g., of oxycodone) are less likely to produce euphoria and thus lead to dependency, but such drugs may be administered in non-indicated ways (e.g., by being crushed and then snorted or injected intravenously) to enhance their euphoric effects.

Use of cocaine and other psychostimulants

Cocaine, including crack, and other classes of psychostimulants such as amphetamines, directly interact with brain dopamine systems to produce psychomotor stimulation and rewarding properties. Amphetamines enhance dopamine release by reversing the action of the *dopamine transporter* (DAT), while cocaine and others block DAT and thus dopamine reuptake. Psychostimulants also interact with noradrenaline and serotonin, which might explain psychostimulant effects in the absence of DAT. Some responses to these drugs can increase with repeated exposure (sensitisation), perhaps due to aberrant glutamate signalling. Psychostimulant effects, which can be exploited for therapeutic effects, include increased wakefulness, attention, increased energy and reduced appetite, while various classes have other properties, for example, cocaine is a local anaesthetic.

Acute psychostimulant toxicity includes hyperthermia, cardiovascular problems, and seizures, especially in vulnerable individuals. Longer-term, high-dose amphetamines in particular are linked with possible neurotoxicity, cognition disorders and psychosis. Long-term toxicity also includes cardiovascular and cerebrovascular disease, including stroke. Use in pregnancy can lead to withdrawal in the neonate and possibly birth defects. Psychostimulants are found more rewarding by people with lower brain levels of dopamine-D2 receptors. Long-term use reduces DAT levels and prefrontal cortex activity, which may enhance risk taking and drug seeking and cause cognitive changes such as reduced decision-making ability. These changes might also increase sexual and other risk taking.

Psychostimulant withdrawal leads to disruptions in mood, motivation and potentially hostility, although the relation with aggression is not straight-forward. These effects may be mediated in part by increased neuronal excitability and neurotoxicity. The potential for dependency is increased with intravenous use or with smokable forms (e.g., crack cocaine and methamphetamine ('ice')). Psychostimulants with slower onset of actions such as oral methylphenidate and atomoxetine are less rewarding, and are used for beneficial cognitive and behavioural effects in disorders such as ADHD. Use of such substances in ADHD is likely to reduce rather than increase subsequent illicit drug use. However, use of such agents in rodents without ADHD-type deficits has variable and sometimes deleterious effects on brain reward systems, emphasising the need to use such treatments only in people who fit ADHD diagnostic criteria. The abuse potential of legal psychostimulants depends on the potency and rapidity of drug action, with dopamine releasers more of a concern than reuptake inhibitors. Increasing diversion of medical supplies with use via non-indicated routes is another concern (e.g., injection or intranasal use of methylphenidate). Newer formulations (e.g., gel-filled methylphenidate capsules) are designed to reduce such problems.

Values of ecstasy (MDMA) and other 'designer drugs'

The many so-called designer drugs related to ecstasy (often grouped together as amphetamine-type substances) are taken for their stimulant and mild hallucinogenic properties. Their effects range from those with mostly psychostimulant properties to those with mostly hallucinogenic properties, presumably dependent on their individual underlying mechanisms. MDMA is most extensively researched, and acts predominantly to block serotonin uptake and enhance release, but also via the dopamine, noradrenaline and possibly other systems. Use of MDMA and some analogues is widespread though few harmful effects are reported, with notable exceptions such as idiosyncratic toxicity, hyperthermia and effects due to excess serotonin, called serotonin syndrome, liver damage and post-intoxication mood instability. Rare deaths have occurred with MDMA, MDEA, MDA, PMA, and 2CT7, although the extensive usage of these drugs with few mishaps plus other data indicate they are less acutely toxic than classic psycho-stimulants, opiates or alcohol. Notably, the majority of deaths related to ecstasy have been associated with other psychoactive drug use, particularly opiates.

Physical harms of ecstasy and other designer drugs are likely to be related to the degree of their psychostimulant effects together with pre-existing vulnerability. Other factors that might increase risk of harms include overheating, extreme exercise, dehydration or, rarely, water intoxication, and concomitant use of other drugs, especially opiates or alcohol. MDMA has been the cause of much debate over its potential for long-term neurotoxicity of brain serotonin neurons, as seen with high doses in animal models. Such studies have implicated hyperthermia, loss of temperature regulation and dosing interval in determining neurotoxicity. Imaging studies in human users have produced varied results, from minor deficits that resolve within weeks to substantial long-term effects, while the psychological symptoms that occur in current and abstinent users often predate MDMA use. Nevertheless, it seems probable that a minority of heavy, frequent users with vulnerability (e.g., women) could experience long-term neurotoxicity, though a substantial public- health risk seems unlikely.

Use in pregnancy might lead to emotional changes in children, if rodent studies are confirmed in humans - this effect is also seen with SSRIs. Other potential harms include immunosuppression and possible increased sexual risk behaviours. MDMA was used in the 1960s as a tool to aid psychotherapy - and, recently, two trials on its use have been approved in the USA - to investigate use in conjunction with psychotherapy for post-traumatic stress disorder, and for anxiety in terminal cancer patients.

Values of Dissociative anaesthetics

Certain substances combine anaesthetic with hallucinogenic properties; eg noncompetitive NMDA antagonists ketamine and PCP, which produce states similar to schizophrenia. The use of ketamine as a club drug may be increasing, especially in combination with other substances such as cocaine.

Dissociative anaesthetics have potential for dependency, while other harms are related to reduced levels of consciousness during acute intoxication such as their use to facilitate sexual assault. PCP intoxication also has a possible relationship to violence. Both agents may be harmful to the developing foetal brain. Ketamine has been used therapeutically to treat alcohol dependence, especially in Russia, and is now used in healthy volunteers as an experimental model of schizophrenia.

Values of gamma-hydroxybutyrate (GHB) and precursors

GHB is an endogenous molecule synthesised from GABA, while exogenous GHB can be produced from solvent precursors 1-butanediol and butyrolactone. Exogenous GHB appears to affect inhibitory brain neurotransmitter systems over a short timescale, probably via GABA production, GABA-B receptor binding, and the putative GHB system. Acute use leads in a dose-dependent fashion to relaxation, euphoria and increased disinhibition, poor co-ordination, amnesia, sleep and then coma. Acute toxicity in overdose is high, especially with concomitant alcohol use. The drug has been utilised, especially with alcohol, to facilitate sexual assault. GHB has dependency potential, leading to a withdrawal syndrome similar to and sometimes more severe than that of alcohol. Sold as sodium oxybate, the agent has been used for anaesthesia and now is used to promote night-time sleep for patients with narcolepsy. GHB has been used as a treatment for alcohol and opiate dependence and might be neuroprotective.

Inhalers of psychoactive effects

Various substances are inhaled for psychoactive effects, including solvents (e.g., lighter fuel, toluene and petrol), gases (e.g., nitrous oxide) and nitrites (e.g., amyl nitrite). Intermittent use may be widespread but dependency occurs mainly in certain subpopulations, including 'antisocial' young men sniffing glue, nitrite use in gay culture, and petrol abuse among marginalised aboriginal populations. These subpopulations may be overlooked yet at risk of psychiatric disorders, abuse of other drugs, and possibly crime. One concern is that inhalant use may increase in situations where other drugs are unavailable or unaffordable. All substances carry risks of unpredictable acute and more predictable long-term toxicity. For example, solvent use carries risks such as accidents, including fires, and liver and neurological damage, while any prolonged exposure might be harmful to the developing foetus. Nitrous oxide is available in 'gas and air' for obstetric use, and has been used in alcohol withdrawal.

Synthetic hallucinogens

Numerous natural products such as peyote and '*magic mushrooms*', and synthetic substances (e.g., LSD) are used traditionally (including for religious purposes) and recreationally to alter psychological functioning in ways described as psychedelic – '*mind- manifesting*' – and hallucinogenic – '*mind-wandering*'. Actions may include the enhancement as well as distortion of perceptions and other cognitive and emotional processes, which vary greatly according to set and setting. The physical toxicity of many of these compounds is very low, while tolerance is often rapid and profound, though short-lived. Anxiety, panic attacks and other psychological distress may occur during intoxication in vulnerable people, though effects rarely persist. Unknown ingestion is almost always a distressing experience.

Other harms can result from loss of self-protective abilities during severe intoxication, and toxicity may occur, most often due to mistaken identification (eg *Amanita* species) or the use of plants from the Deadly Nightshade family such as species of *Datura*. Rarely, use has been linked with intermittent or persistent perceptual changes in the absence of the drug, termed '*flashbacks*,' and hallucinogen persisting perceptual disorder, although their existence is controversial since such state-specific recall can occur in non-drug-related circumstances. Many hallucinogens act as serotonin-

2A receptor partial agonists, while specific cellular molecules and other neurotransmitter systems seem likely to be involved in their effects. Known and novel hallucinogens are a potential source for development of new psychoactive agents for therapeutic, research and recreational purposes. A trial of *psilocybin* for anxiety in terminal cancer patients was recently approved by the FDA while a trial of its use for obsessive-compulsive disorder is ongoing.

Cannabis (products from plants)

Products from plants of the *Cannabis* genus are widely ingested, usually as smoke and sometimes as food products. The status of such products in the Misuse of Drugs Act 1971 was recently down-scheduled to Class C, reflecting the extensive evidence that suggests lower potential for harms with moderate usage than most other recreational psychoactive drugs. Cannabis products have low acute toxicity, although panic and anxiety or acute psychosis can sometimes occur. Cannabis use early in life is linked with subsequent psychotic symptoms and psychosis, especially in heavy users and people with predisposed vulnerability. People with certain psychological symptoms might also self-medicate with cannabis. Young, heavy users of cannabis may experience reduced motivation and achievement, plus mood disorders although the causality in this relation is uncertain.

Cannabis use might increase the likelihood of smoking tobacco in youth but does not seem to be a 'gateway' to other drug use. In humans, a minority of individuals who use cannabis heavily will develop other substance use disorders, probably reflecting a common predisposition plus the fact that cannabis is often sold alongside other illicit drugs. Dependency may occur in around one in ten users, or 1.5 per cent or more of adults and withdrawal is characterised by labile mood and hostility. Many dependent cannabis users do not desire to completely stop use of the drug. The main active principle in cannabis, THC, acts on endogenous cannabinoid receptors – CB1 receptors in the brain, and CB2 receptors in immune cells and other peripheral sites. Endogenous cannabinoids occur naturally in the brain and inhibit neurotransmitter release directly or indirectly – e.g., via hippocampal GABA neurons that also contain the neuropeptide cholecystokinin.

Cannabinoid receptor stimulation probably has reciprocal effects on opioid and monoamine systems. In addition, a study in adolescent rats suggests that regular adolescent cannabinoid use might encourage neuronal adaptation that leads to cross-tolerance to morphine, cocaine and amphetamine. Together such studies could indicate that cannabis use at a vulnerable age might alter pleasure/reward signalling. The therapeutic benefits of cannabis have been documented regularly, and many people with multiple sclerosis (MS) and epilepsy, in particular, use cannabis. A herbal cannabis extract in a sublingual spray is being considered for approval as a medicine for the treatment of MS and chronic pain in the UK.

Current research focuses on the function of individual natural and synthetic cannabinoids, which could include analgesia, anti-inflammatory and anti-allergy effects, sedation, improvement of mood, neuro-protection, and anti-cancer effects. THC is available as the approved drug dronabinol for the treatment of nausea and weight loss, while other natural cannabinoids such as cannabidiol may prove useful for neuroprotection, stroke and seizures. Cannabinoids might prove helpful for selective forgetting. Synthetic cannabinoids aim to isolate therapeutic properties without problematic effects, e.g., ajulemic acid, a novel painkiller.

Effects of benzodiazepine agonists and other sedatives

Benzodiazepines and other drugs that act on GABA-A receptors tend to have both anxiolytic and sedative properties. Prescription of these agents fell after their potential for dependency was revealed and stricter controls introduced. However, long-term prescription use, especially of hypnotics, and problematic use, including self-medication via diversion or online pharmacies, seem likely to be increasing, despite the potential for rapid tolerance and dependency with these agents. Acute toxicity, which can lead to death in overdose, may be linked with intravenous administration and coadministration with other drugs, especially sedatives such as opiates and alcohol.

With long-term use, resurgence of original cognitive and other symptoms can occur despite continued ingestion of the drug, which can prompt dosage increases and other sedative use. Withdrawal is often unpleasant, involving insomnia, exaggerated stress responses, cognitive difficulties, and possibly aggression. The non-benzodiazepine hypnotics (i.e. zaleplon, zolpidem, zopiclone) also act at GABA-A receptors and can lead to tolerance and dependency, though less often than benzodiazepines. Barbiturates are rarely prescribed but are highly toxic in overdose.

Other psychoactive drugs

Various other agents that can be abused have psychoactive properties. One key class is hormone-based agents such as anabolic steroids, which are used widely (e.g., by body-builders) outside of medical contexts. Although not traditionally considered psychoactive, such agents may have problematic psychoactive effects in addition to adverse physical effects. Anabolic steroids may increase aggression and lead to other symptoms, such as fatigue, on withdrawal. A growing diversity of psychoactive plant products are increasingly available, partly because of globalisation. Such 'herbal highs' show a similar variety of properties to other psychoactive substances, e.g., stimulants such as khat (*Catha edulis*) and Ephedra species, psychedelics such as morning glory (Ipomoea species) and *Salvia divinorum*, sedatives such as kratom (*Mitragyna speciosa*) and Valeriana species, and others including some with mood-altering or aphrodisiac properties. Such products may interact with other drugs and can differ widely in safety aspects. For example, plant psychostimulants can induce dependency but range in toxicity and abuse potential from low (e.g., caffeine) to potentially high in vulnerable individuals (e.g., khat). Psychoactive plants are a likely source of novel compounds.

CLINICAL TREATMENTS

Drug treatments for substance use disorders are relatively recent, partly due to concerns over giving drugs to people with drug problems. In any case, few treatments have been available until recently. However, increased understanding of problem drug mechanisms has led to an expansion in potential pharmacological treatments. Current treatments work mainly as agonists, substituting a less hazardous compound that acts similarly to the problem drug, or antagonists, which block the effects of a drug, thus leading to reduced use by a process called extinction. By contrast, disulfiram disrupts a step in the breakdown of the problem drug (alcohol), leading to symptoms of intolerance. Bupropion and buprenorphine act at key receptors, in different ways, to stabilise the tone of the neurotransmitter system and to block the action of the problem drug if administered. Harms of drug treatment include side-effects of agents, plus the diversion and misuse of agents, which occur mainly with benzodiazepines, methadone and buprenorphine. Agonist activity can lead to abuse of the treatment agent whereas antagonist action can precipitate withdrawal from the problem drug. The BAP has published guidelines on the treatment of common substance use disorders. The mainstay of much treatment is psychosocial and behavioural intervention.

Pharmacological treatment is likely to be best delivered in the context of such support since it may improve drug treatment compliance and outcomes. Conversely, drug treatments might facilitate the effectiveness of psychosocial treatments – for example, to provide a window of opportunity from craving or withdrawal to allow psychosocial interventions and behaviour change to take effect. This summary does not reflect the complete guidelines, which need to be interpreted within the context of new evidence, specific services and individual patients. For example, treatment of alcohol withdrawal needs adjustment according to local formularies and estimation of individual liver capacity. For treatment of opiate dependency, patients prefer agonist and partial agonist effects rather than antagonists although these are often beneficial in restricted groups whose members have access to the drug, such as doctors or pharmacists.

Other treatments are used for selected patients, particularly for treatment of comorbid psychiatric disorders, e.g., SSRIs. However, preliminary evidence points to efficacy of injectable heroin (diamorphine) for opiate dependency; disulfiram, amphetamines and bromocriptine for psychostimulant dependency; buprenorphine for polydrug dependency (e.g., opiates and cocaine); antiepileptic agents for various drugs, including sedatives, opiates, cocaine and GHB; SSRIs to block acute effects of MDMA, and possibly treat subsequent mood instability; and serotonin-3 antagonists (e.g., ondansetron) for alcohol dependency. The aims of treatment services usually include:

1. Management of withdrawal symptoms;
2. Reduction of harms associated with drug use promotion and maintenance of absence;
3. Management of complications of drug use.

Since dependency is a chronic, relapsing condition, interventions too usually need to be chronic or repeated, if necessary for use of other agents. Although such treatments can be characterised by poor compliance and high rates of relapse, substantial beneficial outcomes both on drug-taking and related problem behaviours such as crime are now documented. Rates of improvement, remission and relapse are comparable to those seen with other psychiatric disorders. Moreover, poor treatment compliance and relapse are seen in many chronic illnesses (e.g., hypertension, diabetes), and service frameworks for substance use disorders could benefit from being based on those used for other chronic disorders. This recognition may also reduce the stigma associated with dependency and its treatment.

EXPANSION OF TECHNOLOGIES

If current trends continue until 2025, we can expect the expansion of neurosciences and related technologies to be coupled with increasing public interest in altering brain functioning for the treatment of impairments, for pleasure and recreation, and for enhancement. There will probably be more psychoactive substances, while the boundaries between categories of usage – pleasure and recreation, therapeutic use, and enhancement – are likely to become more blurred than they are already. Developments in knowledge will probably be rapid and profound, while its application will be dependent on the context in which this knowledge is applied.

Clinical Characteristics

The following trends are under way and seem likely to continue:

1. Knowledge of neurotransmitter and receptor systems and their correlates in terms of cognition, mood, behaviour etc., will expand
2. Increasing sophistication in understanding actions of drugs on the brain will expand development of the new speciality of psychopharmacology
3. Identification of genes important in drug actions, drug metabolism, substance use disorders and therapeutic responses will offer hope of tailoring therapies to individuals (pharmacogenomics); gaining further insights into biological mechanisms of dependency; and developing new therapies
4. Improved imaging of function, including drug actions, in living human brains will permit research on the brain circuitry and changes in receptor and transporter densities linked with dependency and other symptoms. Such technology could eventually be used to tailor treatments to their associated neurochemical deficits
5. Wider availability of drug testing and psychological screening are already mooted and could permit identification of those engaging in or at risk of problematic use
6. Services for substance use disorders are anticipating improved use of existing interventions, including medications, and the development of new treatments. The hope is that this broadening array of interventions will be available within higher-quality treatment services that can combine the most effective medications with psychosocial support and behavioural therapy regimes.

Management

A broader discussion of social and economic drivers of changing drug use and its management is covered elsewhere. Important trends are related directly to psychoactive substances:

1. Overall, prevalence is rising worldwide, with known abusive drugs remaining the key problem. If current UK trends and policies continue for 20 years, smoking prevalence and deaths may fall somewhat, while alcohol consumption could reach among the highest in Europe, with effects on health and other consequences. Blood- borne infections from intravenous drug use also will continue to rise. Prevalence of other drug use is likely to continue to vary, both up and down in different time periods, depending on the substance, though with current trends, deaths due to other drugs could fall further. However, such trends can change rapidly.
2. Increased online availability of prescription drugs and other substances plus information on the use and manufacture of drugs is making control of supply increasingly difficult. Rising off-label use of prescription drugs and illicit diversion seem likely to continue.
3. Concomitant use of several psychoactive substances is becoming increasingly common and can involve all types of substances. Such polypharmacy may make successful treatment more difficult

and increases the likelihood of adverse interactions. In the future, a widening range of novel psychoactive agents plus increasing polypharmacy will present new challenges to health services.

4. The availability of a range of cognition-enhancing drugs and other enhancement techniques, such as the wider use of performance enhancers in sport, could lead to increased acceptance of enhancement, which could alter attitudes to recreational psychoactive substance use.
5. Psychoactive substance development and production is a growth area that is likely to continue to expand. Markets for herbs and other plant-based products are expanding, while pharmaceutical companies are seeking markets for their products beyond standard therapeutic markets (e.g., with 'lifestyle agents' for baldness, obesity and sexual dysfunction). These industries are likely to want to target the potential market of healthy consumers with recreational and enhancement techniques.

Synthesis of Related Compounds

The number of psychoactive substances in human use has expanded at growing rates throughout history. Thus by 2025, a large number of new psychoactive substances are likely to be discovered. The probable sources of these substances are similar to those of today but there may be unanticipated sources. Approaches to the development of new psychoactive substances will include known and developing technologies, such as the isolation of new natural products, including neurotransmitters; synthesis of analogues or combinations of known compounds; agents selected or designed for specific functions (e.g., receptor binding by rational drug design); and new strategies, such as protein- and peptide-based agents, cellular products, and proteomic and genomic interventions (e.g., gene therapy, anti-sense approaches to inactivate cellular material containing instructions to make proteins).

Developers of psychoactive drugs will continue to look to developments in other fields to inform drug development, a process encouraged by information technology. Future drug development is likely to follow similar models to today. Novel psychoactive plants and other substances will be investigated, including those used by indigenous peoples, new psychoactive compounds will be isolated or synthesised and new analogues derived, which may be combined with known substances or components. Recreational drug manufacture is likely to occur in illicit laboratories, but therapeutic agents will continue to be diverted for recreational purposes from pharmaceutical and other companies, which could begin to manufacture pleasure agents. Illicit manufacture might increasingly occur in small laboratories, with growing diversity of products and quality, and potential for environmental damage. Pharmaceutical companies will expand their portfolios with new targets, new agents for old targets, and whole new areas such as cognition-enhancing drugs. Academic institutions will continue to identify new targets and strategies.

Major developments in gene-related and stem-cell technologies seem likely, which ultimately may be brought to market by 'spin-off' technology companies. New forms of known agents such as anti-depressants, painkillers and pleasure drugs, and novel classes, are likely to be described and used, perhaps according to their effects on neurotransmitter systems. One prominent underground chemist has predicted that around 2000 psychedelic compounds alone will be available by 2050. New drugs will include effective analogues that can be anticipated to some extent from current pharmacology and new psychoactive agents. New pharmaceutical leads already include appetite modulators, for weight loss and enhancement of sexual functioning, and skin-tanning agents; one drug class in development – the melanocortins – seem to promote weight loss, enhance libido, and tan skin simultaneously. The publication pattern of drug-development knowledge means that it is hard to predict what agents will be available by 2025. We can anticipate developments in where data are published, plus some developments from current science on neuropeptides, genomic approaches, stem cells, neurogenesis and nanotechnology. We cannot predict unexpected successes or failures that are likely to occur in development, but we can predict that only a minority of pharmaceutical drug leads will reach the market, as is the norm for drug development. The current aim and thus the future likelihood is that new therapeutic, lifestyle and pleasure drugs will be better designed to minimise toxicity, dependency and abuse potential.

PHARMACOLOGICAL STUDIES

Trends in psychopharmacological research suggest that the next 20 years are likely to bring a large expansion in the types of strategies to prevent and treat substance use disorders, some specific to a problem drug and some generic. Developments in related fields will mean that such therapies are likely to be delivered in the context of an increased understanding of appropriate psychosocial and behavioural interventions and the targeting of treatments to individual symptoms. New classes of drug treatments will be produced, in terms of action (e.g., partial inverse agonists, stem-cell replacement), targets (e.g., neuropeptides, genomic targets) and new formulations (e.g., patches, nasal sprays, nanoparticles). Increased understanding of the 'downstream' events that occur after receptor binding may allow different molecular pathways inside neurons to be exploited. High-throughput technologies, such as genomics and proteomics, promise to revolutionise the pace and breadth of scientific investigation into the causes, consequences and treatment of substance use disorders.

Since individuals vary in their responses to abused drugs and treatments, due in part to genetic variations in key enzymes, pharmacogenomics is likely to prove helpful to tailor certain drug treatments to individuals by 2025. Other techniques developed to manipulate matter at molecular scale and below (nanotechnology) could bring unanticipated developments. Therapeutic strategies will probably include:

1. Agents targeted at specific drug use, including:
   (a) *Alcohol* – e.g., GABA-A and GABA-B enhancing agents,
   (b) *Tobacco* – e.g., nicotinic antagonists,
   (c) *Opiates* – e.g., new formulations of opioid-modulating drugs.
   (d) *Cocaine and other psychostimulants* – e.g., dopamine-D3 receptor antagonists and partial agonists, dopamine-reuptake inhibitors,
   (e) *cannabis*– e.g., CB1 receptor blockers,
   (f) *other approaches* including active vaccination against nicotine, cocaine and possibly other drugs.

2. Interventions with generic actions to modify excess appetitive behaviour, including:
   (a) Learning and anti-learning agents that undo conditioning to external cues (e.g., association of alcohol with pubs, heroin with needles),
   (b) Drugs acting on inhibitory systems, such as GABA and cannabinoids, that might regulate the switch from normal to dependent states,
   (c) Anti-stress/anxiety drugs, which might reduce important factors in relapse,
   (d) Agents that target and modulate brain reward and pleasure circuits,
   (e) Gene-silencing or turn-on agents (e.g., for endorphins, dynorphins and other peptides involved in drug dependency),
   (f) Brain stimulation and psychological approaches.

Treatments

This overview of treatments includes agents in development for specific substance use indications. Many of these may ultimately be found to act on generic circuitry involved in substance use disorders, and thus may cross over for use in other abused drugs.

Use of tobacco

In addition to known medications, three novel strategies are in clinical studies for tobacco smoking cessation – vaccines, rimonabant and varenicline. Nicotine vaccines produce antibodies that bind nicotine in the bloodstream and thus reduce entry to the brain. This 'blockade' is thought to reduce the subjective effects of smoking. The vaccine TA-NIC now is in a phase II trial of 60 smokers, with promising early findings. The nicotine partial agonist varenicline currently is in phase III trials, with about 50 per cent cessation rates in earlier trials. The cannabinoid-1 receptor antagonist rimonabant (SR141716) has shown promise as an anti-smoking treatment in late clinical trials. This strategy is also one of several generic approaches that have been proposed for smoking. Others include dopamine-D3 receptor partial agonists, noradrenaline reuptake inhibitors, GABA-B receptor agonists, metabotropic glutamate-5 receptor (mGluR5) antagonists, and CRF1 receptor antagonists.

Alcohol consumption

Improvements in abstinence from drinking with acamprosate and naltrexone have prompted further investigation of their use, alone and in combination. Since naltrexone seems to block the pleasurable effects of alcohol, its use may be warranted in problem drinkers who continue to use alcohol intermittently. Long-acting formulations of naltrexone that can be administered monthly are in development to improve compliance compared with oral naltrexone. A naltrexone depot injection and sustained-release oral naltrexone are both in phase III trials. Agents that act at NMDA and other glutamate receptors seem to have specific beneficial actions in alcohol dependence to ameliorate the effects of withdrawal, in addition to generic effects on learning and sensitisation.

Acamprosate may act via binding to NMDA and mGluR5 receptors to reduce withdrawal-induced excitation of glutamate neurons. Treatment with acamprosate alone and in combination with naltrexone is effective in maintaining abstinence and preventing relapse, perhaps through effects on craving mediated by excess neuronal excitation. In addition, acamprosate given before withdrawal may prevent glutamate excitation and consequent neurotoxicity. Other NMDA antagonists that selectively target the NR2B receptor subunit also inhibit alcohol-withdrawal neurotoxicity in animal studies. These subunit-selective NMDA antagonists (e.g., CP101606, RG1103) may prove useful to promote withdrawal and abstinence. Potential new GABA-based strategies for alcohol-use disorders have been proposed using selective GABA-A agonists and other agents.

Use of the GABA antagonist flumazenil might be beneficial in certain circumstances. GABA-A binding to the alpha5 receptor subtype is thought to mediate the reinforcing properties of alcohol, so GABA (benzodiazepine) alpha-5-subunit inverse agonists in development might reduce alcohol effects and related pleasure. GABA-B binding induces mesolimbic dopamine release, so potential exists for treatment with agents such as baclofen (GABA-B agonist), tiagabine (GABA uptake blocker) and topiramate, which enhances GABA neurotransmission without affecting GABA receptors and has effects on glutamate signalling. Any efficacy of these compounds might be due to generic inhibitory effects on reward-related circuitry.

Other novel targets implicated in alcohol dependency include the inhibition of neuropeptide Y, protein kinase-C epsilon and adenosine. Strategies that target acute alcohol intoxication are of interest given that the effects of intoxication vary from potentially hazardous to life-threatening. Metadoxine is a coenzyme precursor that alleviates symptoms of acute alcohol intoxication via complex mechanisms that include accelerated breakdown and preservation of brain and liver cellular energy stores. The implications of availability beyond medical contexts encapsulate some wider issues surrounding socially available pharmacology. Off-label use could increase harms by allowing further ingestion of alcohol, or facilitate harm reduction, for example, if it permits safe driving.

Opioid acting drugs

Various opioid-acting drugs are in development to improve withdrawal symptoms and abstinence in heroin dependence. A sublingual buprenorphine tablet with added naloxone to prevent intravenous abuse has been recently approved by the FDA. Injectable or implantable formulations of known treatments seem promising to improve compliance. Monthly depot naltrexone is in phase III trials, while a depot formulation that releases naltrexone for six months is in development. A buprenorphine depot is also in development. Other agents in clinical trials are the oral opiate dihydrocodeine, alpha-noradrenergic agonist lofexidine in phase III trials, and nicotinic antagonist mecamylamine in phase II. An accelerated lofexidine regimen and a naloxone/lofexidine combination both show advantages over the standard lofexidine regimen. The use of nicotinic antagonists is prompted by reports of cross-tolerance between morphine and nicotine and other animal studies. Other potential agents include the NMDA and nicotinic antagonist dextromethorphan and agents developed to act via the neuropeptides galanin and nociceptin.

Cocaine dependency

Cocaine dependency has the highest number of potential treatments in development. Many clinical trials are being conducted by the US National Institute on Drug Abuse Clinical Research Efficacy Screening Trial (NIDA CREST) programme, which rapidly screens existing medications. CREST has five medications in phase II trials that directly or indirectly affect dopamine levels:

cabergoline (dopamine agonist), reserpine (depletes presynaptic monoamine stores), sertraline (SSRI), tiagabine (GABA uptake blocker), and NS2359 (inhibitor of dopamine, noradrenaline, and serotonin transporters). Ondansetron, a serotonin-3 receptor antagonist, is in phase II for combined cocaine and metham-phetamine dependency.

Disulfiram and selegiline, which inhibit dopamine breakdown in different ways, were effective and well tolerated in phase II trials, but selegiline was found ineffective in a recent phase III trial. Agents in NIDA phase I trials include modafinil, cyclazocine (mixed opioid agonist and antagonist), metyrapone (corticosteroid synthesis inhibitor), and vanoxerine (GBR12909, a long-acting dopamine reuptake inhibitor). The rationale for the investigation of these and other agents is their direct or indirect effects on dopamine neurotransmission. Other direct-acting agents include the dopamine-1 receptor agonist DAS431 in phase II trials. Methylphenidate (a dopamine reuptake inhibitor), amphetamine (a dopamine releaser), adrogolide (a selective D1 receptor agonist), and BP897 (a D3 partial agonist) have reached clinical studies.

Further agents that show potential are SB277011 (a D3 antagonist) and various other known and novel compounds that act in different ways on dopamine receptors, transporters or breakdown inhibitors. Such strategies are likely to be refined when the role of different dopamine receptors is more fully defined. Another specific strategy is active immunisation with a vaccine that 'mops up' cocaine before it can enter the brain. Phase II trials found good short-term abstinence rates in abstinent and using cocaine users. Another vaccine strategy in animal studies involves the intranasal delivery of a bacterial virus designed to produce cocaine antibodies within the brain. Other agents showing promise include gabapentin and drugs that target serotonin, noradrenaline, opioid receptors, glutamate, endocannabinoids, neuropeptides and neurohormones. Since alterations in serotonin and dopamine neurotransmission seem to be required for the reinforcing effect of cocaine, agents that affect both these neurotransmitters might show specific benefits for cocaine use.

Amphetamines

Fewer agents are in development for other psychostimulants such as amphetamines. However, in theory, some successful agents for cocaine dependency, especially dopamine modulators, will also show benefits for other psychostimulants. For example, vanoxerine has shown promise in animal studies, though it may be rewarding itself. Currently, NIDA is testing bupropion and selegiline in phase I trials, and ondansetron for cocaine and methamphetamine use in phase II.

Treatment of cannabis

The discovery of cannabinoid receptors and the development of antagonists such as rimonabant offer promise for the treatment of cannabis dependency since cannabinoid-1 receptor antagonists block the effects of cannabis in humans and animals and precipitate withdrawal in dependent animals. Long-acting agonists and partial agonists might provide additional benefit.

Aberrant Learning

In theory, agents that promote learning could either facilitate the extinction of the aberrant learning thought to underlie dependency, or facilitate new learning, eg to associate drug cues with abstinence. D-cycloserine, a partial agonist at glutamate NMDA receptors, has been shown to facilitate the extinction of learned fear in animal studies and to prevent reinstatement of extinguished behaviour. D-cycloserine consolidates human neuroplasticity and has shown potential in combination with naloxone in a small experimental study of opiate-dependent people. Importantly, NMDA receptor binding seems to mediate sensitisation, a possible initial step in dependence, although glutamate is also implicated in tolerance, craving and withdrawal.

The role of glutamate may also partly explain the success of anticonvulsants in substance use disorders since agents that act as agonists and antagonists at NMDA receptors possess anti-convulsant activity. Agents such as memantine, and drugs that modulate AMPA and metabotropic glutamate receptor functioning, seem likely to show benefits over older drugs and some are under study for dependency. Agents shown to play similar roles include the alpha2-adrenoreceptor antagonist yohimbine and a novel serotonin-4 receptor agonist RS67333. Other cognition enhancers might have beneficial properties in this respect, including cholinesterase inhibitors and calcium-channel blockers. Effects on calcium channels might also underlie the potential antidependency mechanism of gabapentin. The timing of use and other effects of drugs that affect learning might be crucial, since yohimbine has also been found to reinstate drug-seeking behaviour while D-cycloserine can facilitate tolerance at least to alcohol.

GABA Neurotransmitter

GABA neurotransmission in particular is suggested to regulate a switch from normal to drug-dependent states. Possible agents that promote GABA include approved drugs, especially vigabatrin, a GABA-transaminase inhibitor, that has already shown promise in reducing animal consumption of cocaine, nicotine, alcohol, and heroin. In animal studies, the GABA-B agonist baclofen has effects on usage of alcohol, nicotine, morphine, cocaine, amphetamine, methamphetamine, and GHB. Clinical trials of baclofen have been promising for opiates and cocaine. Other agents include tiagabine, topiramate, etifoxine, ganoxolone, and retigabine. Analogues of vigabatrin are in development, while new analogues of GHB might target both GABA and GHB inhibitory systems. Endocannabinoids have indirect effects on reinforcing dopamine release, though the link is less well researched than for GABA. However, CB1 receptors have been found to be important in alcohol consumption, including stress-related drinking, and in opiate and amphetamine effects. Moreover, the cannabinoid antagonist rimonabant has shown benefits in trials as an anti-smoking agent and as a weight-loss agent.

Stress and anxiety

Stress is a major factor in drug seeking and relapse, while withdrawal is also considered a stressor. Moreover, anxiety is one of the pre-existing conditions thought to predispose individuals to use both sedatives and stimulants. CRF and related neuropeptides (e.g., neuropeptides Y and S) are known to mediate effects of stress in the brain. So CRF antagonists, such as antalarmin, are being developed as potential dependency treatments. The role of CRF and the potential for CRF-antagonists have been demonstrated in animal models of cocaine (CP154526), cannabis withdrawal, alcohol [D-Phe-CRF(12-41)] and opiates (antalarmin). Recent findings suggest that alcohol directly induces release of CRF. Thus, CRF antagonists may show particular promise in the treatment of alcohol withdrawal, as may other anxiolytics like GABA agonists and serotonin-1A partial agonists such as buspirone. Finally, CRF2-receptor agonists such as urocortin may produce similar effects to CRF-1 receptor antagonists.

Dependency treatment

Many agents now used or in development as dependency treatments act directly or indirectly on the dopamine, opioid, serotonin and/or endocannabinoid systems. Since these systems seem to be involved in most appetitive behaviours, various agents that act on reward/pleasure circuits might prove beneficial in disorders related to excessive appetitive behaviours. These include buprenorphine, naltrexone and other opioid-acting agents such as the antagonist clocinnamox and methoclocinnamox, a partial agonist with delayed, long- lasting antagonist effects, dopamine modulators, serotonergic agents like ondansetron, and cannabinoid antagonists. This generic approach has already borne fruit, although combination treatments might be required for full efficacy. Thus, buprenorphine was not found very effective for cocaine dependency, but might be efficacious in combination with disulfiram. The molecular effects of drugs that activate reward/pleasure circuits is beginning to be elucidated. Such effects range from short-term activation of intracellular signalling pathways after receptor binding, through longer-lasting changes in molecules that underlie learning and memory (e.g., CREB) in brain reward/pleasure regions, to long-term effects mediated by neurotrophic factors (e.g., BDNF) that alter nerve complexity and growth.

Intracellular and other molecular targets represent challenging but important future targets for therapy of substance use disorders. Neuropeptides are another category of agents that might influence reward/pleasure circuits. Little is known about most neuropeptides, but tools for their manipulation are becoming increasingly available. For example, agents that target cholecystokinin might be beneficial for dependency, while CRF antagonists might help alleviate withdrawal. Manipulation of neuropeptide systems is likely to broaden possibilities for altering brain functioning and behaviour, including for substance use disorders. It seems possible that many different types of agents for therapy, pleasure and enhancement will be produced from increasing knowledge of neuropeptide functioning.

Neutralizing effect

Several therapeutic strategies have been proposed that neutralise a drug in the bloodstream before it reaches the brain. Such strategies might be useful for prompting or maintaining abstinence and also to treat overdose. Active vaccination is under study and may have potential against many substances, except alcohol and endogenous molecules like GHB. Other strategies include infusion of drug-binding antibodies (passive immunotherapy), use of enzyme accelerators to speed up drug metabolism, and a combination of the two (catalytic antibody therapy).

Traditional substances

Various medicines based on plant products are traditionally reported to be useful anti-addiction measures, including *Hypericum perforatum, Pueraria lobata, Lobelia inflata, Sceletium tortuosum, Passiflora incarnata* and ibogaine from *Tabernanthe iboga*. All might have actions on several neurotransmitter systems. While virtually no rigorous clinical data exist, ethical investigation of plants with traditional use might inform research on anti-addictive compounds and mechanisms.

CONCLUDING REMARKS

Advances in drug treatments are likely to bring the capability to genuinely tackle problematic drug use and reduce the demand side of the 'drugs problem'. However, very few strategies are in development specifically for substance use disorders. Strategies are often developed for other conditions and companies then may seek to license them as treatments for dependency. Thus far, drug dependency has not been a priority area for pharmaceutical companies, since aside from problematic alcohol use and smoking, markets are small, while people with dependency attract negative perceptions. Currently, up to 10 times as many leads are in development for cognition enhancement as for dependency and addiction.

Thus, responsibility for the conduct of trials to ensure a good evidence base may fall to UK funding bodies, the National Health Service, and academic institutions, to prove and implement the promise of better treatments. Research on problem drugs including alcohol seems to be a low government priority. The success of attempts to reduce the burden of substance use disorders will depend not only on good treatments and services, but also on the social, legal, ethical and personal contexts of psychoactive substance use and treatment over the next 25 years. Few data exist to help estimate the impact that broader social, legal and other changes might have on the prevalence of drug use, while effects of such changes on treatment access and outcomes could be complex. For example, access to treatment could be easier for legal drugs (as with tobacco smoking), except if disordered use of the substance is normalised or stigmatised as with alcohol. Illegal status seems likely to represent a barrier to accessing treatment services.

Enforcement of treatment, as with current drug testing and treatment orders, has not resulted in improved levels of abstinence compared with voluntary services, and raises ethical issues. Most societies are likely to continue to experience a rise in overall psychoactive substance use, together with an increase in the number and type of psychoactive substances available over the next 20 years. The current key problematic drugs are likely to remain in society, while a minority of new drugs will potentially lead to substance use disorders. Development of psychoactive substances is likely to be a growth area in the next 20 years. Future psychoactive substances will probably be developed by pharmaceutical and other industries, academic institutions, plus 'underground' and illicit sources.

The production of new psychoactive substances is likely to include current and novel methods. Advance knowledge of future psychoactive substances is unlikely to be accessible from most commercial and other sources, unless reported in scientific literature or via the Internet. Future psychoactive substances will probably be used for various therapeutic purposes, disease prevention, health maintenance and the enhancement of aspects of functioning such as performance, cognition, mood and pleasure, but may also be abused and lead to dependency. The majority of new psychoactive drugs are likely to have potential benefits and harms – very few future psychoactive drugs will be entirely safe or wholly without benefits. Although the potential for new agents to cause adverse effects or lead to substance use disorders may be largely predictable from their pharmacology, a minority will present unexpected challenges.

The expected increase in polypharmacy could well increase interactions between agents, which may result in additional effects that are useful or harmful. New uses and misuses of psychoactive substances are likely to develop alongside alternate or new classifications of these substances. Future classification will depend on changes in societal attitudes and in legal and regulatory frameworks, but will probably include nutrients and foodstuffs, herbal medicines, therapeutics, legal recreational drugs, other lifestyle and enhancement drugs including cognition enhancers, and illicit drugs. An increased blurring of boundaries is anticipated within current categories. Nutrients may be used in high doses as drugs, cognition enhancers as anti-addiction measures, psychiatric drugs as 'lifestyle' agents, and current illicit drugs used or under study as therapeutics. New types of classification could be based on neurotransmitter actions, actions on gross central nervous system functions, or both.

Psychoactive substances are likely to have an increasing impact on individuals who take them, their communities, and wider society, although the wider effects of increasing psychoactive substance use do not seem predictable at present. The use of psychoactive substances may well become increasingly acceptable and desirable overall. Prevalence and treatment of problematic use will depend on resources invested, wider social perceptions and social policies. Mechanisms used to manage psychoactive substances in the future will almost certainly include education and other measures aimed at prevention and harm reduction, use of treatments and regulatory mechanisms. Through education and experience, a wider understanding will probably arise in society of the mechanisms of psychoactive drugs, dependency and related problems, plus increasing recognition of the non-abusive use of recreational and other 'lifestyle' or enhancement drugs. The prevalent view of dependency as a moral weakness is likely to be abandoned and instead will incorporate other elements (eg neurobiological and genetic influences with psychosocial elements). Individuals will tend to become more informed over drug use, and may also seek greater autonomy in choices over substance use and the treatment and prevention of drug use problems.

Increasingly sophisticated public attitudes and changes in social policy could facilitate better use of psychoactive substances and encourage entry into treatment services for problematic use. The treatment of drug dependency and the minimisation of its consequences are likely to become more effective, with more options that allow targeting of therapy. More effective and earlier intervention in the course of substance use disorders could be possible, with expanded treatments and services, including dedicated drug services and others including peer-led education initiatives, and GP and psychiatric services. Prevention might become a realistic possibility. If voluntary entry to treatment or prevention initiatives becomes more acceptable, more available, and associated with better outcomes, society may be less accepting of continued drug dependency that leads to related problems, such as ill health, crime or antisocial behaviour. Current regulations are likely to be revised to encompass a broader spectrum of psychoactive substances than today, and increased resources will be required to maximise benefits and minimise harms of future agents. More sophisticated regulatory assessments of individual agents could represent the most rational future approach to inform society of novel risks.

Properties of psychoactive drugs that will remain important for assessment include abuse potential, potency and efficacy, other safety aspects, history of use and accessibility. Education and research on new and existing agents will probably remain valuable measures. Regulatory authorities are likely to be most effective if they can respond flexibly, rapidly and scientifically to the introduction of new agents and new research findings on known drugs. Continued and long-term assessment after regulatory approval or other release into society of psychoactive agents will probably be increasingly important. Efforts to eradicate drug abuse seem unlikely to be successful. However, social policy at one extreme could evolve to emphasise prohibition, with coercion as an element of prevention and treatment, or at the other, universal availability with market and other controls. The effects of such policies on the prevalence and treatment of substance use disorders are not predictable since no simple relationship exists between prohibition, availability, usage and harm. A more likely alternative to these extremes is a stratified approach that incorporates different levels of regulatory control for different substances according to potential for harm. Directing resources towards a reduction in demand rather than supply is an under- utilised strategy that may be effective at minimising harm from substance use disorders. Ultimately, fewer resources may be required to enforce avoidance of substance use disorders through legal frameworks if more sophisticated use of psychoactive substances is promoted and more effective treatments for substance use disorders are available.

4

Brain and Addiction

Experimental psychology uses controlled laboratory experiments to study the role of attentional, cognitive and emotional processing in human behaviours, and aims to understand the inner workings of the mind from inferences based on experimental observation. It contributes to, and draws on, a number of related disciplines including clinical psychology, anatomy, pharmacology, neuroimaging and genetics. Experimental psychology uses techniques that have made it possible to experimentally target the reward pathways in the brain in such a way as to provide insight into addictions such as drug taking and gambling. It is now possible to explore the cognitive effects of psychoactive substances in healthy volunteers and, in combination with other techniques such as neurobiological and neuroimaging advances, determine the neural basis of these processes. This report focuses on the contributions of experimental psychology to our understanding of psychotropic drug use, but refers to related areas where these have contributed to advancing our understanding of the brain and addiction. The major strength of experimental psychology is that it provides objective, precise and testable explanations of behaviour by systematically deconstructing complex human behaviours such as drug addiction and eliminating the extraneous uncontrolled factors that are found in more naturalistic observations.

Equally beneficially, clinical observations can be transferred into a laboratory environment to define the processes involved in observed behaviours such as dependence and withdrawal. The identification of the processes involved may facilitate targeting for treatment. Exciting recent developments in the creation of computer-based psychometric tasks and learning procedures allow similar aspects of behaviour in animals and humans to be tested. Paradigms have been developed that allow observations in humans to be modelled in animals, and for animal observations to be tested directly in humans. This cross-species translational approach has been particularly fruitful in the area of cognitive psychopharmacology. For example, there are versions of the Cambridge Neuro-psychological Test Automated Battery (CANTAB) intradimensional-extradimensional (IDED) attentional set shifting test, a computerised test of cognitive flexibility akin to the well-used Wisconsin Card Sorting Test (WCST), for rat, monkey and man. This test is crucial for assessing the ability to shift attention and learn new rules. Ornstein *et al.* (2000) found that chronic amphetamine abusers could be discriminated from heroin users on performance of this test.

The chronic amphetamine abusers were significantly impaired in their performance on the extra-dimensional shift (a core component of the WCST) whereas, in contrast, the heroin abusers were impaired in learning the normally easier intradimensional shift component. Neuroimaging studies in humans and neurobiological studies in animals indicate that the frontal lobes are key areas for the successful performance of tasks requiring cognitive flexibility. The development of brain imaging techniques is an important recent technology that allows the experimenter to draw parallels between a specific behaviour and changes in activation in precise brain areas in healthy volunteers and different patient groups. The use of cognitive psychological tests in combination with functional magnetic resonance imaging has enabled the identification of the neural basis for human reward systems.

These discoveries have enormous potential for identifying abnormal reward circuits in people suffering from drug addiction and for identifying effective mechanisms for correcting these changes. Research using learned behaviours in animals has helped achieve significant progress in determining the processes involved in maintaining drug taking and those aspects that lead to relapse during abstinence. In animals, it has been shown that the attentional processing of drug-related cues can lead to changes in an animal's motivational state and that this can produce the behavioural responses required to obtain the drug. Analogous methods are now available to test these components of drug taking in humans.

There is a need for a translational medicine approach to drug development which uses these experimental psychological measures to determine the effects of drugs, including neurological effects, in man. People traditionally take psychoactive drugs in order to feel better, enhance their mood or improve their memory. However, with repeated use, many of these drugs produce dependence and addiction. The *Diagnostic and statistical manual* (DSM) description of substance dependence refers to continuing drug use despite awareness of the long-term negative consequences, repeated attempts to cut back or quit substance use, and a gradual exacerbation of drug intake over time. It is well recognised that addiction is characterised by a long-lasting risk of relapse (often initiated by exposure to drug-related cues), although good-quality drug treatment can be highly effective.

There is still a large amount of work to be done in understanding the behavioural and neural substrates of compulsive drug use. There are several frameworks within which experimental psychologists can approach the study of addiction. One interesting proposal, which integrates several psychological indices mapped onto neural networks, has been put forward by Volkow and colleagues. Four neural networks, all of them modulated by drugs of abuse, have been proposed to be instrumental in drug abuse and addiction. These include brain circuits of reward (including the nucleus accumbens and ventral pallidum areas of the brain), motivation and drive (including the orbitofrontal cortex and subcallosal cortex), memory and learning (including the amygdala and hippocampus) and control (including the prefrontal cortex and anterior cingulate gyrus).

These four circuits all receive direct innervations from dopamine neurones, but they are also connected to each other through various direct and indirect projections, mostly involving the neurochemical glutamate. This means that a large number of brain regions may be involved in each circuit. In addition, one region may participate in more than one circuit. Patterns of activity in this four-circuit network may be at the core of behavioural choice. Volkow and colleagues have proposed that the decision to take a drug will depend on the expected positive feelings (the reward) to be had from taking it, which in turn will be affected by previous knowledge and memories, as well as a person's internal needs or motivation. The cognitive decision to act (or not) to procure the drug would be processed in part by brain areas such as the prefrontal cortex and the anterior cingulate gyrus.

Addiction could occur when disruptions to these circuits by agents (most likely dopaminergic) produce a greater and longer- lasting activation of the reward circuit than that obtained with non-drug stimuli, and this could overcome the inhibitory control exerted by the prefrontal cortex. This framework is perhaps too simplistic to explain some of the more sophisticated aspects of addiction. Nevertheless, it provides a useful basis for conceptualising some of the processes and structures that are important in understanding addictive behaviour. This report highlights developments in our understanding of the scientific basis of drug addiction and the effects of psychoactive substance use in terms of these core psychological components. This is followed by a description of some of the psychoactive substances that we believe will feature in the future, and potential interventions to manage their use. We conclude with several visions of the future and how experimental psychology will impact on these.

EXPERIMENTAL PSYCHOLOGY

Experimental psychology has approached the analysis of drug addiction with research into several main areas, which are largely incorporated into Volkow's model. Five main areas - reward and reinforcement, motivation, emotion, learning, and cognition and cognitive control - form this part of the report. Advances in neurobiology have enabled neuroscientists to manipulate specific brain molecules, neurons and systems. This has led to major advances in the neuroscience of reward, which is the key to our understanding of addiction. An understanding of how humans respond to reward is central to providing insights into disorders involving psychoactive substances and also into conditions such as obsessive-compulsive disorder, depression and eating disorders. Reward consists of numerous psychological components and is often considered in terms of the processes of emotion, motivation and learning. The challenge is to identify how different brain circuits mediate different psychological components of reward, and how these components interact.

Reward and Reinforcement of Addictive Drugs

Addictive drugs can be both rewarding (interpreted by the brain as intrinsically positive) and reinforcing (behaviours associated with such drugs tend to be repeated). A large amount of work using classical pavlovian conditioning paradigms has shown that it is possible for a number of effects of a drug to become conditioned. This occurs when a previously neutral stimulus becomes an incentive when paired with the acute reinforcing effects of a drug. Another striking phenomenon observed as part of addictive behaviour is the response of the individual to the presence of drug-related cues.

Appetitive pavlovian theories of drug addiction, for example, argue that stimuli associated with drug reinforcement (such as feeling high, and a reduction of anxiety or tension) acquire the capacity to elicit conditioned responses that lead to drug-seeking and drug-taking behaviour and become reinforcers in their own right. One important consequence of such a property is that the drug's initial rewarding effects (such as a desirable taste, or feelings of warmth), or even expectancies related to the drug consumption, become conditioned stimuli that come to control behaviour. This phenomenon, known as the '*priming effect*', has been studied with alcohol in the laboratory. Studies on the priming effects of alcohol in social, nonalcoholic, drinkers showed that priming will lead to an increase in choosing alcohol over an alcohol-free beverage or over money, as well as an increase in the desire to drink and the amount of drinking.

Similar effects have been shown with alcoholics. Since the priming effect is associated with relapse in alcoholics ('one drink, one drunk'), understanding the phenomenon is of major importance. Similarly, data from research on humans and rodents has shown that re-exposure to the drug itself, exposure to stress or the presentation of drug-associated cues are all effective at triggering relapse or causing reinstatement of the behaviour. Molecular changes in the brain can promote continued drug taking that becomes increasingly difficult for the individual to control. Imaging studies in drug abusers have shown that drugs of abuse increase extracellular concentrations of dopamine in the striatum (where the nucleus accumbens is located) and that these increases are associated with the reinforcing effects of the drugs. Studies examining the reinforcing and rewarding effects of drugs have shown that these effects are associated not only with the magnitude, but also with the abruptness, of dopamine increase in the brain.

Thus, for an equivalent increase in dopamine, intravenous administration of methyl-phenidate (Ritalin) – which results in a rapid uptake of the drug and a corresponding rapid alteration in dopamine concentrations – is considerably more reinforcing than oral administration. Indeed, there is strong evidence that the dopaminergic system that projects from the ventral tegmental area of the midbrain to the nucleus accumbens, and to other forebrain sites, is the major substrate of reward and reinforcement for both natural rewards (such as food and sexual opportunities) and addictive drugs. Drugs can stimulate brain reward circuitry with a strength, time course and reliability that exceeds almost any natural stimulus, powerfully consolidating responses to drug-associated stimuli.

Drug-induced synaptic changes in the nucleus accumbens and dorsal striatum may therefore contribute to addiction by consolidating drug-wanting, drug-seeking and drug-taking behaviours. The nucleus accumbens is involved in responding to the motivational significance of stimuli, and the dorsal striatum is involved in the learning and execution of behavioural sequences in order to ensure an efficient response to the cue. Addictive drugs increase the levels of synaptic dopamine in the nucleus accumbens while pharmacological antagonism and lesion studies have shown that dopamine is generally required for reward and reinforcement.

Opiates, such as heroin and morphine, represent a partial exception to the central role of dopamine. Although opiates can produce reinforcement by dopamine release, they can also interact directly with opioid receptors on nucleus accumbens neurones. However, increases in dopamine concentrations during intoxication can occur in both drug-addicted and non-addicted volunteers. This means that addiction might arise not so much from dopamine increases as from repeated marked increases and decreases in dopamine in the reward circuits, which in turn disrupt motivational, memory and higher cognitive circuits. For example, drug-addicted volunteers display longer-lasting decreases in the number of dopamine D2 receptors than controls. It is possible that this results in a decreased sensitivity of the reward circuits, leading to reduced interest in ordinary environmental stimuli, apathy or anhedonia, which could predispose a person to seek drugs to enhance the importance of these stimuli.

Thus dopamine-regulated reward circuitry in tobacco smokers fails to activate in response to monetary reward compared with non-smokers. Decreased sensitivity of reward circuits to acute alcohol administration has also been documented in cocaine abusers compared with control subjects. These findings suggest an overall reduction in the sensitivity of reward circuits in drug-addicted individuals to natural reinforcers and to drugs, including others in addition to the one to which they are addicted.

Role of Orbitofrontal Cortex

Motivation and decision making are also implicated in determining how rewarding a substance is. Drugs have the potential to be much greater reinforcers than natural rewards. One area of the brain involved in rating the relative value of reinforcers is thought to be the orbitofrontal cortex. Chronic amphetamine abusers are known to make suboptimal decisions, which correlates with years of drug abuse, as well as deliberating for far longer before making their choices during tests of decision making. In contrast, in the same study, opiate abusers were shown to have intact decision-making abilities, although they also deliberated far longer than controls. What is particularly interesting is that this study included patients with focal damage to the prefrontal cortex, with the suggestion that chronic amphetamine abusers show similar decision-making deficits to patients with focal damage to the orbitofrontal cortex, although these findings are awaiting replication. Imaging studies support this behavioural evidence of involvement of the orbitofrontal cortex in the deficits associated with addiction.

The orbitofrontal cortex has been shown to be underactive in drug-addicted volunteers tested during withdrawal, possibly as a result of the lack of brain stimulation by salient stimuli during detoxification. In contrast, in active cocaine abusers, the orbitofrontal cortex has been shown to be overactive - in proportion to the intensity of the craving experienced by the volunteers. It is therefore possible that, in the withdrawal state, exposure to the drug or drug-related stimuli reactivates the orbitofrontal cortex, resulting in compulsive drug taking. There are several other avenues of research that implicate the orbitofrontal cortex in motivation. Increased orbitofrontal cortex activation has been associated with compulsive disorders, and it is known that tobacco users show greater impairment on neuropsychological measures sensitive to orbitofrontal function, and that tobacco use is correlated with the degree of impairment. It is also known that nicotine activates orbitofrontal-striatal and limbic circuits, and that cigarette-related cues cause activation of orbitofrontal cortex. This is consistent with the accounts of drug addicts who claim that once they start taking the drug they cannot stop, even when the drug is no longer pleasurable.

Since the orbitofrontal cortex also processes information associated with associative learning and the prediction of reward, its activation during cue exposure could indicate changes in signal reward prediction, which could then be experienced as craving by the addicted subject. Motivated action, or instrumental conditioning, is the process whereby behaviour is altered in the presence of an association between the behaviour and a reinforcing outcome. Motivation to take the drug cannot be dissociated from the behaviour to seek and consume it, that is, 'we will work for what we value'. Rescorla and Solomon's (1967) two-process learning theory proposes that appetitive behaviour is controlled by two associative learning processes. The first involves pavlovian stimulus-reinforcer contingencies resulting in the stimulus acquiring the capacity to elicit physiological conditioned responses that prepare for, and support, the motor activity of seeking and consuming the reinforcer. The second process involves response–reinforcer contingencies, which endow the agent with the repertoire of motor responses to acquire and consume the reinforcer.

These two processes interact so that the stimulus acquires the capacity to elicit instrumental responses trained with the same reinforcer. This final prediction is demonstrated in the pavlovian-to-instrumental-transfer effect, where pavlovian conditioned stimuli augment instrumental responses trained separately with the same reinforcer. All three predictions of Rescorla and Solomon's (1967) two-process learning theory are supported, albeit sometimes indirectly, by the human tobacco addiction literature. For example, conditioned stimuli, paired with tobacco-smoke reinforcement and stimuli related to smoking, have been shown to elicit greater excitatory physiological responses than control stimuli. These excitatory physiological conditioned responses could conceivably prepare for, and support, the motor activity of reinforcer acquisition and consumption. Evidence for a pavlovian-to-instrumental-transfer effect in human tobacco addiction is indirect. Smoking-related cues have been shown to augment aspects of human smoking behaviour relative to control cues including naturalistic cues and arbitrary cues that have been associated with the reinforcer in the laboratory.

Effect on Emotions

In contrast to many other drugs of abuse, ecstasy (MDMA) has been shown to cause permanent reductions in serotonin axons and receptors in rats, non-human primates and humans, in addition to effects on dopamine. Because of the important role of serotonin in modulating mood and emotion, it might be expected that such long-lasting depletions may cause vulnerability to depression in chronic ecstasy users. Consistent with this hypothesis, a number of studies have demonstrated that ecstasy users report higher levels of depression than controls, both in the days following ecstasy use and following a period of abstinence. However, not all studies have found evidence to support a relationship between depression and ecstasy use. This discrepancy was recently clarified using advances in genetic subtyping.

Specific genetic subtypes of drug users were shown to be more susceptible to the emotional and cognitive effects of ecstasy than others. For example, ecstasy users who were homozygous for the *s*-allele of the serotonin transporter gene showed higher subjective depression ratings, increased affective lability and less rational betting behaviour on a decision-making test than others. Patients who suffer from depression show a similar affective bias and respond more rapidly to sad as opposed to happy stimuli compared with healthy control subjects, thus providing validation for this task in mood disorders. This test is sensitive to detecting first episode depression in adolescents and might provide a useful mechanism for detecting vulnerability to mood changes in ecstasy users. This task has also been shown to be sensitive in healthy volunteers to manipulations of the dopamine and serotonin neurotransmitter systems in tyrosine and tryptophan depletion studies respectively.

Formation of Habits

Learning and the formation of habits are enormously important in addiction. Seeing certain people, places and paraphernalia can trigger an intense desire for the drug in addicts, and this can often result in relapse. Drug addiction has been considered in terms of normal learning and memory systems of the brain which, through the actions of chronically self-administered drugs, are pathologically subverted, leading to the establishment of compulsive drug-seeking habits. Many forms of learning are mediated by different brain systems, with multiple memory systems having been implicated in drug addiction. As described above, exposure to a drug may sensitise the incentive-motivational system, producing compulsive drug-seeking behaviour (or '*craving*'), even though the drug has ceased to give pleasure.

Much of this work has been explored using the phenomenon of '*second-order schedules*', in which animals will work for a long period to obtain access to a drug, providing that their operant behaviour is maintained by the response-contingent presentation of conditioned reinforcing stimuli paired with the eventual drug infusions. Similarly, it is known that addicts will work under a '*second-order*' schedule to produce eventual injections of opiates, provided there is intermittent presentation of drug-associated cues. Learnt knowledge is thus extremely useful for making predictions about rewards, anticipating outcomes, and performing goal-directed actions. All of these can be studied in the laboratory in both humans and animals. Humans are exquisitely able to evaluate relationships between events with great accuracy and adapt behaviour rapidly in response to unexpected occurrences.

This ability to learn about whether an action will lead to a rewarding outcome or not, and to use knowledge of causal relationships between our actions and their consequences on the world, is critical for the behavioural expression of cognition. However, the process by which selective attention to drug-associated cues relates to learning stimulus–response associations that support drug acquisition and consumption is not yet fully understood. According to certain attentional theories of both pavlovian and instrumental learning, stimuli that signal that an instrumental response will be drug-reinforced should acquire the capacity to command the focus of attention and bring about the instrumental response. Although these learning theories all agree that attention for discriminative stimuli plays a crucial role in the acquisition of discriminative control, thereafter they assume that attentional processing of the discriminative stimuli varies with knowledge of the stimulus-reinforcer contingencies.

Mackintosh predicts that attention to a rewarded stimulus should increase to the extent that the stimulus is the best predictor of the reinforcer. Similarly, according to incentive salience theories, as the relationship between the stimulus and the drug reinforcer is learned, the amount of attention allocated to the reinforced stimulus increases to asymptote. These theories regard the attentional orienting response elicited by the stimulus as being like other conditioned responses, such as salivation and skin conductance, which increase with training. By contrast, alternative attentional learning theories propose that, although attention for the reinforced stimulus is necessary for learning about the predictive relationship between the stimulus and the reinforcer, once this relationship is well learned the processing of the stimulus becomes efficient. As a consequence, the attentional orienting response towards the reinforced stimulus might well decrease when the stimulus is established as a good predictor of the reinforcer. Dual-process theories predict that, as exposure to a drug reinforcer increases, the mechanisms that control addictive behaviour change from being controlled, cognitively mediated and goal-directed, to being automatic and habitual.

At this stage, the circle of reward–motivation–drive and learning must become self-sufficient and does not require mental representations or control input. A further important consequence of drug- related cues acquiring incentive properties is that the cues can capture the attention of the addict and control the addict's behaviour. There is a vast amount of evidence supporting the presence of attentional bias towards drug-related stimuli in addiction-related disorders, as seen during the performance of tasks of attentional orienting (for example, using MacLeod's dot-probe task; or eye tracking), disruption with a secondary task and adverse interference (e.g. the Stroop task). This type of attentional bias has been demonstrated for dependence on alcohol, nicotine, cocaine and opiates. Changes in the motivational state of the individual, either by priming with the drug or by deprivation from the drug can also alter attentional bias to substance-related stimuli.

A significant number of studies with human volunteers in the laboratory using arbitrary cues have provided evidence that attentional bias is the result of the conditioning of incentive associations between the drug-reinforcing properties and the cues, including alcohol and nicotine. By experimentally modulating the learning of stimulus-reinforcer contingencies either by instruction or by over-training, we can understand how the capture of attention by drug-related cues relates to the response to obtain them. In one such study, explicit knowledge of stimulus-reinforcer contingencies was accompanied by a decrease in attentional orienting towards the stimulus, demonstrating that behaviour, through learning, became more efficient. Eye-tracking techniques that have become available are now applied to investigate the role of attention in the development of addiction. In terms of the neural basis of learning, some theories of reward-dependent learning suggest that learning is driven by the unpredictability of the association between the cue and the reward.

Error-driven learning only advances when prediction errors occur, with single dopamine neurones coding reward prediction error in a signed (positive or negative) bidirectional fashion. Neural manipulation could influence rewarded behaviour by altering one of many forms of learning. The activation of regions linked with memory has been reported during drug intoxication and during craving induced by drug. For example, studies of drug abusers experiencing withdrawal have shown decreased dopamine D2 receptor expression and dopamine release in the dorsal striatum. The dorsal striatum is associated with habit learning and it is possible that, in addiction, the routine associated with drug consumption may be triggered automatically by exposure to the drug or drug-related cues. Drug addiction may be an aberrant form of learning, possibly mediated by the maladaptive recruitment of particular memory systems in the brain.

Neurotransmitter Systems

Effective cognitive functioning typically involves numerous neuronal pathways and neurotransmitter systems, with several distinct neurotransmitters being implicated in the alteration of cognitive function. In particular, executive functions, such as attention, planning, problem solving, adapting behaviour and attending to relevant stimuli while ignoring distracting ones, are crucial for the successful performance of many everyday procedures such as prioritising tasks and remembering important information while performing an action. Frontostriatal neural networks in the brain have been shown to subserve many of these crucial functions and to be modulated by neurochemicals, such as the catecholamines dopamine and noradrenaline. It has been suggested that alterations in brain dopamine activity might lead to a reduced sensitivity to natural reinforcers, such as food and sex, as well as disrupting frontal cortical functions implicated in inhibitory control.

It is hoped that a clearer understanding of the neural mechanisms of substance abuse will help drive effective strategies for treatment. Drug addicts are known to display abnormalities in the prefrontal cortex, including the anterior cingulate gyrus. The prefrontal cortex is involved in decision making and in inhibitory control. Its disruption could lead to inadequate decisions that favour immediate rewards over delayed but more favourable responses. It could also account for impaired control over the intake of the drug, even when the addicted subject expresses a desire to refrain from taking the drug. An extensive literature has pointed to frontal lobe dysfunction in alcoholic patients. Studies in animals and neurological patients have been able to inform human laboratory testing with addicts so that conclusions can be drawn as to the damaged or dysfunctional area of the brain. Studying cognitive dysfunction in addicts in relation to the history of their addiction can indicate the critical factors contributing to its development. For example, repeated detoxifications from alcohol leads to an increased severity of withdrawal often accompanied by seizures.

Evidence has now accumulated that other aspects of brain function can also deteriorate, resulting in altered motivation. For example, cravings have been shown to be increased, together with impairments in emotional processing and executive functions. The latter has also been associated with binge drinking. Feeling emotions and perceiving emotions in others forms the basis of social encounters and everyday relationships. Similarly, executive functions such as the ability to plan, follow goals, be flexible, and inhibit habitual tendencies all reflect virtues that are essential for controlling excessive drinking. If binge drinking, leading abruptly to high amounts of alcohol in the brain, interacts with withdrawal from alcohol to intensify anxiety, emotional and cognitive disturbances, then it becomes clearer how alcoholics become sensitised to experiencing more intensified withdrawal and are susceptible to relapse.

Repeated detoxifications from alcohol or alcohol binges can impair new learning of both aversive and appetitive associations in animals as well as impairing the ability to relearn items such as a new path to escape. More importantly, once aversive learning has been established, stimulus generalisation can also occur. Such a phenomenon (currently only observed in animals) might underlie the increased incidence of panic attacks noted in alcoholics as a result of inappropriate responses to innocuous stimuli. The identification in animals that structures such as the amygdala, hippocampus or cortical structures might be dysfunctional or sensitised as a result of bingeing or withdrawal from alcohol provides the beginning of an understanding of the neuropsychology of withdrawal-induced and binge-drinking-related change in emotional and cognitive functions. With respect to other drugs of abuse, preclinical studies have shown that chronic administration of cocaine or amphetamine results in a significant increase in dendritic branching and the density of dendritic spines in the prefrontal cortex.

These changes could be involved in the changes in decision-making, judgement, and cognitive control that occur during addiction. Imaging studies have shown evidence of changes in prefrontal activation during a working memory task in smokers compared with ex-smokers. It has been proposed that disruption of the prefrontal cortex could lead to loss of self-directed behaviour in favour of automatic sensory-driven behaviour. Moreover, the disruption of self-controlled behaviour is likely to be exacerbated during drug intoxication by the loss of inhibitory control which the prefrontal cortex exerts.

The inhibition of top-down control would release behaviours that are normally kept under close monitoring and would simulate stress-like reactions in which control is inhibited and stimulus-driven behaviour is facilitated. Indeed, a recent review has suggested that, in addition to the brain's reward system, two frontal cortical regions (anterior cingulate and orbitofrontal cortices) that are critical in inhibitory control are dysfunctional in addicted individuals. These same regions have been implicated in other compulsive conditions characterised by deficits in inhibitory control over maladaptive behaviours, such as obsessive–compulsive disorder. It is hoped that a better understanding of the neurochemical basis of prefrontal inhibitory control might enable the development of drugs and psychotherapies that selectively activate this function and prevent the general trend for substance-dependent individuals to relapse a few months after quitting.

PSYCHOACTIVE SUBSTANCES

The increasing availability of potentially beneficial psychoactive substances demands research to establish the full effects of their use. Psychoactive substances can be taken for a wide range of reasons, including mood-altering or cognitive enhancing purposes. Preclinical, cognitive, imaging and genetic studies are furthering our understanding of the potential benefits and harms of these agents. This section describes our current understanding of the effects of certain of the most commonly used recreational drugs, cognitive enhancers and addictive non-chemical behaviours. Many of these effects are an essential feature of the cycle of abuse, and will apply to new substances as they are introduced in the future.

Drugs of Recreational Values

Recreational drug use usually begins because of a drug's beneficial effects on mood or cognition. However, the defining characteristic of addiction is compulsive, out-of-control drug use that persists despite serious negative consequences. Indeed, the high risk of relapse to drug use that persists even in abstinent addicts long after the cessation of withdrawal symptoms has led some to consider addiction as a chronic medical illness. This section focuses on some of the physiological and behavioural effects of drugs of abuse, and in particular highlights findings with respect to some major classes of drugs such as alcohol and the opiates.

Mechanism of dependence

Several effects are commonly seen in drug addiction. Tolerance often develops to the desired pleasurable effects of drugs, leading to dosage increases that can exacerbate the molecular changes that lead to addiction. Depending on the pattern of use, some drugs, such as cocaine and amphetamine, can also produce sensitisation (or enhancement) of some drug responses where a smaller dose produces an equivalent effect. Dependence can also develop. Dependence, narrowly defined, refers to the adapted state of cells, circuits or organ systems that occurs in response to excessive drug stimulation. On drug cessation, this adapted state can result in the production of cognitive, emotional and 'physical' withdrawal symptoms. However, it is possible to experience tolerance, dependence or withdrawal without producing compulsive use, prompting researchers to search for other explanations for addiction. *Mechanisms of dependence*, withdrawal and sensitisation have been studied extensively.

In particular, the neural and molecular basis of physical opiate dependence and withdrawal has been thoroughly investigated as a model of drug action in the brain. By contrast, convincing animal models of addiction - compulsive use despite negative consequences - are lacking. Robinson and Berridge (1993) have proposed an incentive-sensitisation theory of addiction, which suggests that brain systems normally involved in incentive motivation (but not pleasure) become hypersensitive to drugs and drug-associated stimuli, markedly increasing their incentive salience. However, the motivation of drug-induced pleasure does not fully explain compulsive use, because of the presence of effects such as habituation. Some people have argued that compulsive drug taking might continue in an effort to avoid the negative effects of drug withdrawal.

Positron emission tomography (PET)

Clinical evidence in drug addicts has shown that relapse can occur even a long time after withdrawal symptoms have cleared, arguing against drug addiction continuing solely from the desire to avoid negative withdrawal symptoms. Researchers have identified other factors that also impair an addict's ability to remain independent of substance abuse. Cues associated with previous drug use can initiate craving and conditioned emotional responses in addicts, and are associated clinically with relapse, both during active drug use and after a period of abstinence. Cues can elicit drug-related behaviour in animal models, including sensitised responses to psychostimulants. Importantly, sensitisation in animal models has been shown to be context-dependent, raising the possibility of its providing a good model of relapse in humans.

In rodent models, stress has been shown to produce resumption of drug self-administration. Its role in relapse in human addicts seems to be significant and deserves further study. Drug-associated cues can elicit conditioned emotional responses, including drug urges, in human addicts. Studies of responses to drug-conditioned cues using *positron emission tomography* (PET) and functional magnetic resonance imaging (fMRI) have shown activation of prefrontal cortex regions and the amygdala, a brain region involved in the consolidation of stimulus-reward associations. Some of these imaging studies have also found activation of the nucleus accumbens and other regions. The cue-induced brain activations observed with PET and fMRI correlate with the onset of subjective drug urges and physiological responses, such as activation of the autonomic nervous system. Robbins and Everitt (1999) have suggested that conditioning and memory retrieval may represent useful targets for the treatment of addicts and relapse prevention.

Long term drug abuse

Long-term drug abuse is known to induce cognitive and emotional impairments, detectable using psychological methods. Alcohol-related, long-term dysfunction of the frontal lobe, and alcohol-related amnesia, including the devastating Korsakoff's syndrome, are examples of the harm that can result from this most common recreational drug. Even moderate consumption of alcohol has been shown to attenuate the brain's capacity to detect errors in performance, potentially explaining alcohol's detrimental effects on tasks such as driving a car and operating machinery. Similarly, ecstasy has been studied in the laboratory with sophisticated neuropsychological methods and it is now generally accepted that it produces memory impairments.

Amphetamine-induced psychosis, and the recent data on cannabis-induced psychosis, as well as Ecstasy-induced depression, demonstrate the spectrum of harm that recreational drugs can produce. Ornstein *et al.* (2000) reported deficits across a range of cognitive functions including verbal fluency, pattern and spatial recognition memory, and executive functions such as planning. Heavy ecstasy use has been shown to result in selective executive deficits, although it is difficult to ascribe the deficits definitively to ecstasy in this particular study as all participants also consumed a range of other illicit drugs. Robinson and Berridge (2000) have suggested that some drugs of abuse may weaken the *'rational break'* of cognitive regulatory processes, which are necessary to inhibit motivational impulses, and could impact on other psychological factors such as decision-making.

Numerous studies support the observation of impaired decision making or altered neural activation in substance users have suggested that reduced reflection may represent a cognitive marker for substance abuse that may be associated with vulnerability to substance use or the neurotoxic effects of drug exposure. Only a detailed study in the laboratory of the psychological factors that underlie these phenomena will allow us to further explore these cognitive effects and ultimately help prevent or treat them. These studies must control for polydrug use in order that we can better understand the differences in effects between the individual drugs.

Effects of Psychoactive Substances

Any discussion on the effects of psychoactive substances must include both their harms and their benefits. Many of the agents mentioned already, such as amphetamine and methylphenidate, bridge both camps, with recognised harmful side effects in addition to their cognitive-enhancing properties. There is a huge need for effective cognitive enhancers and it is anticipated that an increasing number of compounds will be developed to target cognition while reducing the potential for addiction.

Mental illnesses

Major mental illnesses are extremely common and their effects on behaviour, perception, emotion and cognition are an enormous factor in worldwide disability. Numerous neuropsychiatric disorders, such as attention-deficit hyperactivity disorder (ADHD), schizophrenia, frontal dementia and Parkinson's disease are characterised by cognitive impairments. The potential public health benefit of improving current treatments for cognitive disabilities in patients is largely undisputed. However, cognitive difficulties are not confined to psychiatric disorders, but are also endured by a much wider sector of society, including those whose deficits arise from sleep deprivation, head injury or age-related cognitive decline. With the increase in use of ecstasy and other recreational drug use as well as alcohol binge drinking, there will be additional groups of people who suffer from cognitive decline who might benefit from cognition enhancers. Their possible use raises many considerations in terms of their effects on the individual concerned and the ethical implications of their use. The global impact of cognitive impairment has only recently been recognised. The disorder of schizophrenia provides a particularly good illustration of the potential benefits to be had from exploring new options for the treatment of cognitive dysfunction.

It is estimated that 24 million people worldwide suffer from schizophrenia, which ranks third in terms of the global burden of neuropsychiatric conditions following depression and alcohol dependence. Its economic impact alone is enormous. In 2000 in the United States it was suggested that the direct and indirect costs of schizophrenia could have been as much as $40 billion. In many patients, cognitive difficulties have been shown to be the main factor in limiting full rehabilitation (such as returning to work) and quality of life, even after the clinical symptoms have remitted. Schizophrenia is an important example in the context of drug use. Cannabis use is a risk indicator for schizophrenia, but its causal effects are not yet established.

Stimulant drugs and cannabis can cause transient psychotic symptoms and precipitate relapse of an existing psychotic illness. Andreasson *et al.* (1987) showed (in a study of over 45000 conscripts to the Swedish army) that those who abused cannabis at 18 years of age were more likely to be admitted to hospital with schizophrenia over the next decade and a half than those who did not, and that there was a dose-response relationship so that the more cannabis was consumed, the greater was the likelihood of schizophrenia. The active metabolite of cannabis, delta-9 tetrahydrocannabinol (THC), raises the levels of cerebral dopamine and might precipitate the illness through this mechanism. However, as McDonald and Murray (2000) point out, it is also possible that patients take cannabis as a 'self-treatment' in order to increase dopamine levels to overcome the negative or affective symptoms of the disorder. Through a greater understanding of neurochemical mechanisms, it is increasingly becoming possible to identify drugs that will help patients to lead effective and high-quality lives, unhindered by cognitive impairments.

Several psychopharmacological studies have shown that it is possible, using single doses of pharmacological agents, to improve cognitive functioning in a wide variety of patient groups that suffer from memory impairments. These have included people who have primary disorders of attention such as ADHD as well as patients with secondary cognitive deficits such as schizophrenia and frontotemporal dementia. Patients with cognitive impairments as a result of traumatic brain injury have also been shown to derive some benefit from pharmacological treatment. In patients with brain damage or disabling neuropsychiatric conditions, an important clinical objective is to try and enhance cognition, with the aim of improving functional outcome and quality of life. For example, cholinesterase inhibitors and the ampakines may prove to be beneficial for problems of learning and memory resulting from brain trauma or neuropsychiatric illness. However, an important point of discussion is the increasing use of cognitive enhancement by the general population to improve performance in school and the workplace.

Research in the healthy adults

Our understanding of cognition would not have reached its current state without research in healthy adults. Volunteer studies enable a comprehensive understanding of the effects of cognitive enhancement, free from the confounds of underlying pathology. This work is vitally important in furthering our understanding of cognition, and deepening our knowledge of underlying brain mechanisms. It is also the most contentious owing to the ethical issues inherently surrounding enhancement of the healthy brain. Until recently, psychotropic medications had significant risks and side effects that made them attractive only as an alternative to disorder or illness. However, a growing body of evidence is now demonstrating that it is possible to pharmacologically enhance cognition with minimal side effects in healthy volunteers. For example, in normal individuals, modafinil (a drug recently licensed for the treatment of narcolepsy) induced reliable improvements in short-term memory and planning abilities in the absence of any impairments. These newer drugs, originally designed to treat a medical condition, but which prove to be safe enough for widespread use during healthy volunteer studies, are increasingly being used off-label. Impaired cognition and alertness chronically affect millions of otherwise healthy people (often as a result of ageing, jet lag, shift work or disorders of sleep or the circadian rhythm).

The cognitive performance and alertness deficits that result from monotonous activities or sleep loss are recognised as a considerable threat to productivity and safety in both industrial and military settings. Thousands of normal, healthy children and adults have discovered the benefits of cognitive enhancement on their own, with many people already self-medicating using over-the-counter psychoactive remedies such as herbal stimulants, tonics and caffeine to improve performance at work and school. Increasingly, people will be turning to pharmaceutical cognitive enhancers. Many agents, such as methylphenidate and modafinil, are readily available though the Internet. It is also possible that, in the future, cognitive enhancers will be used to target deficits associated with the use of cognitive impairing drugs and other forms of cognitive dysfunction. Drugs such as modafinil that improve inhibitory control might be a useful treatment to help overcome the difficulties that addicts experience with cognitive control and impulsivity. The application of a cognitive enhancer targeted against impairments that contribute to the maintenance of addictive behaviours could potentially provide the breakthrough needed to enable the effective treatment of addiction.

Addiction of Short Lived Pleasurable Activities

There are many behaviours that can become addictive in excess for vulnerable individuals. Pathological gambling, Internet use, shopping and binge eating have all been reported to be associated with common effects that might be linked to their addictive potential. Effects include short-lived pleasurable feelings, euphoria and anticipation accompanied by a physical 'rush' (sweaty palms, rapid heartbeat, nausea), tolerance (gamblers might bet increasing amounts or Internet users need more and more sessions to achieve the pleasurable feelings), withdrawal symptoms during abstinence (including craving, anxiety, insomnia, dependence and loss of control) and continued involvement despite negative consequences (such as bankruptcy). Neuropsychological support for the addictive potential of these agents has been shown in neuroimaging studies illustrating excessive striatal dopamine release in volunteers playing a 'battlefield' video game. Another example is overeating in obese individuals, which shares some similarities with the loss of control and compulsive drug-taking behaviour observed in drug-addicted volunteers.

A recent study has shown that pathologically obese volunteers also have reductions in striatal dopamine D2 receptors, a result very reminiscent of that seen in drug addicts. Dopamine D2 receptor levels were found to have an inverse relationship to the body mass index of the obese volunteers. The authors postulated that decreased levels of dopamine D2 receptors might predispose people to search for food reinforcers to temporarily compensate for a decreased sensitivity of dopamine D2 regulated reward circuits. These findings are very tantalising but a large amount of work is still to be done in the field of understanding non-chemical addictions, and what prompts these behaviours to deviate from normality, in order to help derive strategies targeted at treatment of these disorders.

BENEFICIAL PSYCHOACTIVE SUBSTANCES

The management of psychoactive drugs can be considered in terms of harm minimisation, and centres on establishing a person's vulnerability or resistance to drug addiction, as well as the implications for treatment of an addiction, and the potential for developing 'safe' psychoactive drugs. After considering these issues, we examine how we should aim to manage the use of beneficial psychoactive substances, such as cognitive enhancers. Studies that aim to differentiate the addictive from the harmful potential of a drug can provide a useful basis for a refinement of the methods of experimental psychology.

Resilience to Addiction

A challenging problem in the neurobiology of drug addiction is to understand why some individuals become addicted to drugs while others do not. Volkow and colleagues (2003) have proposed several scenarios in which a person could become more or less vulnerable to addiction. For example, decreased sensitivity of reward circuits to natural reinforcers, decreased activity of cognitive control circuits, or an increased sensitivity of memory or motivation circuits to drug or drug-related stimuli could all make an individual more vulnerable to addiction. Acute drug intoxication results in a complex and dynamic pattern of activation and deactivation that includes regions neuro-anatomically connected with the dopaminergic system and known to be involved in reward, memory, motivation and control.

In addition, personality type and temperament, which have their roots (in part) in genetic and early environmental factors, age, stress and co-existing psychiatric conditions, may all predispose an individual to drug addiction. An individual may become dependent on drugs because of a single factor or, more likely, several interacting factors. Environmental factors have also been shown to influence these neuronal circuits, which in turn affects behaviour related to drug addiction (e.g. drug consumption) and needs to be considered by experimental psychologists. For example, a study in nonhuman primates showed that social status affects dopamine D2 receptor expression in the brain, which in turn affects the level of cocaine self-administration. Dominant animals possessed higher numbers of dopamine D2 receptors and had much lower levels of cocaine intake than subordinate animals. Studies in humans have also implicated dopamine D2 receptors in sensitivity to stimulant drugs.

Volunteers with low numbers of dopamine D2 receptors tended to describe the effects of methylphenidate (Ritalin) as pleasant, whereas volunteers with high numbers of D2 receptors tended to describe it as unpleasant. Thus, volunteers with low numbers of D2 receptors might be at a higher risk of abusing stimulant drugs than those with higher numbers. In terms of future potentials and the effective collaboration between genetics and experimental psychology, work in animals has shown that insertion of the D2 receptor gene, via a viral vector, into the nucleus accumbens of rats that had previously been trained to self-administer alcohol significantly reduced the animal's alcohol intake. Neuroimaging studies examining the risk factors for addiction have shown that children of alcoholic families that were considered to be at a high risk for alcoholism showed smaller amygdala volumes than control subjects.

Imaging studies have also reported structural changes in the orbitofrontal cortex of cocaine-addicted volunteers, although it is not known whether these changes preceded drug use (making these volunteers vulnerable to drug addiction) or were a result of drug abuse. Access to transgenic and knockout animals now provides a means to evaluate the role that specific genes may play in the vulnerability or protection against drug abuse and addiction. Susceptibility factors are an important consideration in the study of psychoactive substance use. Individuals at risk of becoming alcoholics (usually sons of families with alcoholism over multiple generations) show impairments in executive and working memory functions and reveal reduced activation of prefrontal areas when performing attentional tasks.

Impairments in executive functions and working memory are substrates of impulsive behaviour and might be a general factor predisposing to addiction if they lead to an inability of the subject to suppress automatic responses. We need to understand how the acute effects of the drugs on memory as well as on executive function can contribute to drug taking. It is thought that taking drugs, drinking alcohol or smoking are learned responses that become automatic and difficult to interrupt. By using behavioural fingerprints of groups with high risk for familial alcoholism and drug addiction, as well as identifying the differences in the acute effects of drugs in these populations, it might be possible to develop tests predictive of addiction vulnerability in the general population.

Conceptualised Addiction

Many researchers have *conceptualised addiction* as a state initiated by the qualitatively different and larger reward value of the drug, which triggers a series of adaptations in reward, motivation, memory and cognitive control resulting in an enhanced and permanent saliency value for the drug and a loss of inhibitory control, which favours the emergence of compulsive drug administration. Experimental psychology can offer a route to the development of behavioural treatments. The ability of drug- related cues to attract the attention of the addict is well documented. If research in the human laboratory determines the relationship between attentional, cognitive and emotional processes in the presence of drug- related cues and the maintenance of addictive behaviours, psychological techniques can be developed to break this relationship and protect the individual from maintaining the behaviour.

The cognitive impairments associated with addictive behaviours may also be targeted by psychological techniques directed to train the individual to strengthen the domains of cognitive abilities that can contribute to addiction. Targeted pharmacological treatments can also be developed using a translational medicine approach. Potential avenues for treatment to combat drug addiction could include attempts to decrease the rewarding value of a drug (for example, a pharmacological treatment that interfered with a drug's reinforcing effects) or to increase the value of non-drug reinforcers that might help weaken learned drug responses (for example, behavioural therapies), or interventions that strengthen cognitive control such as cognitive therapy. Targeting many of these cognitive effects with novel cognitive enhancers might also prove an effective treatment strategy. Very little is currently known about how improvements in cognitive function might result in improved clinical outcome. This is largely because there has been a lack of effective cognitive enhancement therapy, making it difficult to examine the extent to which improvements in cognitive functioning can predict improvements in clinical response.

Long-term studies that measure clinical outcomes associated with these cognitive changes are essential. It is also possible, using experimental psychology, to define how individual drugs interact with the different processes involved in addiction. For instance, the knowledge of a role of endogenous opiates, GABA and glutamate, in the development of alcoholism has initiated treatment strategies including naltrexone (an opiate antagonist) and acamprosate (a glutamate modulator). As these drugs affect different neurotransmitter systems, they might influence different processes involved in addiction. An under-standing of the psychological processes with which these drugs interact could allow their use to be tailored to specific addictions. O'Malley *et al.* (2002) have used naltrexone in the laboratory to determine which of the psychological processes involved in the addictive behaviour are most likely to be influenced by the drug.

Pharmacological Survey

As mechanisms to improve mood, performance and memory are identified, safety becomes of particular concern if a pharmacological agent is to be used to enhance, rather than to treat, performance. Patients with severely debilitating symptoms will often tolerate the side effects of drug treatment because the benefits outweigh them. It is always very difficult to be certain about the potential for subtle, rare or long-term side effects, particularly in relatively new pharmaceuticals, and thus a full exploration of the long-term implications of any treatment that might be used by the healthy population is imperative. Children, especially, are at risk if these drugs are found to affect normal brain development. It would be devastating to discover that a dazzling youth of successful mood and cognitive enhancement meant a middle age of premature memory loss and cognitive decline. However, as Sententia (2004) points out, public policy decisions about cognitive enhancement should be guided by the democratic right to cognitive liberty and the principle of safeguarding one's own thought processes, rather than by moralism or paternalism.

Moral and safety precautions will inevitably have a place in determining the appropriate uses of drugs, but each individual should have access to the information necessary to determine for him or herself what is an acceptable personal risk. In the future, with the development of pharmacogenomics, this decision might also be based on information about our genetic make-up. However, even if such drugs are safe, some ethicists will no doubt continue to worry about the impact that their widespread use will have on society. We will inevitably have to embrace the consequences of pharmacological enhancement, and scientists have a moral responsibility to monitor and foresee, as far as possible, the consequences of their research. Novel psychoactive drugs might have the potential to aid millions of individuals who voluntarily take them for cognitive impairments, but their widespread availability also brings with it the more sinister prospect of compulsory or forced drug administration to individuals unwilling or unable to give informed consent. At the moment there are no drugs that can be taken for mood or cognitive enhancing reasons without at least a risk of side effects.

The development of safe and effective compounds, however, is becoming increasingly likely. As the neural mechanisms behind cognition and drug addiction are elucidated, it may become possible to develop pharmacological agents that have very selective mechanisms of action. Agents such as the novel cognitive enhancer modafinil already have minimal side effects. The potential for these types of drugs to be used recreationally by a 24/7 society where people maintain their performance in the absence of sleep is increasing.

NEW TECHNIQUES

New techniques mean that several areas that hold particular promise for increasing our understanding of the brain with respect to addiction and drugs are now within reach. Three main growth areas – the Internet, genetics and pharmacological imaging – will directly enhance the work of experimental psychologists.

Experimental Psychology

Almost all the current research into the neural mechanisms underlying substance abuse focuses on '*bottom-up*', primarily limbic, incentive motivational processes to understand the underlying attraction of taking certain substances. But in the treatment of substance misuse, a more clinically significant question is often why some individuals spontaneously recover from substance dependence and even more importantly, why some individuals relapse and others maintain abstinence. Experimental psychology is well placed to attempt to provide a definitive answer in the next decade, particularly through further research into the neural basis of '*top-down*' cortical inhibitory systems and individual genetic variance.

Challenge for experimental psychology

Another challenge for experimental psychology is to provide a clearer understanding of the assumption that substance dependence reflects pathological learning. This concept has been the basis of considerable cue-exposure research, in an attempt to train substance-dependent individuals to unlearn or 'extinguish' associations between drug- related cues and drug use. This approach has been limited in practice by the time taken to extinguish drug-related conditioning and the huge number of potential drug-related cues operating in a dependent user's environment. It has therefore been proposed that a more radical approach might be to move away from slow and relatively ineffectual extinction and counter-conditioning approaches to treatment, and to exploit a better understanding of the biochemical basis of learning. One approach for the future could be to intervene directly using pharmacological and other targeted amnesia-inducing manipulations to selectively delete drug cue-activated pathological learning – using the same principle as extinguishing drug-related cues, but accelerated by the use of biochemical treatments with the aim of reprogramming the brains of dependent individuals back to their pre-drug state. For such an approach to be viable the intervention must be targeted at drug-related memories rather than memory per se.

Until recently such an approach was not feasible because the only procedures available (such as electroconvulsive therapy) produced general memory impairment. There is, however, some preliminary evidence that such an approach might be realised. Debiec *et al.* (2002) recently demonstrated that intra-hippocampal infusions of the protein synthesis inhibitor anisomycin caused amnesia for a consolidated hippocampal-dependent contextual-fear memory, but only if the memory was reactivated prior to infusion. A similar approach is being developed for a pharmacological deletion of memories related to post-traumatic stress disorder. With further research, such an approach might provide the basis for reactivating drug-related memories and then deleting them as part of a treatment intervention.

Adolescents neurodevelopment

Another possible direction for future experimental psychology research relates to the recent recognition that the brains of adolescents are often more sensitive than, and react to abused substances in a qualitatively different way from, adult brains. For example, Bjork *et al.* (2004) reported that adolescents showed less recruitment of the right ventral striatum and right-extended amygdala while anticipating rewards (in contrast with anticipation of non-rewards) than young adults. Similarly, Chambers *et al.* (2003) portrayed adolescent neurodevelopment as a biologically critical period of greater vulnerability for experimentation with substances and the acquisition of substance-use disorders.

Adolescent impulsivity and novelty seeking can be explained in part by maturational changes in frontal cortical and subcortical monoaminergic systems, which may be adaptive for learning adult roles but may also confer greater vulnerability to the addictive action of drugs. Indeed, there is evidence that taking risky decisions decreases with age. Adriani and colleagues (2004) have reported a similar vulnerability in adolescent rats. Acute nicotine administration had opposite effects on anxiety in adolescent and adult rats. Furthermore, nicotine exposure during mid-adolescence, dose-dependently down-regulated AMPA GluR2/3 subunits (which are thought to be involved in the control of addictive behaviours) in the striatum and hippocampus. Comparable exposure during adulthood had either opposite or no effects. Thus these data identify a nicotine-vulnerable age window, characterised by long-term disruption of locomotor habituation and down-regulation of AMPA receptors that suggests that further research into neurobiological vulnerability to drugs in adolescent humans would be profitable.

New Communications

One current hurdle to the understanding of psychoactive substance use is the study of '*real-life*' drug users. One mechanism might be through the use of new communications and real-time recording technology to move testing out of the laboratory and into the real world. At the moment, experimental approaches to the chronic effects of illicit abused substance are highly constrained by ethical issues. For example, if one wants to know the long-term effects of consumption of an illicit substance on human behaviour, one cannot simply recruit drug-naive participants and repeatedly administer potentially dangerous drugs to one group and placebo to the other for years. Because of these constraints, the majority of current experimental psychological studies into the chronic sequelae of illicit drug use have employed cross-sectional designs, in which existing illicit drug users are assessed. These studies tell us little about the causality of any behavioural problems associated with particular patterns of illicit substance abuse.

The best approach to establishing cause and effect is to employ large, prospective longitudinal studies in which a cohort of people are recruited before they have used illicit drugs and then tested repeatedly before and after some of them 'naturally' progress on to substance abuse. Such studies are expensive and time-consuming. One way of improving data collection and reducing costs would be to exploit developments in the Internet and mobile phone technology to gather data on environmental stimuli, drug intake, cognitive performance, physiological state and behaviour, remotely and in real time outside the laboratory. The development of the Internet will also have an impact on our understanding of brain science and addiction. The ever-increasing availability and influence of the Internet means that more people are able to learn about and obtain psychoactive substances with relative ease. This, in itself, provides a unique mechanism for researchers to determine whether the increased availability of drugs results in greater levels of use in the general public.

The Internet also offers a unique tool for surveying which psychoactive properties are sought by different groups of individuals and whether new patterns of technology-related addictions are emerging, such as dependence on cognitive enhancing substances to improve computer game performance. Perhaps in the future people will play tailored video games instead of taking drugs. We already know that excessive striatal dopamine is released in volunteers playing a 'battlefield' video game. We might also want to deliberately stimulate the neural networks involved in reward in order to reduce the need for more harmful reward-seeking behaviours, although the deleterious effects on society of people solitarily playing computer games are also a consideration. An example of how advances in computational modelling might be used in the future has recently been put forward by Redish (2004). In this study a computational model of addiction was developed using temporal-difference reinforcement learning, in which learning only occurred when the reward was incorrectly predicted and an error signal present.

Fluctuations in dopamine like those which might follow a natural reward were then introduced as the error–reward signal, and, as predicted, the model achieved a stable response to this 'reward'. However, when a dopamine surge mimicking an addictive drug was modelled, the demand for reward increased disproportionately, and although the process still showed some elasticity (or connection between reward and cost), it was still very inelastic when compared to modelling for a natural reward. This type of computational modelling may help us better understand the mechanisms and factors that are involved in addiction, by providing a mechanism to explain and confirm observations and predictions about behaviour.

Cognitive Psychology

An important and emerging area of cognitive psychology is the effect of genetic subtypes on cognitive response. For example, functional polymorphism in the catechol-O-methyltransferase (COMT) gene (Val$^{108/158}$Met), which is important in dopamine metabolism, has been shown to predict cognitive performance. Patients with schizophrenia who have the low-activity Met allele show enhanced cognitive performance on tests of cognitive flexibility and have a more efficient physiological response in the prefrontal cortex (as measured using fMRI), compared with those with the high-activity Val allele. Egan *et al.* (2001) suggest that the COMT Val allele results in increased prefrontal dopamine decomposition and thus impairs prefrontal cognition and physiology, which might even predispose to psychiatric disorders. It would be of interest to determine whether patients who show improved cognitive flexibility following modafinil can be determined by their genetic subtype. With respect to drugs of abuse, Mattay *et al.* (2003) have shown that the effect of amphetamine on working memory performance and brain activity is modulated by genetic influences. Assessing the same COMT gene, they found that individuals with the Val/Val genotype had faster reaction times on a working memory task and less dorsolateral prefrontal cortex activity, when receiving amphetamine compared to placebo. In contrast, volunteers with the Met/Met genotype showed impaired performance on amphetamine and increased dorsolateral prefrontal activity at high working memory loads.

Genetic loading has important implications for the study of cognitive-enhancing agents. An understanding of which genes are specific to a disorder and to aspects of cognitive function is important to determine whether functional polymorphism has an effect on response to treatment, and to establish why people vary in their response to pharmacological intervention. Allelic variants in genes involved in dopamine metabolism have also been shown to influence the amount of tobacco consumed by smokers. These researchers found that certain allelic variants for the enzymes dopamine beta-hydroxylase and monoamine oxidase were associated with whether a person was a heavy smoker and how many cigarettes they consumed. They suggested that these enzymes are critical in determining a smoker's requirement for nicotine and could be used to explain why some people are predisposed to tobacco addiction and why some find it very difficult to stop smoking. In the future, we are likely to know increasingly more about allelic variations and their effects. This is likely to impact on our understanding of genetic predispositions to substance abuse. Kabbaj *et al.* (2004) have reported the use of micro-arrays to discern the 'neural signature' or 'neural phenotype' associated with individual differences in vulnerability to substance abuse, and susceptibility to the effect of psychosocial stress on drug-seeking and drug-taking behaviour, in mice. Better animal genotype models of patterns of human polydrug use, dependence and relapse will form an important part of future work.

Pharmacodynamic Properties

Understanding the precise pharmacokinetic and pharmacodynamic properties of pharmacological agents *in vivo* in humans, and determining their precise locations of action, will help contribute to our understanding of psychiatric illness and treatment. An important potential area for exploration with agents such as methylphenidate and modafinil is the visualisation of their effects *in vivo.* Positron emission tomography can be used to test the mechanisms of drug action and has the potential to provide rationales for further drug development, although a greater understanding of the effects of neurochemical changes downstream of the receptor site will be critical for explaining the mechanisms by which drugs exert their effects. It is nevertheless possible to correlate changes in receptor binding with behavioural measures. It has been shown that changes in raclopride binding potential in the ventral striatum, induced by amphetamine administration in healthy volunteers, correlate with the associated hedonic response of the drug. This suggests that the effects of amphetamine are, in part, mediated through action in the ventral striatum.

In the future, examining the effects of psychotropic drugs such as modafinil on raclopride binding potential would give an indication of whether any of the effects of this drug are mediated through downstream dopaminergic effects. This would help elucidate common neural pathways involved in the enhancement of cognition as opposed to addiction. The role of functional imaging in assessing the impact of treatment regimes is also emerging. It is obviously essential with fMRI that the physiological effects of the drug on brain tissue be understood, as they can have an important influence on the interpretability of results. Nevertheless, the effects of several psychoactive compounds on functional activation have already been measured by fMRI. For example, striatal activity in children with ADHD has been shown to be differentially affected by methylphenidate when compared to healthy children, with healthy children showing decreased activity from their baseline and hyperactive children showing increased activity from their lower baseline. This work showing differential responses to drugs might also have implications for identifying susceptibility to drug addiction.

CONCLUDING REMARKS

This section briefly highlights interesting but hypothetical future scenarios for brain science, addiction and psychotropic substances. For example, are we any nearer developing a 'perfect' reinforcing drug that could be taken for pleasure without any opponent motivational effects? Will cognitive and psychological drugs change in the future? How precise can we make cognitive enhancing drugs? Individual drug preferences depend on a number of factors, including the social status of the user, the desired effect, and ease of manufacture and access. It is perhaps more helpful to consider what types of substances will mark our future than to attempt to answer this question directly. It is likely that polysubstance abuse will continue to emerge as the dominant pattern of consumption, as we are not yet within reach of specialised, targeted single agents. We may also see the emergence of novel routes of administration in humans.

It is conceivable that even intracranial self- administration might become possible if the technology were refined for use in humans. Also possible is an increase in the prevalence of novel delivery methods such as implanted mini-pumps, patches, transcranial magnetic stimulation (TMS) and neural prosthetics. In the future it is likely that we will be able to systematically develop truly selective psychopharmacological agents based on a rational understanding of individual brain mechanisms. However, at present we are still some distance from entirely being able to understand the incredible complexity of the interactions within the brain, and most drugs described in this report have effects on individuals that are neither entirely predictable nor uniform in their actions. Some progress is being made towards determining individual responsiveness through the use of genetics, but the majority of pharmacological studies still use statistical differences between groups as the gold standard. This approach has yielded enormous benefits in the past, and will continue to do so in the future.

However, it is also perhaps worth considering how we might develop our experimental armoury to target the question of individual differences in terms of biological, genetic and environmental make-up in order to gauge what the psychoactive substances of the future will be. If we can develop an accurate mechanism for determining individual susceptibilities, vulnerabilities and responsiveness, it might be possible to envisage a society in which 'customers' were screened and then provided with a perfectly tailored range of treatments to ensure optimal performance and enjoyment, with the guarantee of no negative side effects or addiction. Psychopharmacological informatics would be used to determine the precise contexts in which, for example, cognitive enhancers might be expected to produce beneficial effects and when they might be deleterious. The obvious need for psychopharmacological counselling could result in a revamping of psychological and psychiatric practices. Practitioners might be trained to optimise individual performance in the context of individual genomic information. For example, the Yerkes-Dodson principle of optimal levels of arousal for effective performance has been used to explain differences in response to methylphenidate.

Mehta *et al.* (2000) showed that the beneficial effects of methylphenidate on working memory in normal adult volunteers were greatest in those subjects with lower baseline working memory capacity. However, the opposite effect was noticed in children with ADHD, where this time it was those with the highest scores who demonstrated the greatest improvement in spatial working memory following methylphenidate. However, it is inconceivable at the moment that humans will ever be able to dispense with our need for reward. Perhaps our 'substances of the future' will be non-drug activities that elevate mood and replace the need for drugs. Through advances in graphics technology and virtual reality we might well be able to create computerised environments or games that can be used to enhance mood and cognition in participants without any adverse effects. This is an attractive concept in terms of its potential to reduce drug- and alcohol-fuelled public disturbance and accidental deaths, and to increase positive social interaction. But it may be more likely that, instead of a single scenario, we will see innovation in both drug and non-drug areas.

Interventions

It is likely that we will see considerable refinements in our ability to predict abuse liability in humans. Screening for genetic polymorphisms as risk factors for substance misuse might become widely available in the future. Tailored psychological and pharmacological interventions to combat psychoactive substance misuse, perhaps even based on a person's genetic, psychological and developmental identity, are also likely. In addition, there is considerable debate about the recent emergence of potential 'vaccinations' against drug use, including cocaine and nicotine. The ethical issues are considerable and worthy of debate even at this stage.

Future impacts of psychoactive substances

More work is required to investigate the relationship between the cognitive impairments associated with substance abuse and their effects on future dependence. For example, the heightened impulsivity associated with ecstasy abuse, and the impaired executive function and memory seen with alcohol abuse, may render abusers more liable to dependence on other substances. A clearer understanding of these types of associations may help to improve treatment and management in the future. Future studies should also incorporate measures of personal financial management into surveys of patterns of substance misuse, to determine whether the selective neurocognitive deficits associated with particular patterns of abuse, such as elevated impulsivity in heavy ecstasy users and executive dysfunction in alcohol users, have a significant impact in areas such as credit card debt, as well as occupational, academic and social functioning. Our understanding of the biological and genetic factors driving substance misuse is also likely to grow.

For example, dopamine D2 receptor densities are relatively depleted in individuals at the lower levels of the social dominance hierarchy, and this could potentially result in these people finding substances of abuse more reinforcing. Thus, the extent of future substance abuse may well reflect relative disparities in perceived social status and social inequality. It is clear that our understanding and predictions of the future impact of psychoactive substances must be based on the experiences and knowledge of all interested disciplines if we are to maximise the benefit of agents such as cognitive enhancers, while minimising potential harms, and provide the best possible treatments and preventative strategies to people vulnerable to addiction. This report has considered how we will use experimental psychology to embrace the new opportunities presenting themselves as our insight into psychotropic drugs, addiction and cognitive enhancement deepens. Drug use is extremely prevalent and the age of first drug use is dropping. Many agents have enormous potential to improve the quality of life for numerous individuals and to benefit society. However, it is also important we not be complacent about the harms that many agents can cause. It is thus imperative to use experimental psychology paradigms to screen drugs to ensure the safest possible use of current and future psychotropic effects.

5

Clinical Testing of Psychoactive Drug

The term '*psychoactive drug*' is a loosely applied term used to describe substances that act on the *central nervous system* (CNS) to exert a psychological effect. There are many types of psychoactive drug which, for the purpose of this chapter, are considered to fall into two categories: those that are widely used for medicinal purposes and although they have some abuse potential are largely not abused, and those that have a significant abuse potential. The purpose of drug testing is to determine whether an individual has used any particular substance that might have affected their well-being or behaviour. The most common scenarios where testing is undertaken to detect psychoactive drug use include workplace drug testing, forensic drug testing (e.g. with drugs and driving) and compliance with treatment.

Most drug testing is restricted to these areas and is not widely applied to the analysis of psychoactive drugs used in medical practice unless they pose a threat to the safety of the individual. In workplace drug testing and treatment compliance testing, the objective of the procedure is to extend the window of detection to determine whether or not the subject had been taking drugs over a period of days or possibly months prior to specimen collection. In forensic drug testing, interest is focused more on whether an individual may have been affected by a drug substance at the time of a particular event such as driving a car or at the time of an assault. In order to obtain the most appropriate information, therefore, different types of samples may be used, depending on the reason for testing. Testing for drugs of abuse evolved in the United States during the 1960s within the armed forces and was used extensively to monitor members of the armed forces on active duty in the 1970s.

All early drug testing and most testing today is based on urine samples. As technology and our understanding of how drugs behave in the body have increased, screening tests have become available that may be used on a range of samples. The significant increase in the availability of psychoactive drugs in the mid- to late-1990s stimulated an increased interest in drug testing that resulted in the development of tests suitable for use within the workplace (point-of-care) as well as in the laboratory. At the present time, testing for common drugs of abuse, whether in a clinical, forensic, sport or workplace setting, is usually undertaken as a two-step process. The first step normally involves a simple and inexpensive screening test followed by a more specific and definitive identification test (called a confirmatory analysis) that may also be quantitative. This review outlines the techniques currently in use for the screening and confirmation of commonly encountered psychoactive drugs and then considers how advances in technology and miniaturisation might be applied over the coming years.

Discussion centres on the types of test available for the analysis of common drugs of abuse and medicinal drugs, as it is not possible to predict with any certainty what new psychoactive drugs will be introduced during the forthcoming 20 or more years. It might reasonably be predicted, however, that as advances are made in the miniaturisation of electronics and microfluidics, tests will be developed for near-patient use that will combine both screening and confirmation in a single test, and that the range of tests on offer will be extensive. Most of the psychoactive drugs subject to misuse today have been used for centuries, including alcohol, opium derivatives, cannabis and cocaine, and it is envisaged that these, together with amphetamine and related compounds, will continue to be the most popular substances used for the foreseeable future. The most widely used psychoactive drugs (alcohol, heroin, morphine, cannabis and cocaine) are derived from natural sources or are easily synthesised from widely used industrial reagents in the case of synthetics such as amphetamines.

These substances are easy to produce in large quantities at relatively low cost. Until supplies become limited there is little incentive for users to switch to alternatives. Although methods for the synthesis of numerous synthetic psychoactive amines have been published few of these substances have attracted widespread use and few are regularly encountered other than amphetamine, methylamphetamine (MA), methylenedioxyamphetamine (MDA) and methylenedioxymethyl-amphetamine (MDMA, i.e. Ecstasy). In addition to these common drugs of misuse, there are numerous psychoactive substances used medicinally that can affect performance, including the benzodiazepine anti-anxiety and sleep-inducing agents (e.g. diazepam, temazepam and flunitrazepam), hypnotics (e.g. zopiclone, zolpidem and zaleplon) and antipsychotics and antidepressants (e.g. fluoxetine and amitriptyline). The analytical methods employed today in drug testing for psychoactive substances are determined by the chemistry of the individual substances.

As our understanding increases about how and where different drugs act in the body, drug testing may turn towards the measurement of endogenous markers within the body, for example, the monitoring of neurotransmitters and neurotransmitter binding sites rather than the drugs themselves. But for the next 5-10 years at least, testing is likely to remain orientated towards the determination of the presence of psychoactive substances or their breakdown products. It is already possible to use electromagnetic scanning techniques to monitor changes to receptor binding in certain circumstances, but this technique is costly and therefore unlikely to be used routinely for drug-abuse screening.

The analytical techniques used for drug detection also depend on the chemical and physical properties of each substance or its breakdown products. Consequently, there is no universal test that can be applied to detect the presence of a drug and it is necessary to employ a range of different tests for different drugs.

BIOCHEMICAL TESTS

The chemistry of any particular drug and its biological breakdown products (metabolites) determines how it is distributed within the various body tissues and fluids, and therefore the time-window during which it may be detected in the various body secretions and excretions. Drugs that are lipid (fat) soluble, such as cannabinoids, tend to accumulate in the body's fat depots and may subsequently be released over a period of days or weeks into the general circulation of the body, whereas those that are more water-soluble can be eliminated within a few hours. The time spans over which drugs can be detected in different tissues may be influenced by a number of factors.

These include the state of health of an individual, presence of disease, age, the properties of the drug and its distribution within the body, the quantity of drug taken, whether drug intake involved single or chronic dosing, and the sensitivity of the analytical methods used. Historically, urine has been collected for drug testing as it is easy to collect, may be collected in large volumes and is easy to analyse. However, as a consequence of improvements in analytical technology, blood, oral fluid, sweat and hair are now being used with increasing frequency, depending on the rationale behind the test. Urine, oral fluid, sweat and hair are favoured for use in many drug testing applications because they can be collected non-invasively by suitably trained lay people. The collection of blood, however, is regarded as invasive and requires the use of more highly trained collectors such as doctors, nurses and phlebotomists.

Analyzing Urine

Urine is the most widely used specimen employed for drug testing. Urine is produced in large quantities, is easy to collect and offers a wide window of detection, often facilitating the detection of drug use over a period of 2-4 days after the last dose was consumed. It should be noted that detection times can only be regarded as approximate since they are dependent on a number of factors, including the route by which the drug entered the body, the quantity taken, the frequency of use and the sensitivity and limits of detection of the assay.

Table 5.1 Approximate detection times for commonly used psychoactive drugs in urine.

Drug	*Approximate detection time in urine*
Amphetamines	1–2 days
Benzodiazepines	0.5–7 days (depends on half-life)
Cannabinoids (single use)	Up to 3 days
Cannabinoids (moderate use-four times a week)	4 days
Cannabinoids (chronic use)	21 days or longer
Cocaine metabolites	2-3 days
Gamma-hydroxybutyrate (GHB)	0.5 day
Ketamine	2-4 days
Methadone	7-9 days (maintenance dosing)
Methamphetamine (single use)	1-3 days
Methamphetamine (heavy use)	3-5 days
Opiates	2 days

In many ways urine is an ideal matrix for drug testing since it is relatively easy to collect and analyse and, depending on the elapsed time between drug use and specimen collection, drugs may be present in much greater quantities than in other matrices. But because of the ease with which chemicals that may interfere with the analysis may be added to specimens during collection, observed collection may be necessary to prevent specimen adulteration. Observed collection is costly, unpleasant for both the donor and collector, and is regarded as a potential infringement of personal privacy.

Analyzing Blood

Blood is primarily used in forensic and some areas of clinical testing and is not commonly employed when screening for drugs of abuse. The advantage of blood analysis is that it provides information about how much of the drug is present in the circulating blood and can give an indication of whether the donor of the specimen is likely to be affected by the drug at the time of collection. But blood collection requires specially trained personnel and can be costly, its collection is regarded as invasive, and current analytical methods require extensive sample pretreatment prior to analysis, which is expensive.

Table 5.2 Advantages and disadvantages of blood drug testing.

Advantages	*Disadvantages*
Provides indication of recent use	Requires specially trained personnel to collect specimens
Difficult for donor to adulterate	Low concentrations of drugs present, hence relatively short window of detection
Results can be interpreted	Analysis requires specialised laboratory facilities
	Analysis is complex, therefore tends to be costly
	Current technologies not developed for point-of-care testing

The advantages and disadvantages of using blood for drug testing are summarised in Table 5.2. One of the potentially exciting consequences associated with the developments of miniaturisation combined with microfluidics is the possibility of being able to analyse blood without the requirement for complex and expensive sample preparation procedures. The use of blood for general non-clinical drug testing may not develop, owing to the issues surrounding its collection.

Analyzing Oral Fluid

Recent advances in a technique known as lateral flow immunalysis have resulted in the development of point-of-care drug tests for the analysis of oral fluid, a term that covers fluid collected from the mouth, including saliva and other secretions, together with cellular and food debris. It was originally believed that drugs secreted via saliva reflected the concentration of drug present in the circulating blood and that collection of this matrix would enable results to be interpreted. More recently, it has been demonstrated that the concentrations of some drugs in oral fluid (e.g. mono-acetylmorphine, morphine, codeine, cocaine and benzoylecgonine) may be exceptionally high and would not be representative of concentrations in the circulating blood. The most likely explanation for this phenomenon is buccal absorption following smoking or snorting (inhaling the drug in powder form), since the effect is particularly noticeable in association with drugs that are smoked or snorted, such as heroin, cocaine and crack cocaine. Oral fluid is particularly attractive as a matrix for drug testing as it is inexpensive, easy and non-intrusive to collect, requires no special facilities, is difficult to adulterate and does not require complex sample preparation prior to confirmatory testing. It can be reasonably predicted that methods for oral-fluid drug screening will predominate during the next 5-10 years.

Analyzing Sweat

Many drugs are excreted in sweat and hence this matrix can be used as a source material for drug testing. Few manufacturers have devoted attention to the use of sweat as a routine specimen for drug testing, possibly because it is not easy to collect in large quantities for multiple drug screening and subsequent confirmatory analysis. Two types of test are currently available for testing drugs in sweat. One comprises a small absorbent pad that is wiped across an area of skin such as the forehead of an individual to collect a small quantity of sweat that can be subsequently analysed by lateral flow immunoassay. The other system comprises a small absorbent patch enclosed within a tamper-evident adhesive cover that sticks to the skin rather like an Elastoplast dressing. The patch is applied after cleaning the skin by swabbing and left in place for a number of days, during which excreted drugs are absorbed onto the pad prior to collection. The sweat patch is subsequently sent to the laboratory for analysis.

The principal use of sweat patches is for the screening of prison inmates in the US. The major disadvantages of sweat testing are that only very small amounts of sweat are collected, thus limiting the extent of analysis with the present state of technology, and that skin can potentially become contaminated with drugs via environmental exposure prior to collection of the specimen. If a point-of-care device is used, prior swabbing of the skin to remove external contamination would also remove traces of drugs excreted in sweat. Although sweat testing is possible, these drawbacks have limited its take-up and we envisage that it will have only limited application in the future.

Analysis of Drugs in Hair

The introduction of methods for the analysis of drugs in hair has developed rapidly in recent years because hair theoretically offers a non-intrusive way to monitor an individual's drug-taking history over several months or even years. Drugs enter the hair during its formation within the scalp and also by diffusion into the hair from the outside environment or via sweat and sebaceous secretions. Since drugs can enter hair following environmental exposure it is desirable that testing is designed that can detect drug metabolites, since these can only be formed within the body and allow the analyst to distinguish between drug use and external contamination. Hair may be used to back up disputed urine or oral fluid tests, particularly where an individual claims not to be a regular drug user and that, for example, his or her drink was spiked without their knowledge. In these circumstances, with hair analysis it may be possible to distinguish one-time from chronic drug use.

Drug binding and subsequent retention in hair may be influenced by the types of melanin present (black or blond hair) and is thus subject to potential genetic and racial bias. Hair testing is currently confined to laboratory testing and is relatively expensive. The main uses of hair testing at the present time are in child custody dispute cases where one parent may be alleging the other to be unsuitable to have custody of a child because he or she is a drug user, or in cases where a urine drug test has been disputed.

ROUTINE TECHNIQUES

Screening tests usually employ antibodies directed at a region of a test molecule, which is usually shared with other structurally related compounds. For example, the opiate drugs (heorin, morphine, codeine, dihydrocodeine) are structurally related to morphine, and an antibody raised against morphine will also react with codeine and heroin, among others. This is called cross-reactivity. A positive result in an opiate assay is therefore not specific enough to be of diagnostic value, but differentiates between someone who has taken an opiate and someone who has not. An immunoassay is a flexible technique that can provide a rapid, inexpensive and convenient method to screen large numbers of samples in a variety of matrices and allow the differentiation of negative from positive specimens, thus preventing the need for further, more complicated and expensive processing or investigation. The reaction between antibody and drug molecule is measured by using a label which emits a detectable signal. Labels used in immunoassay include radioisotopes (RIA), enzymes, chemiluminescent molecules, particles such as colloidal gold, latex beads, rare earth metals, and fluorescent molecules.

Immunoassay formats vary from systems designed for rapid point-of-care testing of a single sample through to fully automated systems that can process thousands of samples per day. *Enzyme immunoassay* (EIA) and *enzyme-linked immunosorbent assay* (ELISA) are the most frequently used drug-screening techniques, and are often used interchangeably to describe non-radioisotopic assays. EIA and ELISA have largely replaced RIA as the assay system of choice because radioisotopes have more health and safety constraints on their use than enzyme-based reactions. Immunoassay is the preferred screening technique in many areas, e.g. therapeutic drug monitoring, toxicological and workplace drug testing.

The main difference between these applications is the degree of quantification required, and the selection of antibodies with the required cross-reactivity profiles. For example, drug-monitoring assays may require highly specific antibodies that are reactive only with a single chemical entity, whereas immunoassays used in toxicology or drug-abuse screening may be designed to be broadly cross-reactive with a whole drug group. A further criterion for assay selection is whether the result is required immediately, or whether the collected sample can be sent to a laboratory for analysis. Most immunoassays are performed in the laboratory, either using plates with 96 or more reaction wells, allowing many samples to be analysed simultaneously, or on large autoanalysers can process many hundreds of samples per hour.

Homogeneous immunoassays do not require the separation of bound and free fractions, and as a result have been automated successfully in a number of different systems. Examples of such assays include *enzyme multiplied immunoassay technique* (EMIT), *cloned enzyme donor immunoassay* (CEDIA), *fluorescence polarisation immunoassay* (FPIA) and *kinetic interaction of microparticles in solution* (KIMS). Point-of-care single-use cartridges are basically variants of a competitive ELISA assay run on a solid-phase strip rather than inside a well or tube. These tests employ a heterogeneous lateral flow mechanism using antibodies labelled with colloidal gold or other reactant particles which must be separated from the matrix before development of the result. A drug-protein derivative is bound to a nitrocellulose membrane in a band at a defined position on the strip. An antibody labelled with colloidal gold, coloured latex or an alternative visualisation probe is located in a pad that overlaps with the nitrocellulose membrane close to the point of the test strip where the sample is introduced.

When urine, diluted oral fluid or some other liquid that might contain drugs is added to the pad, the labelled antibody migrates from the pad onto the cellulose membrane by capillary action. As the sample and rehydrated gold-labelled antibody flow along the membrane, any drug in the sample binds to the labelled antibody and inhibits it from binding to the immobilised drug on the membrane. Separation of bound and free drug continues with the movement of liquid along the membrane strip. Where a specimen contains no drug, the antibody binds to the immobilised drug forming a visible line. In specimens containing drugs, the antibody binding sites are taken up by the drug in the specimen resulting in no antibody binding to the immobilised drugs on the membrane. A positive result is therefore indicated by no coloured band being formed on the strip, whereas a line will be observed if the donor's specimen does not contain drugs. It is possible to add a number of drug-derivative bands in different positions along the length of the nitrocellulose membrane together.

With suitable matching labelled antibodies this allows the analysis of multiple drugs to be undertaken simultaneously in a single test cartridge. Most commercially available test strips are interpreted visually by the test operator. However, the criticism has been offered that visual interpretation may be subjective. A number of point-of-care test strips are now being developed by the manufacturers to be read using small portable readers based on digital imaging to facilitate more accurate and consistent results. Lateral-flow immunoassay test strips based on colloidal gold-labelled antibodies are currently limited to performing up to a maximum of around five or six tests per strip. This is usually sufficient for a drug-testing clinic, or for an employer who is mainly interested in determining whether a subject has used commonly available drugs of abuse. The up-converting phosphor system developed by OraSure Technologies, however, has the potential to offer up to 15 or more tests per strip. This could in theory offer clinicians much more information, perhaps including tests to determine whether individuals were infected with HIV, hepatitis and other drug-related diseases. The future of immunoassay-based screening tests for drugs other than those currently subject to misuse is limited by the availability of suitable antibodies, and by market forces.

The cost involved in the development and production of suitable antibodies is high and unless substantial commercial sales are predicted, companies involved with the production of test kits are reluctant to risk investment. The result is that there is always a significant time lag between a new drug being encountered and the availability of an immunoassay test for the new drug. Past experience has shown that drug test manufacturers have been reluctant to develop screening tests for the European market alone as this was perceived to be too small to be commercially viable. Given the assumption, however, that a suitable commercial market is available, the next predicted development in immunoassay screening is likely to involve the miniaturisation of tests so that 20 or more assays can be performed on a microchip platform. At least one company is currently developing an automated laboratory-based microchip system capable of offering the rapid screening of several hundred urine or oral-fluid samples per day for common drugs of abuse. Within a few years, such a system could be miniaturised to facilitate portability. The limitations associated with immunoassay testing might stimulate the development of drug testing based on different technologies in the future as miniaturisation is developed further.

Fuel Cell Technology

Fuel cell technology has been in use since the 1980s for measuring the amount of volatile substances, and in particular alcohol, in breath. Because fuel cells are small and relatively inexpensive, they are ideally suited for use in small portable or hand-held screening devices and have found widespread application for use by police officers at the roadside and also in workplace drug testing. It is now possible to link alcohol-detecting fuel cells to electronic locks that can only be opened when an alcohol-free specimen of breath has been supplied. Such devices have been proposed to control entry onto the bridge of a ship or to prevent drivers from being able to start their car engines if they have more than a predetermined concentration of alcohol in their breath. In principle this offers a means of preventing an intoxicated person from driving while under the influence of alcohol, but is not totally secure and has not been widely implemented at the time of writing. The use of devices as a means of harm reduction is still rather limited. Fuel cell technology is particularly suitable for the analysis of volatile solvents and is unlikely to be applicable to the analysis of non-volatile substances or solid-dose formulations.

Screening Test

Any positive result from a screening test must be subjected to further, more specific, confirmatory analysis to identify and quantify the substance taken. The acknowledged '*gold standard*' technique for drug identification is *Mass Spectrometry* (MS), coupled either to a *gas chromatograph* (GC-MS) or *liquid chromatograph* (LC-MS) to separate the compounds in a complex mixture. The mass spectrometer is a detector that produces a signal based on the structure of the analyte itself, and so is compound-specific. There are several types of mass spectrometer, but the most common for drug analysis use the quadrupole, ion trap and *time-of-flight* (TOF) mass analysers. In some clinical and toxicological drug testing laboratories, LC-MS, or more specifically its powerful sibling technique LC-MS-MS, is being used to screen and confirm medicinal drugs and drugs of abuse. The cost of these systems can be prohibitive, but high-throughput applications make it more cost-effective. Mass spectrometry in its various configurations (e.g. GC-MS, or LC-MS, or TOF) is highly versatile and can be used to analyse common drugs of abuse and most drugs used in medical practice.

Liquid Chromatography-Mass Spectrometry (LC-MS)

This is one of many hyphenated techniques that combine the separation of components in complex mixtures such as biofluids with a sensitive means of detection and identification. In this case, the separation is facilitated by a flow of liquid under pressure through a short (10-20 cm length, 2-5 mm diameter) metal column packed with a silica-based medium. The components in the mixture will ideally interact with the medium to differing degrees and so will separate spatially. As the individual components leave the column, they pass into the mass spectrometer, where controlled fragmentation occurs. Measurement of the abundance and masses of the fragments produced gives a very specific '*mass spectrum*' of the compound, which is then searchable against standard reference libraries of spectra to assist in obtaining an absolute identification of the compound. The separation in this case is usually done at room temperature, making it ideally suited to the analysis of thermally sensitive compounds. It is also suited to the analysis of large molecules (e.g. small peptides), parent drugs and their conjugates (especially glucuronides and sulphates), which can provide information on the time since ingestion or administration. One example is the simultaneous measurement of free (non-conjugated) morphine and its more water-soluble glucuronide metabolites.

A further benefit of this technique is that sample pretreatment and derivatisation are often unnecessary (unlike with GC-MS), which saves time and money. LC-MS is routinely used in toxicology, research and pharmaceutical laboratories, and many methods are available in the scientific literature for the separation, detection and quantification of many drugs and their metabolites. An enhancement of this technique, LC-MS-MS, involves the addition of a second fragmentation facility and a second mass spectrometer. This can be used to generate structural information on individual fragments produced from the first fragmentation. These 'daughter' mass spectra provide additional discrimination where structurally similar compounds produce the same mass fragments.

The use of MS-MS also improves sensitivity and is suitable for the analysis of a wide range of drug compounds that exert psychoactive effects. There are published methods which can screen a single sample extract for several compounds simultaneously using LC-MS-MS, demonstrating its analytical capability and value. It is likely that LC-MS-MS will soon replace GC-MS as the routine laboratory analytical tool for drug analysis. Modern LC columns are now being developed to provide greater resolution, or separating power, but require higher pressures to push the liquid through the columns. It is envisaged that these higher resolution systems will become routine in the next 10 years.

Gas Chromatography-Mass Spectrometry (GC-MS)

This is probably the most commonly used analytical technique in all fields of drug testing. Often referred to as the 'gold standard', this technique effects separation using the flow of a gas (usually helium) to carry the vaporised sample through a long (15-60 m length, 0.2-0.5 mm diameter) column whose inner surface is coated with a silica-based medium. The entire column is held in an oven, and heat is used to effect separation of the compounds from the medium. The sample must be soluble in a volatile organic solvent to allow vaporisation upon introduction into the gas flow, and must be thermally stable to withstand the high temperatures (up to 300°C) required to achieve separation. In order to produce satisfactory separation and resolution, it is frequently necessary to perform sample clean-up (either liquid/liquid or solid-phase extraction) and then derivatisation, the addition of a chemical 'cap' to cover up potentially active functional groups. This takes longer, but the result is better chromatographic separation than LC because the derivative has a higher mass and so produces more discriminatory mass spectra.

The mass spectrometer is the same as that used in LC-MS, but there are very different methods of sample introduction to accommodate either a high pressure flow of liquid solvent, or helium gas. Many toxicology laboratories will use a combination of LC- and GC-MS to enable the identification and quantification of any drug or metabolite known to have been ingested, or indicated from a preliminary screening test. Instrumentation is moderately expensive but, with automated sample introduction, large cost-effective batches can be processed unattended. As with LC-MS, it is possible to add further fragmentation capability to give GC-MS-MS, which has the same benefits of enhanced specificity and discrimination.

Nuclear Magnetic Resonance (NMR) Spectroscopy

This technique, like mass spectrometry, measures a physical characteristic of a molecule (or more precisely, the nuclei of its component atoms), producing specific information about the structure of the molecule. The property in question is '*spin*', with an energy change due to the spin of an atomic nucleus being measurable in a strong magnetic field. It is especially useful for detecting molecules containing isotopes of carbon, hydrogen, nitrogen, fluorine, silicon and phosphorus, which covers the vast majority of therapeutic and abused drugs and their metabolites. There are functional groups that do not produce NMR spectra, and so this technique is sometimes used to complement other analyses such as MS. It is possible to couple a liquid chromatograph to the NMR instrument, allowing spectra to be obtained from complex biofluids, although low sensitivity is a problem for trace analysis. NMR spectroscopy is commonly used on pure compounds, solid dosage forms or where a high concentration of analyte is present.

This instrumentation is not routinely used in analytical laboratories, but is used in the pharmaceutical industry to provide structural information capable of distinguishing polymorphic forms of the same drug. A development of NMR technology is used in the diagnosis of disease states *in vivo*, with the aid of magnetic resonance imaging (MRI) scanning instruments. This points towards the future possibility of noninvasive drug detection using the same or similar techniques. We predict that this technique will be increasingly applied to the analysis of psychoactive drugs and their metabolites at their site of action in the body (body scanning) and to study their interactions with receptors.

Capillary Electrophoresis (CE)

This is purely a separation technique and must be coupled to a suitable detector to produce a measurable response. Instead of using a flow of gas or liquid to transport the sample through a separation medium, the innate electrical charge on the drug molecule is used to separate it from other molecules of differing size and charge. In classic electrophoresis, the sample is introduced into a channel, usually in a gel, and a strong electric field is applied that effects separation based on size as well as charge, small molecules travelling further and faster than heavy ones. Traditionally, this technique has been used to separate high molecular weight molecules such as proteins and DNA fragments, but smaller molecules can also be separated in this way.

In principle, any molecule which is naturally charged or can be made charged in a buffered solution can be analysed using CE. The discovery that the same principles could be applied to molecules travelling through a narrow silica capillary, and produce more focused results, has led to considerable interest in CE for drug analysis. CE offers the ability to analyse most types of psychoactive drug and their metabolites including peptides, but suffers from the fact that it requires a very sensitive device to detect the small concentration of components submitted to this technique. Capillary electrophoresis is seen as complementary to LC, but with the advantage of utilising smaller sample volumes (~10 μl).

More analytes are detectable than in LC because negatively charged anions, positively charged cations and non-charged neutrals can be detected simultaneously. Although it is less sensitive than LC, CE lends itself to miniaturisation, and the potential for portable (or even handheld) devices is evident. Instrument costs are similar to those for LC, but CE is much cheaper to run, with much lower solvent requirements. Possible detectors include ultraviolet (UV) absorption, laser-induced fluorescence or even MS. CE as an analytical technique is becoming more popular, many drugs and their metabolites have been analysed by CE and the number of published methods is growing, but it has not become widespread or routine in many laboratories because of the popularity of LC.

Thin-layer Chromatography (TLC)

Thin-layer chromatography is probably the oldest analytical technique still in use today, although its use is very much diminished with the advent of GC and LC techniques. Separation is effected by the flow of solvent up a silica-coated plate standing in the solvent (or mobile phase) causing the migration of components in the same direction; much like LC, only in one dimension. Once the solvent has reached the top of the plate, the run has finished and a visualisation reagent must be added to show up the position and quantity of the components. There are no general methods for drugs as a class, but there are a large number of methods for individual drugs defined by their therapeutic or chemical categories. Most use colour development as the detection method, while some rely on induced UV fluorescence, for example, for LSD and psilocin. Thin-layer chromatography methods are most effective for the low-cost analysis of a large number of samples (e.g. drug screening in biological fluids and tissues, and in herbal preparations), for example, the Toxi-lab TLC system, which is widely used in clinical toxicology laboratories for qualitative screening of tissue and fluid samples.

Supercritical Fluid Chromatography (SFC)

This is another separation technique that must be coupled to an appropriate detector. It uses carbon dioxide at relatively high temperature and very high pressure, producing the supercritical fluid, and this is the carrier medium in the column. This technique is not widely used in drug testing or toxicology, although it offers the potential for analysing a wide range of psychoactive drugs.

Electrochemical Detection (ECD)

This is a very sensitive LC detector which measures the current produced by the electrolytic oxidation or reduction of analytes at the surface of an electrode. These detectors are quite sensitive (down to 10^{-15} mol) and also quite selective. Two types of detector are available. The coulometric detector has a large electrode surface at which the electrochemical reaction is taken to completion. The amperometric detector has a small electrode with a low degree of conversion. Despite the difference in conversion rate, in practice these two types have approximately the same sensitivity. Samples for ECD must be electrically conductive, and this is accomplished by the addition of inert electrolytes. These detectors are difficult to maintain, and so do not lend themselves to routine analyses and are not used in high-throughput drug testing laboratories.

ADVANCED METHODS

Miniature Instrumentation

It is likely that there will be a large increase in *miniature instrumentation* that does not require skilled operators available for use at point of care, by the roadside, or in the office. These systems rely on a variety of physico-chemical principles that are currently being refined. A recent example uses a disposable potentiometric sensor. The speed of development and hence implementation of these systems will depend on the level of investment, which is relatively large since innovation in this field involves an interdisciplinary approach including physicists, analytical scientists and engineers. Commercial investment is limited because the market is insufficiently large. The NanoChip microelectronic array is the inner component of the NanoChip cartridge and is a 100-site electronically powered microarray fluidic device for DNA and RNA analysis. Nanometre-sized low-dimensional semiconductors known as quantum dots (QDs) are also being investigated for use in drug and poison detection. One recent application of this technology has been in the development of a sensitive test for cyanide.

Time-of-flight (TOF) Mass Spectrometry

Time-of-flight mass spectrometry provides an example of where mass spectrometry is already beginning to be miniaturised to provide portable testing devices for the detection of solid-dosage forms of drugs (e.g. tablets, powders, and drugs on banknotes) and explosives. Advances in this field have received considerable attention as a response to the increase in drug trafficking (e.g. their use by Customs and Excise) and the potential threat from terrorism, as they can be used to screen for explosives at airports and sea terminals. TOF technology will play an increasing role in the laboratory for the analysis of a wide range of psychoactive drugs in biological fluids and tissues.

Isotope Ratio Mass Spectrometry (IRMS)

Although most psychoactive substances are foreign to the human body and their presence in body fluids is evidence of administration, some substances are virtually identical to those produced naturally in the body. Examples of these include testosterone, misused by sports competitors at least partly because of its behavioural effect, and *gamma-hydroxybutyrate* (GHB) allegedly used in drug-facilitated sexual assaults. Since the source of these administered '*pseudoendogenous*' substances is different from the naturally produced endogenous material, a subtle difference in their natural isotope composition is often present. The IRMS technique has been used to detect sports cheats misusing testosterone, and work is underway to use the technique to provide evidence of GHB use. Unfortunately, the technique has relatively limited sensitivity at present, requiring about 20 ng or more at the point of analysis, but is likely to improve by at least one order of magnitude over the next 10 years. The main use of this technique in the near future (2-8 years) will be to distinguish between endogenous and exogenous forms of the same substance in a biological specimen.

Coated Microsphere Technology

This technique is suitable for the quantitative measurement of up to 100 different analytes in each sample and generally relies on immunoprocedures. Thus, any substance for which suitable antibodies have been prepared should be amenable to analysis using this technique. The principle of operation is by the use of polystyrene microspheres (5.6 μm diameter in xMAP technology) that have been internally dyed with two spectrally distinct (red and infrared) fluorophores in appropriate precise proportions, enabling the production of up to 100 distinct sets of microspherical beads. Each of the bead sets may be coated with the appropriate antibodies to capture the analytes of interest, with a third fluorophore acting as a reporter of antibody binding. Multiple sets of the appropriately coated beads may be placed in 96-well microtitre plates and the samples of interest added to each well.

The contents of each well may then be sampled with a suitable analyser (e.g. Luminex 100). Using a system of microfluidics, the beads in the sampled microtitre plate are sequentially excited via two different wavelength laser beams to identify each microsphere (in order to determine which antibody is present) and also to detect the presence of any reporter dye captured during the assay. With suitable automation such as using the Gilson 215 Multiprobe fluidics system, up to 32 000 wells per day have been analysed with, in theory, up to 320 000 different measurements of analytes. This technique has been proven for proteins and DNA. To date, it has not been fully explored for small molecule analysis but it has been used to measure the thyroid hormone T4.

Chemiluminescent Nitrogen Detector (CLND)

The chemiluminescent nitrogen detector (CLND) is specific for nitrogen-containing compounds, although it can also be set up to detect sulphur containing compounds. It is based on the chemiluminescent gas-phase reaction between ozone and nitric oxide which produces nitrogen dioxide formed in an electronically excited state. Relaxation of this excited state gives rise to light emission of 600-900 nm. The intensity of the emission is proportional to the mass flow of the nitrogen dioxide through the detector. The technique has been successfully coupled with gas chromatography, liquid chromatography and SFC. The total nitrogen content can be determined quantitatively whatever the state of the nitrogen (except for diatomic nitrogen N_2) and most drugs of abuse contain at least one nitrogen atom.

Sensitivity is of the order of 5pmol. Since SFC shows great promise for the separation of optical isomers of many substances, it is likely to be useful for the ready distinction of certain drugs of abuse, especially amphetamine, that are marketed legally in a single isomer (dexamphetamine), whereas much of the substance sold on the black market contains both forms of the molecule. Similarly, dextromethorphan is sold in cough mixtures and is not a controlled drug, whereas levorphanol is controlled under the Misuse of Drugs Act. Since levorphanol might need to be distinguished from dextrorphanol (the N-demethylated metabolite of dextromethorphan), SFC with CLND is likely to be a technique of the future.

Surface Plasmon Resonance

This is a detection technique that allows the quantitative estimation of one or more analytes concurrently within one assay, relying on the use of suitable immunoprocedures. Suitable antibodies are bound onto a thin metal (typically gold) surface that is coated onto a quartz prism. Total internal reflection occurs when light travelling in the quartz, which has a greater refractive index than at the interface, being in an aqueous solution, arrives at an angle of incidence above a critical angle. However, the light may be maximally coupled into the metal surface through the prism in such a way that, at the appropriate incident angle, the reflectivity is minimised.

The incident light then feeds plasmons (the collective resonance of electrons near the surface of metal islands) in the metal and an enhanced evanescent wave is produced. Biospecific interactions occurring on the metal layer may be observed using this phenomenon since the interaction will change the solute concentration and hence its refractive index. Thus, suitable antibodies may be bound to the metal surface and the concentration of substance bound to the antibody quantified. No tagging or chemical derivatisation is required. The technique is extremely sensitive and currently achieving quantitative sensitivities as low as 2×10^9 molecules, i.e. femtomoles of molecules of molecular weight of about 1000. At least one further order of magnitude of sensitivity is anticipated. The technique can readily be adapted for multi-sample parallel analysis, for example, by using a beam spreader to focus the beam of light on a line of samples and using a charge-coupled device to measure multiple reflectivity responses simultaneously. More than 100 parallel channels could be analysed simultaneously in the near future.

Surface-enhanced Resonance Raman Spectroscopy

Surface plasmons can also enhance Raman scattering. In contrast to surface-enhanced Raman spectroscopy, this allows the use of a smooth metal or single-crystal metal surface. Thus, in a similar manner to SPR, the interactions of biomolecules at the surface may be observed.

Total Internal Reflectance Fluorescence (TIRF)

Total internal reflection occurs when light travelling in a medium of higher refractive index such as quartz reaches an interface with a medium of lower refractive index such as an aqueous solution at an angle of incidence above a critical angle. Although the fully reflected light does not lose any net energy across the interface, an electrical field wave known as an evanescent field wave is produced that extends beyond the interface surface into the medium of lower refractive index, typically to a depth of about half of the wavelength of the light. The technique may be applied to fluorescent or fluorescently labelled compounds of interest. Only fluorophores adsorbed or otherwise in intimate contact with the interface will be excited and hence fluoresce, whereas those in bulk solution will not. The surface may be made biologically active, for example, with appropriate antibodies to substances of interest, so that fluorescently labelled compounds may be trapped at the surface. Since the excitation light is totally reflected away from the interface by total internal reflection, discrimination of the fluorescence signal from the excitation light is relatively effective, and so very good sensitivity and low limits of detection are achievable. The technique is non-destructive and rapid and more than 10 000 times more sensitive than other biosensor systems based on SPR.

CONCLUDING REMARKS

It is predicted that new developments in analytical drug testing will be driven by commercial forces dictated by the potential market size. Although many of the technologies discussed above have the potential to be developed for drug testing, development will be dependent on market demand and manufacturers are unlikely to invest in the high cost of research and development without a clear indication of demand. So it is not possible to predict how fast, if at all, some of the techniques discussed in this review will be developed for drug testing.

It is envisaged that emphasis will be directed towards producing more specific and reliable point-of-care tests to enable employers, those involved in law enforcement, and clinic-based practitioners to determine whether an individual may have used drugs. Oral fluid offers many potential advantages over urine and is becoming a method of choice in certain areas. Technical advances should make a strong impact in developing the scope for screening drugs in oral fluid still further within the next 5-10 years. The application of pharmacogenomics and our ability to prescribe drugs based on knowledge of an individual's genetic profile could well become a reality within five years, and some of this knowledge will have applicability to interpretative toxicology. Following the completion of the human genome project, microarray tests are currently being developed that will enable genetic variations between individuals to be determined. Such tests will, for example, enable physicians to analyse genes encoding drug-metabolising enzymes of the cytochrome P450 complex.

Variations in these genes influence the metabolism of several drugs including antidepressants, psychotics, beta blockers, analgesics and some anticancer compounds. By enabling physicians to access information concerning an individual's genetic make-up, it should be possible to help prevent harmful drug interactions and to assure that drugs are used optimally. Near-patient testing should be developed in this area within the next 5-10 years. Most drugs interact with specific sites (receptors) in the body, e.g. in the brain. Receptors may be isolated to provide the basis for drug assays depending on the specific interaction that occurs between drug and receptor. This use of receptor assays should be one of the most exciting innovations for the future. The use of DNA technology with yeast cells has enabled the production of specific receptors. For example, the use of glucocorticosteroid receptor assays has already been exploited.

The significant benefit of this approach should be that it will be effective for most of the yet undiscovered psychoactive drugs. Other future techniques are likely to include the use of neural networks to deal with the complex analytical data produced by mass spectrometers and other instruments to simplify the detection of substances from a complex biological background. This approach is likely to become more important if metabolomics (the study of the small molecules, or metabolites, contained in a human cell, tissue or organ (including fluids) and involved in primary and intermediary metabolism) is used to indicate that a psychoactive substance has been used. The three most likely significant advances will almost certainly be miniaturisation, enabling point-of-care analysis; high-throughput screening using enhanced spectrophotometric detection of antibody binding of drugs of interest; and enhanced separation techniques for the discrimination of complex mixtures. The futuristic concept of a scenario where a simple probing device is placed on an area of the body may not, after all, be so farfetched that it could not become a reality for the detection of common drugs within the next 25 years.

6

Economic Values of Addiction

There are many misconceptions about the consumption and control of illicit drugs, including that drug users are unresponsive to price; that controlling licit substance use may encourage young people to switch to illegal drugs; and that availability controls are likely to be more effective than measures to reduce demand. Economics provides both theories and empirical evidence to explore factors influencing current drug use, the costs to society of the harm they create and the cost-effectiveness of different policy options. This summary sets out the results of this review in addressing the Foresight project's questions. Economists do not have the answer to this question but can provide insight into how the markets for different drugs may ebb and flow. In particular, the importance of social factors is considered alongside economic incentives, such as price and income, to yield more comprehensive and complex models of the current and future behaviours of consumers and suppliers of drugs.

An important aspect of this work is how different groups in society may make different demands for policies, the richer groups potentially demanding more restrictions, while the poor bear more of the costs of drug use. The analysis of drug adoption also provides a way to see the characteristics of future substances in terms of their impact, whether on the individual, social groupings or markets. The models can also predict whether new drugs will drive out existing ones or find a niche in separated or polydrug-using clusters.

EFFECTS OF USING PSYCHOACTIVE SUBSTANCES

The economic impact of problem drug use is sizeable, with Class A drug users estimated to generate social costs of £12 billion in 2000. These effects are spread across individuals, families, communities and society more generally. The problematic rather than the occasional user seems to be responsible for the majority of these costs, but this may partly reflect data deficiencies. Also, for novel drugs, longer-term health and dependency impacts associated with problematic use may take time to emerge. Governments may need to adopt a precautionary approach to any drug with a dependence potential. The theoretical demand models also shed light on how social groupings, information, economic constraints and policy affect drug-use choices by affecting perceived costs and benefits, and by triggering simplified or myopic decision making.

MANAGEMENT OF PSYCHOACTIVE SUBSTANCES

In a departure from conventional wisdom, both economic theory and empirical evidence suggest that drug prices influence demand. Also, there are a number of complementary relationships between licit and illicit substances. This suggests that policies can be devised to affect individual behaviour. Controlling supply has proved harder, and theoretical models highlight a number of reasons for the sometimes seemingly contradictory outcomes of the application of narrow policy options without an analysis of complete market interactions. Policy effects may well be localised and insignificant in the short run, but much more substantial after network effects and social change have taken hold. The analysis suggests that gateway and cross-drug effects are significant, but are also potentially reversible, so that policy should be both targeted on the most cost-effective drugs and patterns of use and be adaptive to emerging evidence on changing prices and habits.

Economics also suggests that drug use follows punctuated and/or cyclical paths, so that policies should 'lead' or 'follow' the cycle. Finally, the analysis suggests that policy should take account of and try to influence structures on both the supply and demand sides of the market, if only to avoid such 'rebound' effects as the social isolation of drug-using clusters or inadvertent support for suppliers' attempts to limit entry and enforce market discipline. Where effective treatments exist, their use saves costs to society. But the incentives for industry to develop effective treatments for novel drugs may be lacking until serious and sizeable problems have begun to emerge. The current consumption of both licit and illicit drugs is seen to have a range of economic impacts, and there is general concern about costs arising both as a direct result of their use and from measures such as enforcement, education and treatment put in place in an attempt to reduce these problems.

Economics has a set of tools whose application goes beyond simply quantifying the problem. This review aims to explore the role of economics in explaining changes in drug use, predicting expected and unexpected impacts of policy changes and evaluating the overall worth of different policy options. A wide range of psychoactive substances are used in society and most are subject to some legal controls. The focus of this review are the currently controlled psychoactive substances in the UK. Good data on illegal markets where such controlled substances are traded are hard to obtain. But useful insights can be drawn from theory, and from the empirical analysis of legal addictive goods such as alcohol and tobacco. Also, some studies have explored interactions between licit and illicit markets, in particular whether changes in restrictions on the use of legal drugs such as alcohol and tobacco affect the demand for or supply of illicit ones.

MODELS AND THEORIES

Economics is built on demand and supply, determined by rational, informed (but constrained) choice and reconciled by markets. In making choices, rational consumers are expected to take all current and future costs and benefits of their actions into account, but generally are not expected to take account of the impact of their actions on others, nor of the external costs (or benefits) of their choices. External costs make market-clearing consumption higher than is socially optimal.

There is potential for government intervention to improve overall welfare, providing its costs do not exceed its benefits, by reducing external social impacts or by helping economic agents to internalise them. Addictive goods challenge this simple model in two ways. In an extreme model of addiction, users would be incapable of effective choices, making them insensitive to price or their own income.

The rational addiction model, explored below, is an attempt to find a middle ground, suggesting how consumers can rationally pursue their welfare, even if they are aware of the nature of dependence and how this may influence their future choices. Another issue is the role of information about the effects of existing and novel addictive substances. The risks of some licit drugs such as cigarettes may be relatively well known, but just how well informed are consumers about risks of other licit drugs such as alcohol, or of illicit drugs that have been available for a number of years, or of novel drugs with an unknown risk potential? Lack of information about adverse consequences creates a need to include some private costs in decisions about government intervention. It could be argued that '*non-rational*' addicted consumers are not getting optimal utility from their consumption and are therefore diverting productive resources that could increase overall welfare. Clearly one issue raised by novel substances, or indeed novel consumption patterns, is the difficulty of estimating potential longer-term effects, particularly on individual health but also on dependence potential.

The continuing debate about the potential effects of marijuana illustrates these difficulties. In a recent overview, Pacula (2005) argues that accumulating evidence of the impact of marijuana on educational achievement translates into reduced future earnings in addition to the costs of treating health-related consequences for those, albeit a small proportion, developing a dependency problem. Within the economic model, costs falling on families and carers are often treated as 'private' but the nature of dependence - particularly severe dependence - would suggest that normal models of agreed intra-household choice do not apply. A number of other issues influence the distinction between private and external costs.

The basic economic model takes an individualistic approach to consumption decisions. However, economic models of collective behaviour take account of the interactions between individuals. This has some parallel in the debate about whose costs count. If individuals 'care' about the impacts of drug use on others, this may affect social welfare and how societies value the consequences of addictive behaviour. In particular, what value is given by society to the risks of death from drug taking? Should this and other individual values form part of the 'costs' of drugs to society that prompt government action? These considerations determine the choice and level of government policies in drug markets. They can help make explicit some assumptions being made in often simplistic debates about the legal status of different drugs. They also provide a framework for scenario analyses to explore the potential impact on overall social welfare.

Rational Choice

Rational choice means that each actor has a feasible set of options - determined by constraints - that can be ranked according to preferences. Personal choices involve such matters as consumption (of licit and illicit goods and services), investment (in the form of savings, education, etc.) and supply (particularly labour supply and criminal activity). The points of analysis - where economics touches issues relating to psychoactive substances - involve:

1. The preferences themselves - how past consumption affects preferences, the accuracy with which choices are assessed, the rate at which the future is discounted, preferences for investment and supply as well as consumption, externalities, etc.;
2. The feasible set - prices and income (returns on investment and supply activities);
3. Income-generating activities;
4. The choice process - this includes time-consistency, the influence of context and cues, alternative models of '*addictive choice*' and the handling of risk, uncertainty and incomplete information.

A variety of models explore individual choices to consume drugs and undertake other actions associated with drug taking, such as crime. All these models establish connections between behaviour and economic influences (price, purity, etc.). But they differ in the nature of these connections, and thus in their predictions regarding new drugs, and importantly, policy impacts. They also differ in their consistency with neuroscience, psychology and the empirical evidence. Among the characteristics that such models should explain are:

1. Users do not follow monotone behaviour patterns - they frequently abstain for short periods but relapse in the long run;
2. Users often develop compulsive use patterns, even when prolonged use reduces or reverses the hedonic benefits of drug taking;
3. Consumption, quitting and relapse are often triggered by contextual cues;
4. Users often recognise their behaviour as a mistake, but see themselves as lacking self-control. In other words, they believe they would have been better off had they not taken drugs;
5. Users often resort to precommitment strategies to control their behaviour, for example, by consuming other drugs that block or interfere with the pleasurable aspects of their drug of choice, or entering treatment or rehabilitation centres;
6. Drug use responds to economic factors, though the extent and timing of this response may be problematic. There is evidence of forecasting, with users reducing their current consumption to limit vulnerability to anticipated future price increases;
7. Drug use is affected by the attention paid to consumption choices. For instance, recovering addicts can overcome temporary cravings if they are reminded of adverse effects - even when they are well aware of these effects. Conversely, distractions that mask such knowledge (e.g., rave environments) can increase the likelihood of use. (This suggests that attention-management strategies may be a useful adjunct to addiction therapies, and also that there is value in information-exchange activities even when the message is not 'new'. Of course, an overly familiar message may not attract attention.);
8. There are strong and systematic differences in patterns of use and response to environmental changes between different drugs and methods of administration and between individuals.

The rational addict model asserts that drug use, like other behaviour, forms part of a solution to a global expected lifetime-utility maximisation. Preferences (even time preference) may change in complex ways as a result of past behaviour, but the rational agent takes this into account by working backwards from all possible ends of life. In this view, drug taking is an individual problem in the absence of prohibition and the only societal problems arise from externalities. Drug taking thus enhances welfare (at least ex ante) and externality problems can be dealt with by suitable government action. This model does not imply that '*rational*' addicts are happy with the choices they've made and the life they lead.

Welfare improvements are only expectations, and many may well 'lose the gamble' they've undertaken. Nevertheless, on aggregate, the 'winners' could compensate the losers and still have enough left over for a risk premium. The empirical predictions are quite strong. Users should seek out the lowest-cost way of meeting their demand discount the future geometrically at the same rate for drugs and for all other goods be risk-averse (because expected utility is concave) have no problems of self-command (doing things that they expect to regret) and thus have no need to 'play games against their future selves' as Schelling (1980) puts it. The model has been extended to account for observed behaviour. This adds learning and regret, and endogenous determination of time preference. Experiments show that heroin produces higher rates of time preference (discounting) than other reinforcers.

Ex-addicts report a fear of relapse so that treatment by managing contingencies is effective even when the monetary rewards are far smaller than the costs of the drug habit. Other models invoke time-inconsistent preferences – typically by biasing utility towards the present, for instance, via hyperbolic discounting. In such models, small immediate rewards may be preferred to later, more substantial rewards for only a short period before they are available. Choices made in advance will tend to favour deferred rewards. Under these conditions, precommitment strategies that remove the power to revisit choices or which mask cues that might raise the possibility of choice are in the individual's long-term interest. Individual strategies may be consciously economic (e.g., by limiting the discretionary spending power of vulnerable persons), social (by choosing a different group of friends or living situation) or cognitive (e.g., by making consumption choices more deliberate, and thus sensitive to scientific information, and limiting impulse purchases).

Consumers may rationally demand government action to reduce their choices. Becker *et al.* (2004) recently extended this welfare analysis to suggest that restrictive policies, such as prohibition, could be preferred by the rich and that such policies' costs fall disproportionately on the poor. These models are all based on changing individual preferences. A different approach considers changes in the way individuals perceive choices and translate preferences into actions. A simple and tractable approach, largely consistent with neurological science and the stylised facts above, incorporates different cognitive states, triggered by cues in the environment in ways that reflect past experience.

In 'cold' states, individuals take a wide range of information into account and behave in 'rational' ways, using attention, foresight and memory. In 'hot' cognitive states, individuals' decisions are simplified by ignoring alternatives and consequences. Such 'shortcuts' are not inherently irrational or suboptimal. The brain is a finite mechanism, and cognitive decisions can have substantial opportunity costs. So many repeated behaviours are reduced to routines chosen by rough-and-ready rules of thumb, backed by 'wake-up rules' that identify situations where explicit decisions are needed. Melioration models attempt to reconcile compulsive aspects of drug taking (consistent with a disease model of addiction) and its responsiveness to prices and adverse effects (consistent with learning and rational addiction).

The analysis is based on four observations. First, repeated consumption decreases future values of both the drug and competing activities. Second, the frequency of an activity depends on its relative attractiveness, so drug taking that makes alternative behaviours less attractive can increase, even if its own utility declines. Third, psychological experiments suggest that reinforcement is relative to a frame of reference, which can change to favour suboptimal choices. Fourth, while local frames of reference favour excessive drug use, the reverse is true of global frames of reference. Of further interest is the relationship between different drugs. Some substitutability within classes of drugs (e.g., stimulants or depressants) may be expected. If prices of one depressive drug rose, consumers would switch to a substitute. However, the empirical literature suggests more complementarity, and a rise in polydrug use has been observed.

Complementarity among drugs creates analytical problems for microeconomic analysis and can lead to equilibria that are inefficient - dominated by one or a few drugs - or that fail to exist at all. A second phenomenon is the so-called '*gateway effect*' whereby initiation into one substance leads to the use of another. This suggests targeting patterns for enforcement activity. It may be more efficient to target the gateway drug (or activity) than the eventual drug. However, 'rational addiction' models of polydrug abuse suggest that this effect could be reversed, depending on the relative attractiveness of either drug to novice users as well as any asymmetric spillover from using one drug to using another. Gateway effects are not simple and gateways do not always work in a single direction. Therefore, changes in relative alcohol and cannabis prices, for example, could lead to their changing places as gateway drugs to one another or to '*hard drugs*'. Similarly, where peer group effects are important, gateway trajectories may differ from group to group.

Commercial Values

Supply-side points of analysis involve:

1. Underlying motivation - usually assumed to be profit maximisation, but others - e.g., political or ideological - may require different treatment;
2. Available '*technology*' for turning inputs into output. Inputs include raw materials, labour, capital, etc., but also complementary activities such as (for illicit substances) money-laundering, intimidation of rivals, corrupt services and other means of managing law enforcement and other risk;
3. Investment;
4. Financial constraints.

The '*supply chain*' stretches from international or source-country supply, processing, transhipment and money-laundering markets to domestic wholesale and retail (even street-level) markets. At street level, the separation of supply and demand breaks down; user and seller activity may best be modelled as a way to reduce the cost of obtaining drugs. The supply chain can be analysed in terms of:

1. Structure – in particular, the degree and distribution of market power, both upstream and down. This measures the extent to which market players are able to control their environment, and thus the degree to which deep pockets or other forms of influence may immunise them from policy impacts or even allow them to exploit policy to limit competition;
2. Conduct – suppliers of illicit substances have a range of strategies at their disposal. Many of these (e.g., predatory behaviour, use of violence, bribery, etc.) are of direct policy concern.

Drug markets have different supply-side and market dynamics from conventional markets and other illicit markets in terms of turnover, price and quality variability, violence, corruption and the extent of rents and social costs. The unavailability of legal channels for dispute resolution may predispose illicit drug markets to violence and corruption. This is reinforced on both sides: law enforcement pressure may reduce the expected costs of violence while the cost of violent market regulation may make corruption more attractive. The general tendency to concentration has some important drivers. Law enforcement pressure may control market entry and the accumulation of '*learning economies*'.

In addition, law enforcement limits consumer search and can make punishments more credible and thus reduces the temptation to renege on cartel agreements. On the other hand, the threat of disruption at any level of the market makes one-to-one connections risky, so monopoly is, on the whole, less likely and less resilient than oligopoly. The legal risks that inhibit price search also reduce the effectiveness of consumer recourse and reputations in promoting quality, so competitive forces may well produce low prices and variable quality. The conventions of supply (e.g., fixed prices for unobservable 'doses') mean that supply shocks are transmitted to those least able to bear them. This may magnify adverse health consequences.

Dynamics of Drug-taking Behavior

There is a growing literature on the dynamics of drug-taking behaviour and associated harms and optimal policies. It involves both representations of individual drug-taking decisions and models of collective behaviour. Individual economic models provide predictions about the impact of changes in prices and provide some guidance as to the economic legitimacy of restrictive government policies. They also help explain some of the observed unanticipated consequences of policy actions. The potential to analyse the ranges and interactions of policy changes suggests an important role for economic modelling in future scenario analyses. Existing collective models recognise the crucial importance of social interactions for intentions with regard to drugs and the eventual results of acting on those intentions.

The prevalent form of collective modelling likens drug use to epidemic disease. Choices about drugs are strongly influenced by information and incentives from interactions with peer groups, educators, medical personnel, mass media, authority figures, etc. These influence all-important inputs to the demand models: price expectations, beliefs about the likelihood and severity of risks, tastes for particular experiences, coping strategies for managing adverse effects or future behaviour and even the '*framing*' of drug-use decisions: what consequences are considered, how they are evaluated and weighted, and how uncertainty (as well as risk) is incorporated. But beyond this focus on social determinants of drug use lies a deeper issue; that drug use itself as a social ritual determines the social environment as much as it is determined by it.

Future impact of psychoactive drugs

Current economic models highlight uncertainty about the future impact of different illicit markets. The inherent instability of these markets can lead to health and other consequences for users, beyond the direct harms to individuals, families, communities and society in general from dependent drug use. However, chronic problems may only become apparent after consumption has been prevalent in a large enough group of the population for a long enough period of time. The emerging evidence on marijuana may or may not parallel the case of tobacco, where the full health impact has only been fully measured some 50 years after the mass adoption of cigarette smoking. Economic theory gives some guidance on a precautionary principle for governments to follow, in that individuals may *'rationally'* demand some constraints on their behaviour. The predictions from economic models are that consumers are likely to respond to changes in market incentives and be responsive to policy options, but social factors are also important.

COMPUTERIZED DATA

Finding data on illegal activities will always be challenging. In addition, it has not been a tradition of economic research to test and retest models. New theoretical models and empirical estimates are sometimes presented without rigorous statistical testing, and without refuting previous theoretical models. This makes empirical economic evidence difficult to review systematically. Specific definition and measurement issues arise around prices and levels of use. Measuring prices higher up the value chain is complicated by complex contingent payment arrangements, money-laundering costs, etc. A recent analysis taking account of some of these complexities suggests that black-market prices of cocaine are 2-4 times the likely legal market prices in the US and that the corresponding multiple is 6-19 for heroin. However, care should be used in extrapolating these figures because, for both drugs, prices and purity differ between Europe and the US and among European countries. Certainly, the prices of illicit substances are high. This means that elasticities cannot safely ignore income effects. Also, an addict who responds to price increases by committing more crime is not showing inelastic demand but increasing discretionary income to pay the difference.

Price changes should also significantly affect heavy users' behaviour, as has been found for alcohol users. While light users may respond to price increases by reducing consumption, heavy users may reduce other consumption (eg housing, food) or commit more crime. Price rises may move some across the boundary from light to heavy use and criminal activity. Many components – especially farmgate prices – are so small that policies targeting them are unlikely to achieve much impact. Most costs are unmonetised risk compensation – so accounting profits are likely greatly to exceed economic profits. Accounting losses will almost never induce exit. Also, differences in tolerance for non-monetary costs favour market domination by the less risk-averse (or even risk-seeking). Thanks to this adverse selection, supply-side policies that impose risk may have attenuated or perverse effects. Young, poorly educated and violence-prone males are most likely to tolerate such costs.

So the association of violence and drugs may not wholly be due to the struggle for profit. The second problem is determining consumption patterns. Except for tobacco, almost all drugs have highly asymmetric consumption following 'Pareto's 80:20 law' – 80 per cent of consumption accounted for by 20 per cent of the user population. Heavy users who account for most demand and its attendant problems are likely to be under-sampled in population surveys, and data and analysis are almost certain to be dominated by light or occasional users. However, convenient samples of problem drug users, such as those arrested or seeking treatment, are not representative either. Such samples may *'over-represent'* problems and consumption.

Emperical Estimation

An obvious first empirical estimate is the size of the UK illicit market. However, few research studies are available. Bramley-Harker (2001) took data from arrestees. The NEW-ADAM data measured problem drug use prevalence and estimated expenditure from a sample of arrestees with data collected by means of structured interviews and urine samples. Population surveys were used to estimate numbers of occasional users. The results suggest that in 1998 a regular heroin user spent about £16500 per year on all drugs and a regular crack user £21000. Occasional users' estimated expenditure (£1.9 billion) was much lower than regular users' (£4.7 billion). The total estimate for the UK market (£6.6 billion), while sizeable, is still smaller than alcohol expenditure for the same year (£32 billion). The NEW-ADAM data point to a surprisingly high number of crack cocaine users but the overall expenditure figure was similar to a previous Office for National Statistics estimate based on seizure data.

Calculating Costs

Economic models distinguish between private costs of drug use and externalities borne by others in society. There are major gaps in the UK information. Some large international studies have attempted to estimate the external costs of illicit drug use in comparison to those from licit drugs. In general, the costs of illicit substances, while substantial, are smaller than for alcohol and tobacco. They are generally dominated by crime, health and productivity costs.

Epidemiological data

There is a general lack of UK data on personal impacts (good or bad) and their valuation. Some data are available on mortality, morbidity and loss of earnings, including epidemiological data on premature mortality from different licit and illicit drugs. Tobacco accounts for the largest number of premature deaths at 106 000 per year in the UK, alcohol accounts for 22000 in England and illicit drugs between 1000 and 3000. However, tobacco deaths tend to occur at a later age than those from alcohol or illicit drugs. In terms of life years, the percentages lost to alcohol and tobacco are approximately equal and far greater than from illicit substances. More robust evidence is emerging on the impact of substance use, especially problem use, on employment. US research also highlights the impact of early substance misuse on educational attainment. These two factors could suggest substantial impacts on the lifetime legitimate earnings of substance misusers. Pacula (2005), for example, suggests that early marijuana use may reduce future earnings by 2 per cent.

Outdoor estimation

Few studies attempt empirical estimates of families' costs such as the lost earnings of carers. Rather, empirical studies include some public sector resource costs. For example, care of problem Class A drug users' children was estimated to cost £63 million and healthcare for neonates £4.3

million per year in England and Wales for 2000. Workplace costs of all drugs are of considerable interest but hard to quantify. Some drugs consumed in or close to working time have a direct impact on productivity. Some workers may feel stimulants help productivity, although the long-term productivity impact of most drugs remains unproven. Previous excess use can also lead to workplace problems.

These potential effects can alter drug users' behaviour as they attempt to get desired effects in leisure time while ensuring they are fit for work. The long-term impact of such polydrug use is largely unknown. Most empirical attention has been given to the public sector impacts of different substances. The total shown of £12 billion per year is based on the medium estimate of users and costs and taken from Godfrey *et al.* (2002). Estimates for premature death were calculated using willingness to pay. However, premature death is not a large component of Class A drug-use costs, which are dominated by crime costs (88 per cent of the total). Also, in contrast to alcohol, problem drug users account for almost all estimated cost (99 per cent) of Class A drug use. However, this study could not find data to estimate the potential impacts of drug driving or workplace drug use. Costs for different types of drug user varied substantially, with young recreational users estimated to cost £36 per user and older regular users only £3 per user, in contrast to the much larger estimate of £35 500 for each problem drug user.

Factors Affecting Illicit Drug Demand

Basic price

While limited in number and mainly US-based, most published studies find significant relationships between the price and consumption of marijuana, heroin and cocaine. Available studies suggest that raising the prices of any of these drugs will reduce drug use, but there are differences in the size of the effects between studies. For example, heroin price elasticity ranged from -0.27 to -1.5 and cocaine from -0.44 to -2.5, although estimates for most drugs centre around -1.0. Most studies use population survey data. The research gives support to the rational addiction model. Two more recent studies used drug arrestec and hospital admission data. Again, price was found to significantly affect heroin and cocaine use, although estimated price elasticities were lower (between -0.1 and -0.3) than the population studies.

Similarly Bretteville-Jensen and Sutton (1996) and Bretteville-Jensen (2003) used data from more problematic users and found significant price impacts for Norwegian data. Interesting in these studies was a division between users and user/dealers, with user/dealers having lower price elasticity. Animal and human experimental studies confirm these findings on the impact of price for both occasional and regular users. It is difficult to draw conclusions from the available literature about the relative price effects across different groups of the drug-using population. Recent work suggests that pregnant users are price-sensitive, with a $10 increase in pure cocaine price estimated to reduce use by 12-15 per cent. There are some parallels to the study of licit drug demand, where more data has led to a focus on the impact of price on different aspects of behaviour, particularly the starting and stopping of different drugs. With licit drugs, prices across different brands or types of the same drug can be explored. Patterns tend to be complex, with brand-conscious consumers tending to adopt patterns of behaviour within age cohorts but also proving price-sensitive within these patterns.

Impact of other drugs

Of particular policy interest is the impact of changes in the price of one drug on the consumption of others, particularly across the licit-illicit divide. Several studies attempt to explore such cross-price effects. In particular, studies have considered the relationship between prices and restrictive policies affecting alcohol and marijuana consumption. Studies have generally been based on US youth population surveys. The majority of these studies found complementarity between these two drugs, suggesting that restrictions or higher prices for alcohol will reduce both drinking and marijuana consumption.

There have been similar findings for Australia. However, these studies have some inconsistencies, the complementarity being stronger in younger age groups than in ethnic minorities or women, although the latest study found a similar relationship for college students. Other studies examine the relationships between cigarettes and marijuana; cocaine and marijuana; alcohol and heroin; and alcohol and cocaine. All found complementarities. A related issue is the gateway hypothesis of a set pathway from one drug to another. The evidence is currently inconclusive.

Relationship between drug use and income

As discussed above, the relationship between drug use and income is complex, especially for problem users who may finance use through criminal and drug- dealing activities. Bretteville-Jensen and Sutton (1996) estimated an income elasticity of 0.47 from their survey of non-dealing heroin users in Norway, suggesting income increases will raise drug use. They also found that drug users reduced their illegal income when legitimate income (from work or social benefits) rose.

Penalities or decriminalisation

Many demand studies also explore the impact of street-level enforcement on drug demand, to test whether changing use or low-level-dealing penalties or decriminalisation of possession also affects use by changing the effective price. The results have been mixed, partly due to data difficulties. Policies themselves may be influenced by use levels. There also seems to be some difference between older populations where decriminalisation has an effect and younger populations where no such effects have been found.

More recent studies, which have attempted to control for such effects, tend to find more positive policy impacts. Desimone and Farrelly (2003), for example, found that increases in either cocaine or marijuana arrest probabilities decreased demand for both drugs for both adult and juveniles in the US. But Williams (2004) found that Australian marijuana decriminalisation only increased use among males over 25, and did not affect young males or females.

Awareness

Pacula *et al.* (2000) investigated the impact of perceptions of drug-related harm on marijuana consumption in the US. They found that the perception of harm had increased and these changed perceptions reduced demand. In Sweden it was found that young people may overestimate longer-term health risks of licit drugs such as alcohol and tobacco but can be unaware of some shorter-term impacts, consistent with some theoretical models. Most empirical models are based on individual data and models. There have been attempts to explore some peer effects and an interesting study of college drinking in the US suggests that changing the demographic mix of a college has a direct impact on the rate of problem drinking.

Literature on Supply Behaviour

There is a much smaller economic literature on suppliers' behaviour. As suggested above, it may be difficult to influence retail prices by untargeted interdiction policies. Theory would suggest considerable scope for profit and market concentration. Caulkins *et al.* (1999) found profit margins in US markets varying from 50 per cent to less than 10 per cent. In exploring potential supply-demand interactions, Saffer and Chaloupka (1999 b) did find that police drug-control spending affects supply as well as demand.

Future Impact and Interventions

What can be learnt from the estimates of illicit drug market size and the value of associated problems? Regular (and often problematic) users dominate spending and problems created. However, this may simply reflect current drugs of misuse or a lack of knowledge of longer-term impacts. Even with marijuana, the emerging evidence is that costs are associated with problem use, either impeding schooling or later, when dependence emerges. However, not enough is known about regular users and the impact of their use in the workplace or in situations such as driving that are risky both to users and the rest of society.

Tobacco costs are more evenly spread and patterns of alcohol drinking are related to a complex set of problems connected with both acute and chronic use. Economic evidence on demand suggests that consumers of illicit drugs respond to incentives in a similar way to users of other goods and services. Prices and regulations affect demand for both licit and illicit drugs. In general, drugs are complementary, so relaxing policies on any one drug may increase the consumption of all substances. In particular, concerted policy to change sentiment among emerging cohorts does seem to have the potential to shift demand. While these economic and policy factors shouldn't be ignored, social factors influencing preferences are likely to play a large part in determining overall consumption patterns.

Demographic shifts may influence consumption in the short term, although globalisation and the diffusion of habits mean that trends in use spread across populations. We have less evidence about suppliers' behaviour but markets seem to adapt to changing preferences. Demand for interventions across income groups and generational cohorts raises interesting future scenarios. Current evidence points to similar demand and supply factors across different drugs. Indeed, there has been a growth of polydrug use among consumers and some evidence of suppliers becoming more concentrated on this demand.

ECONOMIC EVALUATION

Economists use economic evaluation techniques to compare the costs and consequences that occur in different policy scenarios. Current evidence is briefly reviewed for enforcement; treatment approaches; workplace policies and school-based information relevant to the UK. The brief nature of the review partly reflects the lack of good quality research worldwide on the cost-effectiveness of these interventions. It is too early to draw definitive conclusions about the detailed aspects of current drug policy but important to note that evidence will not be accumulated without well designed evaluative studies designed to include economic elements.

Drug Policies

In the UK, as in most other countries, interventions designed to reduce the availability of drugs take a substantial share of drug policy expenditure. Recent years have seen a sizeable increase in expenditure on treatment and currently the share for enforcement has fallen from over 50 per cent in 2001 to below 40 per cent. Demand studies suggest that increasing the probability of arrest may impact on demand. The supply literature suggests that availability measures only have a limited impact on the overall size of the market. For example, no evidence was found to suggest a major police initiative had an impact on drug use or prices in London. Models of the cost-effectiveness of availability compared to treatment approaches have consistently suggested that treatment would provide better value for money than enforcement. However, combining more sophisticated modelling techniques using social models of demand and taking account of the temporal patterns of adoption of new drug consumption patterns suggest that the relative cost-effectiveness of policies may vary over this diffusion process.

Treatment and Rehabilitation

Reviews of drug treatment and rehabilitation, mainly for heroin users, suggests that treatment saves society far more than it costs. Most of the empirical studies have been undertaken in the US but a number of UK studies, such as the National Treatment Outcome Research Study (NTORS), have suggested similar benefits with a reduction in social costs of about 9-15 times the spending on drug treatment in the two years following treatment. A simulation model estimating the impact of increases in the numbers of problem drug misusers in community-based methadone treatment of 10000 a year over the next five years predicted a reduction in the social costs of drug misuse of £3-4 billion. Current policy does involve increasing the numbers in treatments of all types, not just community-based methadone treatment, through a number of different schemes, such as arrest referral and other components of the Drug Interventions Programme. Given the potential savings from reduced crime and other benefits from treatment such programmes could be cost-effective and indeed cost-saving.

In the absence of primary evaluations of such policies, simulation models do provide one means of evaluating these different components of current programmes. Unfortunately the research on the costs and therefore the cost-effectiveness of different means of attracting problem drug users into the treatment system is very limited. While studies of the economic costs and benefits of treatment have provided evidence on the overall worth of increasing levels of treatment, far less is known about the comparative worth of different treatment approaches. While a number of studies have recently been completed or are under way, this is an important area for future research. For the future, there is considerable scope for the development of pharmacotherapies and psychosocial approaches. Effective treatments are likely to be cost-effective, given the high social costs associated with problem drug users.

Novel approaches that reduced the potential development of dependence could also be worthwhile. However, the history of securing resources for heroin addiction suggests that it may be difficult to secure investment in such treatments. Similarly it is important to conduct research into the different means of attracting and retaining problem drug misusers into the treatment system.

Effectiveness of Programmes

Systematic reviews of the effectiveness of school-based programmes directed at reduced alcohol consumption suggest that most have limited effectiveness and are costly. Caulkins *et al.* (1999) indicate, however, that if effective school-based interventions can be devised, they may also prove to be cost effective despite their high cost. Other types of study on the demand for illicit drugs reviewed above do suggest that young people are influenced by perceptions of harm. While school-based interventions may remain an important source of information, but are not necessarily expected to change drug- related behaviour, other approaches may prove more promising. However, measuring the impact of any prevention programme on behaviour rather than just awareness is challenging and needs to take account of the other factors influencing demand. There is fertile grounds for new research building upon theoretical insights reviewed in the previous section and taking into account potential models of drug adoption.

Importance of Workplace

Drug misuse could have various impacts in the workplace, as suggested above. This is also an area where there are commercial interests in devising tests and policies for employers. However, the evidence of the cost-effectiveness of such policies in reducing drug use is very limited. In particular, drug testing of employees may be a useful selection device for employers, selecting employees with a range of characteristics such as risk aversion, rather than controlling the use itself. As a setting, it does provide an opportunity for treatment interventions, which could provide some payback for employers.

There are no simple policy solutions to reduce demand and, given that this demand will respond to changing market conditions, policy initiatives may have to be commensurate with these changes and external factors such as globalisation. Prevention programmes will need to be evaluated taking such factors into account and realistic goals for the impact of such programmes on population level data set. Dependent individuals have generally been shown to have the highest social costs and treatment for such individuals has been shown to be cost-saving for society. There is considerable scope for devising novel treatments but the commercial incentives for developing this business are currently small. The commercial response to drug taking has focused on devising interventions for employers.

COLLECTIVE BEHAVIOUR

An important part of the growing literature on the dynamics of drug taking is a theoretical treatment of collective behaviour, calibrated to empirical data. These models of collective behaviour display common phenomena, with greater or lesser fidelity, including:

1. S-shaped adoption paths,
2. Cascade or 'herd' behaviour,
3. 'Punctuated' paths of long periods of slow and localised change separated by brief periods of profound or discontinuous adjustment,

4. Path-dependence or hysteresis (locally irreversible change),
5. Cycles.

Everingham and Rydell (1994) present a simple model of market dynamics based on a division between poor, heavy users of a drug, and light users who may be poor or rich. These groups have potentially different elasticities with respect to price, information and other factors. Poor, heavy users can finance their purchases through legitimate work, property crime or drug sales to rich users. Considering only the third of these, a fall in rich users' demand will constrain the income - and consumption - of poor heavy users. But if their drug use reduces their legitimate income, their need to increase their own drug sales may bid down drug prices. The same analysis suggests that legal sanctions may increase the attractiveness of short-term employment in drug markets relative to activities such as education. More recent models take account of the twin forces of contagion (promoting drug use) and prior observation (deterring it). Endogenously influenced forces pulling in opposite directions can easily lead to cycles.

It is possible to extend this view in several directions. First, 'contagion' from non-users can discourage drug use, as policies aimed at decreasing the social acceptability of drug use recognise. Second, contagion (in either direction) is opposed by cohesion with social contacts. A third factor, as Behrens *et al.* (2002) point out, is that drug reputations reflect both current use and memories or prior observation of heavy users. This is analogous to the '*stock of addiction*' variable in rational addiction models - with the significant difference that its present value may respond negatively to past increases. In addition, recent experiences that depress the '*reputation*' of a drug may switch to a positive impact as the experience recedes, consistent with psychological evidence that compulsive addicts try to recreate 'that first fine careless rapture'. (A similar effect could be obtained simply by attaching a positive weight to the first experience and using negative discounting.) Individual preferences for drug use may also display unobserved - or partially observed - heterogeneity. Individuals observe each other's actions, not their underlying tastes. This by itself can generate multiple equilibria and cycles.

A series of control-theoretic models that take account of age distributions and intertemporal drug-supply decision also produce cycles and multiplicity. Moreover, people differ in proximity and the salience or persuasiveness of their experience. Changes in collective behaviour reflect the structure, strength and direction of societal ties. Indeed, networks depend on behaviour. One justification for standard epidemiological models of drug adoption is that addicts gradually form networks consisting almost entirely of other drug users. In the beginning, this involves some recruitment of existing social contacts, marking the contagious phase. Eventually there are few left to infect. This slows propagation by limiting 'outside' contacts, but - by embedding users in like-minded groups - reduces '*benign contagion*' from non-users and may increase the '*infectiousness*' of contact with non-users. This is consistent with evidence that successful quitting typically involves a change of acquaintances.

Drug Adoption

Treating drug adoption as a social phenomenon sheds light on the progress and likely outcomes of both new-drug introduction and policy efforts to control use or adverse effects. The models in this section extend the epidemiological literature, in ways consistent with individual addiction models and evidence on the growth, spread and decline of drug '*epidemics*'.

The underlying ideas are that drug decisions are influenced by:

1. Individual tastes,
2. Drug-taking prevalence and the strength of '*peer pressure*',
3. Perceptions of risks and costs.

The models take account of the impact of direct and indirect observation on peer pressure and risk assessment as well as '*objective*' information. The structure and strength of those connections - and even the underlying taste for the drug - are influenced by drug use. A simple chain of models captures these phenomena - at least qualitatively - and many features, while remaining tractable enough for scenario studies and policy simulations.

Basic Model

The first model incorporates unobserved and heterogeneous tastes; price, risk information, legal and societal disincentives; '*peer pressure*'; and drug use among the peer groups. There is uncertainty about others' tastes but people make the choices they intend. This essentially static model shows multiple equilibria and can be used as the basis for dynamic analysis. We begin with the decision to consume a single drug; the individual's relative utility for drug use is:

$$U_i(\theta_i, \overset{\omega}{\pi}, \nu, \rho, \Gamma) = \theta_i[1 + \nu\pi_i(\Gamma)] - \rho \quad \text{...(1)}$$

where:

θ_i = the individual's intrinsic taste for the drug

ν = peer-pressure effect: social pressure or social drug consumption

ρ = anticipated price/risk: health, legal, workplace, social, purchase price, etc.

$\overset{\omega}{\pi} = (\pi_1, \ldots, \pi_n)$ the vector of drug use prevalence across the population

Γ the social network: a set of connected pairs (i, j)

$\pi_i(\Gamma)$ drug use prevalence among one's network neighbours (i.e $\{j : (i, j) \in \Gamma\}$)

This model makes very strong symmetry assumptions, uses a particularly simple reduced form and is limited to 'rational addiction' in the sense that only preferences change - though it departs from the standard model in taking account of others' consumption. The purpose is to demonstrate that some unexpected qualitative features can arise in a simple setting.

In the simplest model, everyone is connected to everyone else and the population is large enough so that each experiences the general level of prevalence π. Because tastes differ, all do not make the same choice; the drug will be consumed by those with sufficient taste for the experience: $\theta_i \geq \theta^*$, where the critical value of taste satisfies:

$$\theta^*[1 + \nu\pi] = \rho \quad \text{...(2)}$$

If taste is distributed between q- and q+ according to cumulative distribution $F(F(\theta)$ is the proportion whose taste does not exceed $\theta)$, $\pi = 1 - F(\theta^*)$, so:

$$1-F^{-1}(\pi)=\frac{\rho}{[1+\nu\pi]} \quad ...(3)$$

In a second version of this model, expected price/risk is influenced by learning from others. This utility function is:

$$U_i(\theta_i,\overset{\omega}{\pi},\nu,\rho,\Gamma)=\theta_i[1+\nu\pi_i(\Gamma)]-\rho(1-\beta\pi_i(\Gamma)) \quad ...(4)$$

where β measures peer learning. The equilibrium condition is:

$$1-F^{-1}(\pi)=\frac{\rho(1+\beta\pi)}{[1+\nu\pi]} \quad ...(5)$$

Figure 6.1 shows the (common) left-hand side of equations (3) and (5), plotted with their right-hand sides (for the case F = cumulative normal with mean = –5 and variance = 5; ρ = 20; ν = 6). This shows multiple equilibria and that *stable* equilibrium prevalence falls as peer learning about price/risk becomes more important.

For some values, there is a unique equilibrium. The actual structure of the results also depends on the range of tastes:

$$\rho > (1+\pi)\theta \quad ...(6)$$

$$\theta_+ > \rho \quad ...(7)$$

Condition 6 says that the lowest-preference person would not take the drug even if all their peers did; 7 says that the highest-preference person would consume even if none of their peers did. For the (unbounded) normal distribution, both hold trivially.

Stronger peer effects produce multiple equilibria; catastrophic jumps and S-shaped paths. The model becomes more interesting with endogenous changes in the peer and price/risk effects. Drug use may reinforce an individual's tendency to pay attention to their peer group. Although network structure is fixed, an individual's ties to users may strengthen as a result of their own use or if increasing social isolation introduces social considerations into primarily hedonic or utilitarian drug-use decisions. This can move the individual into the multiple-equilibrium part of the diagram and reinforce choices.

The behaviour of individuals with similar tastes and price/risk perceptions might diverge as peer effects strengthen. More striking effects come from endogenous changes in perceived price/risk. If users' experience and close acquaintance with heavy users increase these perceptions (eg through increased expenditure needs and negative observations) and vice versa for non-users, the model implies anticlockwise cycles in drug prevalence. As

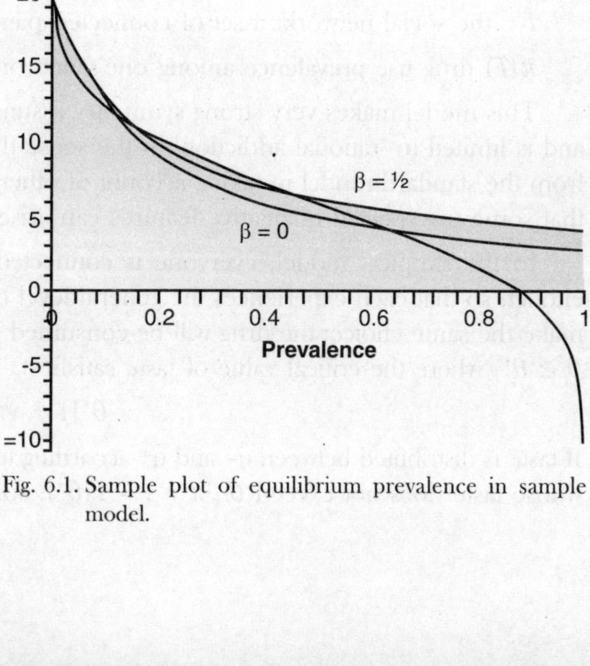

Fig. 6.1. Sample plot of equilibrium prevalence in sample model.

with memory-contagion models, these cycles may be localised and optimally addressed by policy that 'follows the cycle'.

Total welfare

Total welfare can be expressed as:

$$W = (1+\nu\pi)\int_{\theta^*}^{\theta} \theta f(\theta)d\theta - \rho(1-F(\theta^*)) \quad ...(8)$$

Using $\pi = 1 - F(\theta^*)$ to eliminate the critical taste parameter θ^*, we get:

$$W = (1+\nu\pi)\int_{1-F^{-1}(\pi)}^{\theta+} \theta f(\theta)d\theta - \rho\pi \quad ...(9)$$

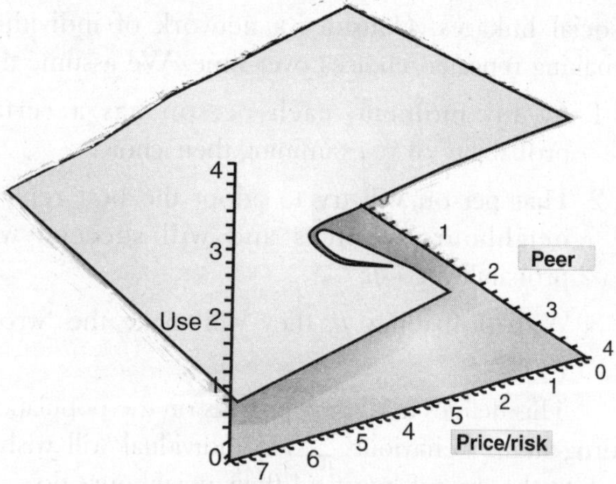

Fig. 6.2. Multiple equilibria in the drug-adoption model.

This implicitly normalises the welfare of non-users to 0, takes no account of welfare decreases (mistakes) associated with drug-taking beyond those modelled with price/risk, and does not model different levels of individual consumption - only population prevalence. The model only implies multiple equilibria for given values of price/risk and peer effects, so we can plot welfare as a function of these parameters and drug taking. A few other phenomena can tractably be handled by simple variants of this model. One concerns information about prevalence and price/risk.

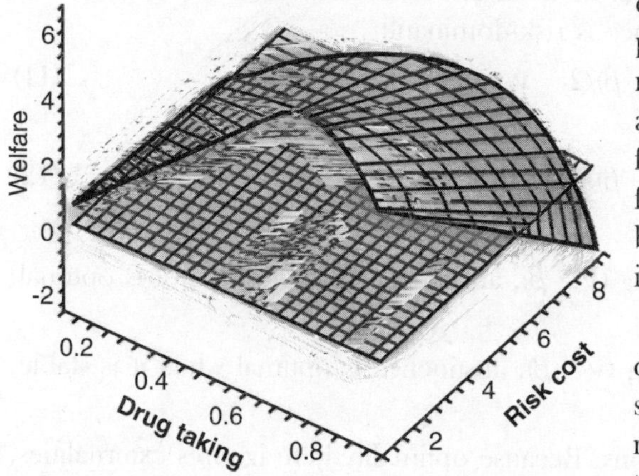

Fig. 6.3. Prevalence, price/risk and welfare.

If individuals rely on general rather than direct experience, users will be a representative sample of the taste distribution. In one sense, the results are broadly similar - roughly S-shaped time-paths of adoption or remission, with '*catastrophic*' jumps when peer effects are large. However, the behavioural response to policy changes may be different - in the absence of peer effects, an increase in perceived price/risk will have much less effect than in the 'strategic' model, but if peer effects are large, increases in price/risk will always reduce prevalence, while in the strategic model modest price/risk increases may increase prevalence.

Conventional Behaviour

Treating drug use as *conventional behaviour* sheds light on the interaction of drug-use choices and social contacts. This dynamic model is based on the twin phenomena of cohesion with one's reference group and contagious '*heterodox*' behaviour. The underlying possibilities are based on the previous model. Choice dynamics reflect, in addition, the stylised facts of addictive behaviour including the possibility of mistakes. A first model examines formation and welfare properties of '*conventional*' behaviour for a given social pattern; the next subsection combines this with evolving social linkages. Consider a network of individuals making repeated choices over time. We assume that:

1. At any moment, each person has a certain probability of re-examining their choice;
2. That person will try to adopt the best reply to neighbours' choices and will succeed with probability $1 - \mu$;
3. With probability μ, they will make the 'wrong' choice.

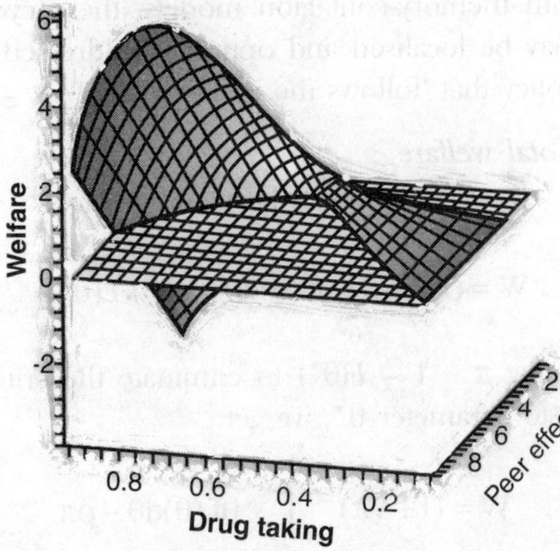

Fig. 6.4. Prevalence, peer pressure and welfare.

This defines a Markov process on the population's drug-taking behaviour $\overset{\omega}{\pi}$. An individual will wish to adopt the drug if 'most' of their neighbours do:

$$\pi_i(\Gamma) \geq (\rho - \theta_i) / (\theta_i \nu - \rho\beta) \qquad ...(10)$$

To begin, we assume that all people have the same tastes (θ) and that every person is connected to everyone else. The limit of this distribution will be a single behaviour, which will involve only experimental drug taking if, and only if, abstinence is risk-dominant:

$$\theta < \rho(2 + \beta)/2 + \nu \qquad ...(11)$$

Abstinence is socially efficient if:

$$\theta < \rho(1 + \beta)/1 + \nu \qquad ...(12)$$

In this model:

1. If peer pressure is stronger than peer learning ($\nu > \beta$), abstinence is stable when it is optimal, but may be stable even when suboptimal;
2. If peer pressure is weaker than peer learning ($\nu < \beta$), abstinence is optimal when it is stable, but may be optimal even when unstable;

This provides some simple policy prescriptions. Because optimality here ignores externalities, stabilising abstinence may be desirable in broader circumstances. Policy can affect perceived price/risk by: conventional law-enforcement cost- or risk- imposing strategies that target users, action against dealers that makes price, quality or availability more uncertain, and education, especially peer reinforcement of risk information. Policy, for example youth policy, can also affect peer pressure. It is even possible – through provision of localised healthcare, law enforcement or price/purity information – to alter the strength of the peer-learning effect (β). Ignoring the dynamics of social interaction, a '*rational addiction*' policy would have to alter these parameters to the point where

$$\theta < \rho \min\left\{1, \frac{1+\beta}{1+\nu}\right\} \qquad ...(13)$$

This involves strictly more policy intervention unless the peer-pressure effect (ν) exactly equals the peer-learning effect (β) – but these are at least partly susceptible to policy intervention!

If peer pressure dominates peer learning, it is easier to make abstinence stable, achieving the desired result through noncooperative evolution. Otherwise, it is easier to make abstinence optimal (among users), achieving the desired result through cooperative (or bargaining) means. Speed of convergence to the stable outcome depends – in this model at least – on the frequency with which choices are revisited and the fidelity with which they are realised. This directly reflects the cue-conditioned cognition model. The dynamics can also be adapted for asymmetry between quitting and relapse – changing rates of onset or recovery but not the ultimate stable state. Risk-dominant behaviour will eventually prevail in any symmetric network. Other 'geometries' can have clusters or pockets of drug use in an abstinent population – or vice versa.

Clustering (small worlds of individuals all of whose friends are likely to be friends of each other) is more likely in certain groups, and has two effects on adoption. It slows propagation, but, within groups, peer effects are likely to be very strong. These impacts are borne out by empirical experience and suggest further forms of educational and youth policy intervention. When tastes differ, the analysis becomes more complicated, but some general observations can be made. Even with relatively few 'high-taste' individuals, a cluster can 'tip' into drug taking if the proportion is large enough (it may well take more than one 'rotten apple').

Vulnerability is reduced if those with middling tastes for the drug have many outside connections to others with low to middle tastes or in clusters with relatively few high-taste individuals. The same holds for abstinence in populations where high taste is prevalent. Adoption or remission are, in general, neither uniform nor monotone. Beyond the stochastic aspects, a given cluster may 'get stuck' for a long period until external changes trigger rapid change. Policy interventions, whether global or localised, do not produce a constant stream of effects but instead have tipping points and delayed impacts that must work through the network as a whole before producing observable changes. On the other hand, the impacts, when they come, may well be profound.

Network Analysis

The previous sections treated tastes (θ_i) and structure (Γ) as fixed. Neither assumption is likely to survive contact with the real world. Network analysis has already developed a treatment of endogenous structures based on chance encounters: unlinked parties may form a link if each benefit by doing so. Links dissolve if either party wishes to end the association. This defines a Markov process – this time on societal structures – whose limiting outcomes may be considered stable. One simple application would hold drug-taking behaviour constant; stable and efficient networks are likely to differ sharply. This may be desirable if individuals take insufficient account of externalities (involving other people, times or random consequences). It also suggests that even well-informed and foresightedly rational individuals might fail to form 'optimal' social contacts. To gain deeper insight, it is desirable to combine these models – to let drug-taking behaviour and social interactions change on the same timescale. Such models can, under the right conditions, reconcile efficiency and stability and produce lower drug use. But this is not inevitable.

One would expect drug adoption to increase the proportion of high-taste individuals among a given person's contacts, and ultimately to increase clustering as individuals sharing a common link are brought into contact. At the same time, links with low-taste individuals – at least non-users – may fall away. This fits heroin users' lifecycle contagion patterns, with simple implications for crime rates, treatment effectiveness and external costs. Policies that enhance clustering - for example, by driving users underground - can have two opposed effects. The first is to reduce the chances and depth of contact between experienced drug users and non-users. The second is to make such contacts more 'dangerous' because the individual making contact with such a cluster is more likely to be influenced by the drug user than vice versa.

In addition, under policy and societal conditions that knit drug-using clusters tightly together, drug use provides a ready-made entree to a tightly-knit cluster, which may be attractive to members of more scattered groupings. A second useful extension is endogeneity of tastes. It is perfectly feasible to make tastes a function of past consumption behaviour. This is consistent with the individual models of addiction in which preferences change (e.g., the rational addiction models), but has not been explored here.

Combined Drug Models

The model given above concentrates on a single drug. There is no difficulty in principle in extending this to the case of multiple or combined drug use. While space does not permit a comprehensive treatment, two interesting phenomena can be treated. Consider different 'drugs' ($k = 1, \ldots, K$), where some may involve combinations (to handle complementarities easily). The utility function generalises to:

$$U_i(\theta_i, \overset{\omega}{\pi}, \upsilon, \rho, \Gamma) = \sum_k \left[\theta_i^{k'} \left[1 + \sum_{k'} \nu_k^{k'} \pi_i^{k'}(\Gamma) \right] - \rho^k \left(1 - \sum_{k'} \beta_k^{k'} \pi_i^{k'}(\Gamma) \right) \right] \quad \ldots(14)$$

where:

θ_i^k individual *i*'s intrinsic taste for drug *k*

$\nu_k^{k'}$ peer-pressure effect of prevalence of drug *k′* use on taste for drug *k*

ρ^k anticipated price/risk for drug *k*

$\overset{\omega}{\pi} = (\pi_1, \cdots, \pi_n)$ the vector of drug use prevalence across drugs and population

$\beta_k^{k'}$ peer learning effect of observing drug *k′* use on price/risk of drug *k*

$\pi_i^k(\Gamma)$ prevalence of drug *k* use among *i*'s network neighbours

Under suitable assumptions, '*diagonal*' outcomes will survive - if peer effects are strong enough then cohesion will stabilise homogeneous behaviour. The Markov analysis can be adapted as follows. As before, individuals have random chances to re-evaluate their behaviour and succeed with some probability, resulting in a Markov process whose limiting distribution defines the stable outcomes. In the single-drug model, the population would drift away from one pattern if enough people made '*mistakes*' - by choosing the other. Here, there are more possibilities, and thus various ways in which a group could change from one state (say abstinence) to another (say heroin use):

1. A sufficient proportion could experiment with heroin use to cause the others to change by peer pressure;
2. A sufficient proportion could experiment with another drug to make heroin a '*best reply*' to the mixed behaviour of the group;

3. A sufficient proportion could experiment with another drug (say crack) to tip the group into its use, from which heroin use could be reached by a smaller step.

In the formal analysis, the stability of a transition from one outcome to another is measured by the smallest number of 'mistakes' required to follow one of the above paths. The stable outcomes are those which are easiest to fall into and hardest to fall out of. There are situations where the transition from abstinence to stable drug use runs through another drug - this is an example of a gateway effect, and depends as much on the network structure of users as on physiological complementarities between the intermediate and eventual drugs. There are situations of stable polydrug abuse - again, these can arise purely through societal interactions. The trajectories and stable states depend on cross-drug peer and learning effects, so that they can be reversed by policies (or information) that change relative prices, risks or social acceptability, such as 'fashion' cascades, as with heroin chic. In addition, policy effects can be attenuated: a policy that inhibits one route into adoption of a particular drug may simply displace the dynamics to the next 'shortest' route.

CONCLUDING REMARKS

The models discussed in this section show that a wide range of policies and types of drugs and drug use can be analysed in a unified way. They capture societal interactions in a flexible but transparent way and thus help bridge the gap between individual models of conventional microeconomics (and some psychological and physiological models) and collective models coming from sociology and criminology. The models do not, by themselves, shed light on the psychoactive substances of the future, but they do shed light on how the characteristics of such substances and incumbent patterns of use will influence which potential substances are taken up and what the impacts are likely to be.

However, the abstract nature of the models means that they are more suited to scenario exploration than prediction. They are best used in conjunction with scenarios to put flesh on the narrative bones, add colour and detail and provide a link between 'rounds' of scenario interaction. In addition to the impacts associated with the prevalence of drug use, these models can shed light on induced changes in social structures. Such changes - in particular the formation of 'small worlds' around drug use - can be profoundly important in determining a range of other societal outcomes, including implications for non- drug crime, health, education and workplace productivity. The models also suggest a wide range of interventions such as:

1. Those aimed directly at the evolution and spread of drug behaviours;
2. The interactive effects of law enforcement, informative and treatment interventions.

They can also lead to a dynamic view of policy intervention that shows how to make policies 'follow the cycle' and how to achieve better results by exploiting network dynamics (for example, by making optimal outcomes stable) and the potential for cooperative progress within the affected community (by making stable outcomes optimal). The economic evidence and modelling surveyed in this section is more useful for exploring the future than for predicting it. In particular, it offers a range of inputs for scenario-based exploration. A scenario, in this context, is a partially-specified description of a possible future which can be used to draw out hidden implications of current knowledge and data, explore the logical consistency of findings based on different data, perspectives and methods, integrate a range of different policy options and objectives, explore how key stakeholders are likely to respond to unfolding challenges, and increase the validity and clarity with which future options and developments are understood.

A scenario is based around four main elements:

1. Key uncertainties – important aspects of the future that are either unknown at present or unknowable in the sense that they have not yet happened;
2. Relationships between key variables and parts of the system;
3. Levers – the policy and strategic options that will act through these relationships;
4. Measures – the terms in which system behaviour is described and evaluated.

Further aspects include descriptive details and a narrative thread. There are so many possible futures that descriptions tend to be either confusing or banal. A scenario should encapsulate a single idea or possibility and convey it clearly and convincingly. This idea may involve a specific mechanism (such as the importance of learning or social interaction to drug adoption, the implications of cue-conditioning or self-strategising or gateway and polydrug interactions), a novel policy lever such as the conscious use of uncertainty-generating strategies to 'steer' individual decisions, or new measures (for example, of the structural characteristics of drug supplies, estimates of rents and value added, estimates of the distribution of use across groups). Scenario exploration can follow one or both of two routes: 'forecast' analysis, which starts from the present and tries to identify possible '*landing places*' and their likelihood and desirability; and '*backcasting*' analysis, which starts from a desired (or feared) future and works out what must have happened in order to realise it.

The useful results are the identification and prioritisation of key areas of uncertainty, highlighting key relationships whose operation might lead to efficiencies or unintended consequences, policy sets – policy combinations, priorities and contingent strategies, and relevant measures – indicators or measurements derived from existing or fresh data, together with an understanding of what they do and do not tell us. In this setting, the empirical results surveyed above can be used to produce coherent starting points for constructing the basic elements and to add convincing colour to the narrative thread that explains the main idea. The theoretical models, by contrast, can be used to illustrate qualitative features or to make quantitative predictions. In scenario exploration, they can be used in the initial description to produce graphic or other illustrations of spillovers and other effects to help link the various parts of the narrative and draw attention to holistic consistency. They can also be used in an iterative exploration of the scenario involving alternative policies or resolutions of key uncertainties.

Finally, they can be used to produce quantitative measures to aid in policy evaluation – though not (given data and econometric limitations and the profound gaps between the measured past and the explored future) in specific detail. More concrete observations can be made of the demand-and-supply aspects of drug use. On the demand side, economic evidence supports the importance of choice and thus the relevance of 'economic' variables relating to individual beliefs about risks attached to drug use, price and income effects, the impact of supply, price and quality uncertainty, the interaction of drug use with education and health, and the importance of peer effects and learning in conveying and appropriately reinforcing good decisions. On the supply side, the evidence suggests that targeting of specific drugs, market locations and levels, key players and supplier strategies (e.g., money-laundering or the use of violence) can play a key role in changing the performance of drugs markets. It also suggests reasons why the possibility of users becoming dealers at the local level may be important for the ability of policy to discourage drugs trafficking by imposing costs, how supply uncertainty may affect prices, qualities and market efficiency and whether the prevalence of minimally refined agricultural products (eg the 'big three' of heroin, cocaine and cannabis) is likely to be threatened by 'high-tech' alternatives.

7

Hereditary Alcoholism

The familial occurrence of alcoholism has been known for many years. Many twin, adoption, and family studies now concur that this familial pattern is to a great extent conferred by genes transmitted to biological offspring. Approximately 50-60% of individual differences in risk for alcoholism is genetic, and this proportion is approximately equal in men and women. Thus, it is an easy task to predict that a close biological relative of an alcoholic is at higher risk for alcoholism. However, risk is not inherited-alleles at specific risk-promoting or -protective genes are inherited. To date, there are only two specific genes known to confer substantial protection against alcoholism, variants at the ALDH2*2 and ADH2*2 metabolic enzymes.

The variant alleles lead to the accumulation of alcohol's metabolite, acetaldehyde, when susceptible individuals drink alcohol. This toxic compound produces nausea, flushing, dizziness, and other unpleasant effects, and slow alcohol metabolizers avoid excessive drinking. Therefore, progress from assigning risk statistically to ascertaining whether specific individuals possess risk-promoting or -protective alleles will require the identification of the specific genes underlying risk. The goal of this chapter is to review the current status in the search for such genes. There are three basic approaches to identifying important genes. First, one can target genes based on their presumed importance in influencing alcohol sensitivity. The targeted gene can be overexpressed, underexpressed, or disrupted to the extent that its function is ablated (a gene knockout). Second, one can seek genes that are identified as important because they are differentially expressed. Third, one can seek variations in the sequence of genes that are associated with alcohol sensitivity.

The first approach can only be attempted using genetic animal models. Each approach will be reviewed, with an emphasis on the second and third. Alcohol, like any drug, produces a wide spectrum of behavioral effects, and the genetic sensitivity to one effect is to a large degree distinct from that to another. For drugs of abuse, the spectrum of responses is typically conceptualized as representing the following: initial sensitivity to the drug in a naive organism; tolerance or sensitization of responses following chronic administration; dependence on the drug, as indicated by the occurrence of withdrawal signs when the drug is discontinued; and reinforcing effects of the drug, inferred from self-administration or from drug-seeking behavior. In addition, differences in drug metabolism are in part genetically mediated. Another chapter in this volume discusses ethanol reward extensively. I will therefore concentrate my discussions on other aspects of drug sensitivity. Finally, I will concentrate on pharmacodynamic differences, that is, those not attributable to drug metabolism.

TRANSGENIC TECHNOLOGIES

The literature on transgenic technologies to alter specific genes' functions in laboratory mice is extensive. Several chapters in this volume review studies exploring transgenics' responses to most classes of abused drugs. In their chapter, Cunningham and Phillips discuss more than two dozen studies testing behaviors related to alcohol reward in a like number of genetically engineered mutants. The genes targeted range from neurotransmitter receptors to protein kinases to other enzymes. It is evident from their review that a substantial number of disrupted genes have been shown to affect alcohol self-administration or other traits relevant to reward.

Many of the same mutants have been tested for alcohol sensitivity, tolerance, or withdrawal using a variety of behavioral assays, and many display altered responses when compared to their wild-type controls. Cunningham and Phillips also present a lucid analysis of the strengths, weaknesses, and future promise of gene targeting technologies. In particular, they note the need for stringent behavioral controls to enable straightforward interpretation of the results from such studies as relevant specifically to alcohol.

One aspect of such control is the selection of the phenotype tested. As they note, virtually all studies of alcohol reward chose a single assay of "*reinforcement*," two-bottle preference drinking, to compare knockouts and wild types. However, they note that this paradigm (like all others) has several problems that hinder its interpretation. Multiple assays that target a putative behavioral domain (e.g., "*reinforcement*," "*anxiety*") should be studied before inferring a genetic effect on that domain. One example is studies of sensitivity to "*alcohol-induced motor incoordination*" conducted in mice lacking the serotonin 1B receptor.

The 5-HT_{1B} null mutants were less sensitive to alcohol intoxication than wild types when tested on a balance beam or in an apparatus called the grid test. However, the genotypes did not differ in sensitivity when tested on a fixed-speed or an accelerating rotarod, the screen test, a static dowel, or when tested for grip strength or sensitivity to loss of righting reflex. Even though these (and several other) tasks are used interchangeably in many laboratories, the effect of altering even one gene is not uniform across tests.

Good examples of how to probe a null mutant are on the increase in the literature. For example, Homanics reported no effect of deletion of the γ-aminobutyric acid A ($GABA_A$) receptor subunit 6 gene on alcohol withdrawal, and tested both acute withdrawal from a single injection and chronic withdrawal from a period of ethanol vapor inhalation. Coste et al. found no effect of deletion of the corticotropin receptor 2 on three measures thought to reflect anxiety. Knockout technology, particularly when the altered gene construct can be expressed in brain with both temporal and anatomical precision, will certainly provide much additional information regarding the genetics of alcohol sensitivity, but new methods will need to be developed to realize this potential.

HEREDITARY CONTROL

Until very recently, studies of gene expression related to alcohol were limited to animal studies and were driven by a candidate gene approach. Because such studies depended on laborious assays (e.g., Northern analysis) and were conducted one gene at a time, expression differences were sought only in neural systems where much prior evidence implicated changes as important for alcohol responses. Substantial evidence has been gathered over the years implicating gene expression in *N*-methyl-D-aspartic acid (NMDA), GABA, and glycine receptors, G-protein-coupled receptors, neuro-transmitter transporters, second messenger systems, and transcription factors, including immediate early genes, among others. Now that it is possible to manufacture gene chips, expression array profiling has become a very hot area in the biological sciences.

By affixing short-sequence probes of mRNA complementary to the DNA sequences of many thousands of genes to a glass chip and exposing the chip to a sample of DNA, those genes currently expressed can be captured by hybridization. This and related technologies allows studies with tissue from humans as well as nonhuman animal models. When expression patterns are compared between alcohol-treated groups and controls, for example, a number of genes appear to be differentially expressed. Only three such studies for alcohol-related gene expression have been reported to date. Lewohl and her colleagues compared postmortem frontal cortical tissue from human alcoholics vs controls. They examined expression patterns for >4000 genes and found that expression of 163 differed by more than 40% between alcoholics and controls. Several myelin-related genes were expressed less in alcoholic tissue, and they posit that this may suggest mechanisms underlying the loss of white matter in brains of alcoholics. Thibault et al. used oligonucleotide arrays with 6000 human gene sequences to compare human neuroblastoma cell lines chronically exposed to ethanol with untreated lines. Forty-two genes showed either increased or decreased expression after 3 d of alcohol treatment.

Of particular interest was the increase in expression of the dopamine hydroxylase (DBH) gene, which was accompanied by releasable norepinephrine in the cultures. Increased DBH expression was also seen in adrenal tissue 24 h after a single high-dose injection of ethanol to DBA/2J mice. Finally, the large differences in behavioral sensitivity to ethanol characterizing ILS and ISS mice were exploited in a expression profiling experiment with whole brain tissue. These mice were selectively bred to display high or low sensitivity to ethanol-induced loss of righting reflex. Basal expression of 6000 genes or expressed sequence tags was studied, and 41 genes were differentially expressed. Gene expression profiling will feature very prominently in advances in understanding genetic determinants for many complex traits, including responses to alcohol. The number of genes affected in such experiments is daunting. Much like the situation with gene targeting experiments, there will soon be so much provocative data that making sense of it will be the next problem. Already, research groups are struggling to establish the bioinformatics capacity for collecting and managing such data, which surely is the first step toward sensibly interpreting it.

SCREENING AND MAPPING

The third approach to finding specific genes depends on screening for allelic differences among individuals. Such approaches are essentially mapping genes by finding polymorphisms either in the gene itself or in a region of the chromosome that is physically closely linked to the gene. When the polymorphism (an altered sequence of bases in the DNA) is physically linked, it can be considered a genetic marker, and it escapes the Mendelian laws of segregation and independent assortment. The closer the linkage, the less likely that an individual will inherit the marker from one parent and the mapped gene from the other, which occurs if there is a recombination during meiosis in the region between the marker and the mapped gene. Responses to alcohol are not all or none.

When a population of individuals is studied, the majority tend to show an intermediate response while few show either extremely high or low sensitivity. That is, alcohol responses are quantitative, not qualitative, traits. This implies that multiple genes contribute to that portion of individual differences that is genetic. As mapping experiments seek such genes, they first establish rough chromosomal locations, and then gradually narrow the focus to smaller and smaller regions surrounding the gene. Hence, the search for alcohol response genes, as for genes affecting all complex traits, begins with the identification of *quantitative trait loci*, or QTLs. There are numerous methods for performing QTL mapping experiments, and they are conceptually similar whether one is using humans or nonhuman animal models.

Good descriptions of the various methods and their relative strengths and weaknesses can be found elsewhere, and technical description will be avoided in this chapter. All methods depend on the existence of many genetic markers, scattered throughout the genome, and a densely populated map that locates each such marker on a specific chromosome relative to other markers. The utility of these markers (initially restriction fragment length polymorphisms, but now typically DNA microsatellites, or short repeated sequences of base pairs or triplets) for gene mapping was first realized in the late 1980s.

Now that the Human Genome Project has essentially finished the task of capturing each human gene and placing it on the map, the number of markers for mapping is greatly amplified. Because humans and mice share an ancient evolutionary ancestor, we also share large strings of DNA that have remained linked. This means that mapping QTLs in mice (and finding the responsible gene) tells us where to look in the human genome more than 80% of the time. While the rough mouse genome map was completed only in 2002, the new generation of markers (*single nucleotide polymorphisms*, or SNPs) is rapidly filling in the remaining spaces.

Multistage Mapping Strategy

Mice have been used in QTL mapping studies for alcohol-related traits since the early 1990s. Most laboratories have employed a multistage mapping strategy. One such popular strategy starts with recombinant inbred (RI) strains. For example, there are 26 RI strains that have been created from the intercross of two standard inbred strains, C57BL/6J (B6) and DBA/2J (D2). Each of these represents a more or less random reshuffling of the chromosomal segments (and, therefore, the genetic maps) derived from the B6 and D2 strains. Each BXD RI strain has been genotyped at more than 1500 genetic markers differing between B6 and D2 and the map of these markers assembled. By ascertaining the sensitivity to alcohol of each BXD RI strain and then comparing those values with the marker maps, it is possible to identify clusters of markers from one strain that are associated with high (or low) sensitivity. When such an association is seen, it suggests the presence of a QTL in that chromosomal region. An early review of QTL mapping studies in RI strains reported associations for 19 different responses to alcohol in approximately 50 different chromosomal regions. However, mapping could not be accomplished using only these 26 genotypes, because the resolving power of the analysis was insufficient.

Many hundreds of phenotype– marker correlations are calculated in such an analysis, and after correcting for the multiple comparisons involved, many such possible linkages are actually false-positive associations. To achieve the statistical certainty generally accepted by the genetic mapping community, it is necessary to study additional populations of animals. For many alcohol traits, such additional data have now been gathered and several QTLs have now been mapped with certainty. A more recent review of progress reported that 15 QTLs had been reliably mapped in mouse and one in rat; the number now is more than 40. These QTLs have one interesting feature–there is occasional evidence for sex-specific QTLs for drinking. This is somewhat unusual, as sex differences have not featured as prominently in QTLs for the other alcohol phenotypes.

Recent Findings from Locus to Gene

The goal of any QTL mapping project is to progress from locus to gene. Two clear examples of such progress are now available for alcohol traits. The first is an ongoing project in the laboratory of Kari Buck. In collaboration with John Belknap, Buck is pursuing several QTLs that harbor genes influencing the severity of acute withdrawal from alcohol. When 4 g/kg ethanol is administered intraperitoneally to a mouse, it is sedated for 2-3 h. However, between 3 and 24 h after injection, an acute withdrawal reaction can be seen, where mice display handling-induced convulsions (HICs) that wax and then wane, peaking at about 7 h after injection. This reaction, first noted in the early 1970s, indicates a state of modest physical dependence on alcohol. The reaction becomes more pronounced when alcohol is administered chronically. Goldstein demonstrated through a short breeding project that the severity of alcohol withdrawal from vapor inhalation was heritable, which we later confirmed, and inbred strains of mice are known to differ markedly in acute and chronic ethanol withdrawal severity.

The D2 strain had the highest acute (and chronic) withdrawal score of any strain tested, while the B6 strain was a low-withdrawal strain, so these strains are a good starting population for genetic mapping. The BXD RI strains differ markedly in acute and chronic withdrawal HIC scores, and comparison of RI strain values with the BXD RI marker database identified several provisional QTLs for acute withdrawal. We then tested 400 mice from the F2 cross between B6 and D2 inbred strains. Each individual F2 mouse possessed a unique pattern of markers due to recombinations during meiosis. By comparing the allelic status of the highest-scoring F2 mice with those from the lowest-scoring mice in the chromosomal regions nominated by the BXD RI analysis, we obtained further evidence for some of the provisional QTLs. Finally, starting with the F2 population, we selectively bred for four generations, mating high scorers together and low scorers together. As selection led to divergence in withdrawal severity between the two populations, the allele frequencies for the D2 (high-scoring) strain increased in the "*high alcohol withdrawal*" selected line for markers in some, but not all of the QTL regions. Together, this evidence allowed us to map three significant QTLs contributing to acute alcohol withdrawal on mouse chromosomes 1, 4, and 11.

QTLs in the same three regions were also mapped using the same methods for acute pentobarbital withdrawal. Although each QTL was now known with certainty to harbor a gene or genes affecting withdrawal severity, the major problem to be overcome was the size of the confidence interval surrounding the map location. Even after combining data from three mapping populations, the confidence intervals were large, containing hundreds of genes, any one (or group) of which could actually be responsible. We chose, as other groups have also, to pursue fine mapping by developing *congenic strains.* We started with B6D2F1 hybrid mice, backcrossed them to D2, and genotyped the backcross generation for four markers surrounding the QTL on chromosome 4, called *Alcw2.* We retained only those mice with B6 markers surrounding *Alcw2.* We then repeated this process for 10 generations to create a congenic strain. This strain was about 98% D2 genotype, and possessed DNA from the B6 strain only in the small region surrounding the QTL, about 35 centiMorgans (cM) in length. Nonetheless, this region still contained 600-700 genes. When the congenics were tested for acute alcohol withdrawal vs the D2 background strain, the congenics had less severe withdrawal HICs, showing that the gene responsible for the QTL effect was captured in the B6 piece of DNA.

Next, three further generations of backcrossing to D2 were pursued, and additional congenic strains were created. These *interval-specific congenic strains* (ISCS) possessed smaller regions of B6 DNA in the chromosome 4 QTL region. Testing of the five ISCS showed that four still retained the responsible gene, whereas one did not. The pattern of markers allowed us to exclude all but a very small region of chromosome 4, less than 1 cM. This region probably contains only 18 genes, and is captured in the ISCS5 strain. Combined with the previous mapping data, this QTL's gene significantly affects withdrawal with high probability (LOD = 8.8, $p = 2.2 \times 10^{-10}$). It similarly captures a gene affecting acute pentobarbital withdrawal severity.

After sequencing 10 of the 18 genes or gene-related sequences in the QTL interval, only one with a substantial number of polymorphisms has yet been found. This gene, *Mpdz,* codes for a multiple PDZ-domain protein. We examined *Mpdz* in 15 standard inbred mouse strains (including B6 and D2) that were known to vary substantially in acute ethanol withdrawal. Eight haplotypes yielding three protein variants were identified. There was a substantial correlation between the severity of withdrawal and the particular protein variant each strain possessed. *Mpdz* participates as a cellular scaffolding protein in the collocalization of receptors such as several serotonin receptor subtypes, a tyrosine kinase receptor, and a nerve growth factor receptor with their reaction partners. Thus, it is an attractive candidate gene for the QTL effect.

Alcohol-induced Losses

The other long-standing gene mapping project in alcohol research that has made substantial progress toward gene identification has been studying the effects of a high dose of alcohol to cause mice to lose their ability to right when placed on their backs. The duration of suppression of this reflex has been used as an index of sensitivity to alcohol and numerous depressant drugs. Many years ago, mice were bred to be sensitive to this effect by breeding together those with a "*long sleep*" (LS) response and those with a "*short sleep*" (SS) response. Although the mice are not actually sleeping, the LS and SS mouse lines after many generations of selective breeding differ markedly in duration of loss of righting reflex. LS and SS strains were crossed to create LSXSS

RI strains, and these strains were phenotyped. LS and SS strains were subsequently inbred (to form ILS and ISS strains, respectively), and a new set of ILSXISS RI strains has also been created. Following logic parallel to that described in the previous section for the acute withdrawal mapping project, Tom Johnson, Beth Bennett and their colleagues have been pursuing several QTLs for sensitivity to the loss of righting reflex after ethanol.

Early studies proposed several provisional QTLs and gradually proved the existence of four of them. Numerous congenic and ISCS strains have been developed to isolate each QTL within smaller regions of the genome, and the confidence interval for some is approaching the subcentiMorgan level. A unique direction pursued in search of the loss of righting reflex QTLs is the search for gene coding variants within the QTL regions. Using high-throughput comparative DNA sequencing technologies to compare DNA from ILS and ISS mice, >1.7 million bases of comparative DNA sequences were generated from 68 candidate genes located within the four QTLs. Eight central nervous system genes that together displayed 36 changes in protein coding were identified from this search.

Congenic Strains

The *congenic strains* derived from QTL mapping studies are useful beyond the scope of the project for which they are prepared. For example, one congenic useful in pursuit of *Alcw1*, the acute withdrawal QTL on chromosome 1, was obtained from Thomas Ferraro, who constructed it to search for QTLs for kainate-induced seizures. A compendium of congenics available from various QTL mapping projects has been published. In addition, a series of more than 60 congenics spanning the genome with D2 genotype introduced onto a B6 background is in progress, and other specialized populations for mapping also exist. Mapping can of course be performed in highly polymorphic, genetically segregating populations, where more than two alleles are present for each gene.

A QTL for alcohol- induced activity has been successfully mapped using such a strategy, and other behavioral traits have been mapped in similar populations. A technique developed many years ago but recently repopularized is the generation of mutations thorough random mutagenesis. By treating male mice with a toxic chemical, N-ethyl-N-nitrosourea (ENU), many mutations are induced at random throughout the genome. The offspring of treated mice are then tested and phenotypic outliers sought on the assumption that they are extreme scorers because of an induced mutation. If such a mutation can be isolated by breeding, it can then be mapped and eventually identified. (It may seem a little odd to consider ENU mutagenesis under the rubric of mapping rather than as variant of candidate gene targeting, but it combines both approaches and the genes targeted are not identified *a priori*).

Many ENU projects are in progress, and their utility for mapping complex trait genes is highly touted. It has been argued that QTL mapping methods are inferior to ENU-based approaches. However, neither approach has as yet identified a gene, although progress is continuing on both fronts. We have argued that it is more prudent to see what success is actually achieved by each approach before deciding which is a superior methodology. Another new technique depends on the proliferation of SNP markers. Using 15 standard inbred strains, Grupe et al. surveyed them for SNPs and identified 500 useful for mapping. This mapping database was combined with preexisting information about nearly 3000 other SNPs among the strains. By examining published phenotypic data for the strains and associating that data with the mapping information, they identified several regions with potential QTLs. These included the three QTLs for acute alcohol withdrawal mapped by Buck et al., and QTLs for alcohol-preference drinking. The method currently does not allow assignment of statistical confidence to such predictions, but it may prove to be a useful alternative to the laborious process of genotyping individual mice as more phenotypic data on inbred strains are accumulated in the literature.

GENE MAPPING IN HUMAN POPULATION

As noted, studies of gene expression differences in humans are in their infancy. Gene mapping efforts in human populations uses variants of the methods described for mice. However, the problem of mapping in humans is markedly more difficult. Mouse populations are characterized by controlled breeding and the existence of inbred strains (each of which is essentially a collection of genetic clones). Humans breed with whomever they want, and are much more polymorphic than mice typically are. In addition, humans frequently form genetic subpopulations with characteristic gene frequencies. Comparison of the gene frequencies between a patient group and a "*control*" is apt to fall victim to stratification, meaning that controls may tend to be drawn from one population while alcoholics (for example) are drawn from another.

The apparent gene frequency differences (or marker associations) are then likely not to be replicated in additional samples. Added to this is the fact that studies with humans tend to look for markers associated with a diagnosis. A diagnosis of alcoholism can be achieved with many different clusters of symptoms, so any group of alcoholics in a genetic study is likely to be diagnostically (and potentially etiologically) heterogeneous. The statistical and methodological difficulties with human genetic mapping for complex traits are discussed more authoritatively elsewhere. Consequently, the search for markers associated with alcoholism, like similar searches for other complex traits, has been a succession of findings of linkage followed by a failure to replicate.

Most such efforts have attempted to search intensively for markers in the vicinity of a candidate gene of interest (a serotonin receptor, for example). When considered collectively, the literature offers mixed evidence for several potential associations. One large association study has been ongoing for more than 10 yr. The Collaborative Study on the Genetics of Alcoholism (COGA) has accumulated information and genotypes from more than 10,000 subjects from hundreds of multigeneration families. A few suggestive and even signficant linkages have been reported. The best hope for such studies is currently thought to reside in shifting the focus to "*endophenotypes.*" These are heritable traits associated with the real trait of interest. For example, alcoholics (and their close genetic relatives) have lower-amplitude sensory event-related potentials, called P3 or P300 waves, than controls. QTLs for P300 amplitude have been mapped in the COGA data set. As the mouse and human mapping projects continue to develop, it will be of increasing interest to compare results directly, given the evolutionary conservation of the genetic maps. Thus far, the most encouraging possibility of convergent QTL identification comes from the domain of alcohol withdrawal.

The chromosome 11 QTL in mice falls in a region spanning genes for several $GABA_A$ receptor subtypes. One variant form of the GABAα2 receptor subtype has been shown to be significantly associated with withdrawal severity across a number of inbred mouse strains. Human QTL studies have also reported some evidence for an association of alcohol dependence with markers in the region of these subunit genes. This is far from proof that the same gene has been mapped in humans and mice. In neither species has the 2-subunit gene been proven to underlie the QTL association. Nonetheless, it offers hope that the efforts to find genes in mice and humans will be increasingly informative and facilitate progress in both species. The search for alcohol-related genes has only been underway for about 10 yr. During that time, significant progress has been made, but no gene has yet been captured. This chapter has largely ignored the formidable complexities that must be addressed for further progress to occur. These include the highly specific nature of phenotypes under investigation, an important point because any gene-finding study is only as good as the phenotyping on which it rests, and the genes eventually found will certainly have a restricted range of phenotypic influence.

Genes interact with other genes (epistasis), and both human and animal studies are faced with extracting genes' presence from this backgound of noise caused by other genes. Genes' effects are not necessarily the same even when identical genotypes are compared in multiple environmental settings. Such gene *X* environment interactions are also of importance. Many of these issues are discussed elsewhere. Nonetheless, the progress in 10 years has been marked, and the tools are improving daily. Soon we will need to deal with the issue of what use to make of this knowledge of specific risk-promoting and -protective genes, and this must be an important future consideration for all geneticists.

ETHANOL REWARD

Substantial evidence supports the suggestion that development of alcoholism is strongly influenced by family history and that genes underlie a significant portion of that influence. Despite recognition of a genetic influence for many decades, it is only within the last 10 years that scientists have had the sophisticated research tools needed to begin the task of identifying specific genes involved in alcoholism. These tools have been provided by advances in molecular biology and by improvements in the animal genetic models and behavioral models used to study the processes that contribute to excessive ethanol consumption. Use of animal models has enabled examination of a wide range of ethanol-related behavioral and neurobiological processes. Although many different behavioral processes have been considered, theorists place strong emphasis on ethanol's rewarding and aversive motivational effects.

These effects are believed to be critical in determining whether individuals exposed to ethanol will continue to consume ethanol and increase their intake over time. Moreover, genetic differences in sensitivity to ethanol's rewarding and aversive effects are hypothesized to contribute importantly to development of the excessive drinking patterns characteristic of ethanol abuse and alcoholism. This section focuses on recent animal studies that address the contribution of genotype to ethanol's rewarding and aversive effects. We begin with a brief overview of the major behavioral models currently used to draw inferences about ethanol's motivational consequences. This overview is followed by more detailed consideration of genetic influences in these behavioral models using a variety of different strategies. Finally, we conclude with a summary of the current status of research in this area along with suggestions for possible future directions.

GENETIC INFLUENCE OF ETHANOL

Most studies examining genetic influences on ethanol reward have used behavioral procedures that fall into one of two categories: (a) self-administration models and (b) conditioning models. Self-administration models typically involve procedures in which animals have substantial control over their ethanol intake, including control over the amount (dose) and temporal pattern of intake. Within this category, it is useful to distinguish between relatively simple home-cage drinking procedures and procedures conducted in operant chambers that require an explicit "seeking" response (e.g., bar pressing) to obtain ethanol.

In contrast, conditioning models are characterized by experimenter administration of a fixed drug dose in combination with exposure to a gustatory (e.g., taste conditioning) or exteroceptive (e.g., place conditioning) stimulus. The value attached to this drug-paired stimulus is typically assessed in a test conducted without ethanol. Each of these models is briefly reviewed in the following subsections, with emphasis on issues of potential importance to interpretation of genetic studies described later. More comprehensive discussions of these and related behavioral models can be found elsewhere.

Regular and Continuous Availability

The oldest and most commonly used technique for studying genetic differences in ethanol reward is simply to place a drinking tube on the home cage and measure the amount/dose of ethanol consumed or the preference for ethanol relative to a concurrently available alternative solution such as water. The principal advantages of this procedure are its simplicity and high face validity. Although it is recognized that many variables influence ethanol drinking in such procedures, rewarding postingestive pharmacological effects are assumed to play a significant role. In home-cage procedures, ethanol is usually available continuously and intakes/preferences are measured over 24-h (or longer) periods of time.

A potential problem with this approach is that 24-h measures may hide important variations in the temporal pattern of ethanol exposure. For example, a strain that drinks a large number of relatively small ethanol bouts each day could achieve the same total daily intake as a strain that drinks a small number of relatively large bouts. Although analysis of total daily intakes would suggest no strain difference, consideration of bout size suggests a difference in the pharmacological effects experienced by each strain. One solution to this problem, of course, is to measure drinking patterns within each day of continuous access (e.g., using a drinkometer). Alternatively, one can measure intakes over smaller periods of time. Another solution is to actually reduce the total amount of time per day that ethanol is available (e.g., 30-60 min), so that animals have the opportunity to engage in only one or a relatively small number of drinking bouts. Such "*limited-access*" procedures promote relatively high ethanol intakes in short time periods, making it more likely that ethanol's pharmacological effects influence intake. Although interpretation of genetic differences in home-cage ethanol drinking often focuses on hypothesized differences in sensitivity to postabsorptive rewarding pharmacological effects, ethanol drinking/preference can also be influenced by other factors such as taste or sensitivity to aversive orosensory or pharmacological effects.

To address the role of taste, investigators have sometimes examined intakes/preferences for primary tastants (e.g., sweet, bitter) in the same strains. When strains differ in ethanol intake/ preference, but not in intake/preference for the primary tastants, interpretations based on differences in sensitivity to postabsorptive pharmacological effects become more compelling. Testing animals in operant or conditioning procedures in which ethanol is administered by a nonoral route offers another strategy for eliminating interpretations based on taste differences. Conditioning procedures are also useful for addressing the possibility of genetic differences in sensitivity to aversive drug effects.

Home-cage Self-administration

In home-cage self-administration procedures, animals simply approach and consume ethanol. In contrast, operant procedures require animals to engage in an explicit "*seeking*" response (e.g., bar press or nose poke) to gain access to ethanol. Although ethanol is usually consumed orally, operant models allow for the possibility of administering ethanol directly into the stomach, blood, or brain via surgically implanted cannulae. Thus, in contrast to home-cage procedures, operant models allow one to separate ethanol "*consumption*" from ethanol "*seeking*" and, if desired, to eliminate the oral route entirely. The value of the latter feature is well illustrated by a recent study showing that two mouse strains known to differ dramatically in home-cage ethanol intake (DBA/2 and C57BL/6) showed little difference when nose poking produced intravenous ethanol injections in an operant procedure. This finding supports the suggestion that aversive oro sensory (preabsorptive) effects of orally administered ethanol contribute to the normally low home-cage intakes of DBA/2 mice. Interpretive issues raised in the study of home-cage oral self-administration generally also apply to operant studies using the oral route (e.g., concern over the temporal pattern of ethanol intake).

In both models, investigators will sometimes add a sweetener or other flavor to the ethanol solution in an effort to increase overall intakes. Although these flavor additives may be "*faded out*" over time, use of this strategy in genetic studies raises the possibility that strain differences are caused by differences in sensitivity to the added flavor rather than to postingestive effects of ethanol. Again, as suggested earlier, this hypothesis can be addressed by examining responding for the flavor additive in the absence of ethanol. One difficulty in the interpretation of both home-cage and operant self-administration procedures is that increases (or decreases) in the target behavior produced by a genetic manipulation do not unambiguously reflect increases (or decreases) in ethanol reward. This ambiguity arises due to the inverted U-shaped relationship between ethanol intake and variables that presumably affect ethanol's reinforcing efficacy, such as dose or concentration.

Thus, a genetic manipulation that reduced ethanol reward (e.g., a null mutation) could either increase or decrease ethanol intake, depending on where control intakes fell on the concentration/ dose-response function. Strategies for addressing this problem include examining a range of ethanol doses/concentrations and using conditioning procedures to provide converging evidence on effects of the genetic manipulation. A unique feature of the operant model is its potential to separate genetic differences in appetitive processes underlying ethanol-seeking behavior from consummatory processes involved in the regulation of ethanol intake. This feature could be important for determining whether different genes influence appetitive and consummatory processes. However, as recently noted, much of the literature on operant self-administration of ethanol has not allowed a clear separation of these processes, due to frequent alternation between "*seeking*" and "*consuming*" within self-administration sessions. Although promising alternative procedures involving chain schedules have recently been introduced, these schedules have not yet been used in the study of genetic differences.

Pavlovian Conditioning

Conditioning models are based on the premise, derived from *Pavlovian conditioning*, that stimuli paired with drug exposure acquire the ability to elicit affective states similar to those evoked by the drug itself. Thus, approach or contact with a drug-associated *conditioned stimulus* (CS) is interpreted as evidence of a drug's rewarding effect, whereas avoidance or withdrawal from a drug-paired CS is viewed as indicating the drug's aversive effect. Early proponents of these techniques sometimes implied that conditioning models could substitute for self-administration models in the analysis of drug reward. Although there is certainly overlap in the information provided by these models, theorists now recognize that conditioning models measure a learning process that is fundamentally distinct from that measured by self-administration models. Consequently, genetic differences observed in self-administration models may not always match those obtained in conditioning models. Place and taste conditioning are the two conditioning procedures that have been used most often to examine genetic differences in ethanol's rewarding and aversive effects.

These procedures differ primarily in the nature of the CS paired with ethanol, and the response used to index learning. In place conditioning, the experimenter initially pairs drug with distinctive environmental cues (e.g., tactile, visual) and later measures approach toward or withdrawal from those cues in a choice test. In taste conditioning, the experimenter pairs drug with novel taste or flavor cues (e.g., saccharin, sodium chloride) and measures changes in subsequent intake or preference for the flavored food or fluid. Although avoidance of foods or fluids previously paired with abused drugs has often been attributed to presumed aversive properties of such drugs, some theorists have argued that taste avoidance may actually be caused by rewarding drug effects. On the basis of genetic correlational data from standard inbred mouse strains, however, it seems likely that taste avoidance induced by ethanol reflects an aversive drug effect. One of the main advantages of conditioning procedures is their sensitivity to both drug reward and aversion.

In the case of ethanol, both the taste and place conditioning procedures have yielded evidence of conditioned preference and conditioned aversion. Although a detailed discussion of variables producing these diverse outcomes is beyond the scope of this chapter, it appears that prior ethanol experience, species, and the temporal relationship between CS and ethanol are among the variables determining whether preference or aversion is obtained. Other advantages of conditioning models include the ability to test animals in a drug-free state and the fact that dose-effect curves are typically monophasic rather than biphasic. However, these models also have certain disadvantages that must be considered in the analysis of genetic differences. For example, initial biases (i.e., unconditioned preference or aversion) for the target CS may interfere with measurement of conditioned preference or aversion. Moreover, genetic differences in those unconditioned biases will complicate interpretation of genetic differences in conditioned effects. Interpretation of genetic differences in a target phenotype may be further complicated by genetic differences in other phenotypes that have an important influence on the target phenotype. For example, genetic differences in initial fluid intake may hinder interpretation of intake changes produced by taste conditioning. Similarly, genetic differences in basal locomotor activity may affect detection of conditioned changes produced by place conditioning.

ETHANOL'S REWARDING AND AVERSIVE EFFECTS

Several different genetic strategies have been applied to understanding the role of genes that influence ethanol's rewarding and aversive effects. These approaches include using traditional animal genetic models such as inbred strains and selectively bred lines, as well as more recently developed techniques that involve the mapping of quantitative trait loci (QTLs), the insertion or inactivation of targeted genes, or the measurement of changes in gene expression. Recent advances based on the application of these strategies to the study of ethanol reward and aversion are summarized below.

Recombinant Inbred Strains

The heading, "*Standard Inbred Strains,*" is used to distinguish this panel of strains from specialized groups such as recombinant inbred strains. Standard inbred strains differ genetically from each other, unsystematically. They are unique, relative to a population of heterogeneous individuals, in that each individual is homozygous at every genetic locus. Further, each individual of an inbred strain is genetically identical to all other individuals of its strain. Therefore, differences among individuals must arise from epigenetic influences. The laboratory is ideal for holding environmental conditions such as temperature, housing density, and handling procedures constant. When several inbred strains are examined under controlled conditions, differences among individuals within each strain can be used to estimate epigenetic sources of trait variation, and differences among strains to estimate genetic influences. Inbred strains have been widely used to demonstrate genetic influences on motivational traits, including those thought to reflect sensitivity to the rewarding and aversive effects of ethanol. Some of the earliest examples are studies that characterized strain differences for home-cage ethanol drinking.

Larger numbers of strains have been included in more recent examinations of this trait. In addition, standard inbred strains have been used to demonstrate genetic regulation of ethanol-induced *conditioned place preference* (CPP) and *conditioned taste aversion* (CTA). The many two-strain comparison studies, which also pepper the literature, suggest genetic regulation, but are inconclusive due to their lack of power for estimating heritability. Panels of 12 or more strains provide reasonable power for estimating genetic contribution, and for examining genetic correlations between traits. However, they have typically fallen short of providing specific genetic information with regard to chromosomal location of genes influencing these complex motivational traits. As marker and sequence information expands for specific inbred strains, their utilization for gene mapping purposes is likely to grow.

A recent publication by Grupe et al. demonstrated that single nucleotide polymorphism data from standard inbred strains could be utilized to rapidly predict the chromosomal regions associated with trait variation. Furthermore, the Mouse Phenome Project, initiated with the goal of promoting systematic characterization of many standard inbred mouse strains for a wide range of phenotypes of potential importance to biomedicine and biological science, may provide a rich phenotypic database, including ethanol reward traits, to which genetic information can be applied. In addition, the characteristics of inbred strains make them invaluable for QTL mapping via the production of recombinant inbred (RI) strains. They also provide a fixed genetic background for assessing effects of targeted mutations and transgene insertions, a characteristic that is proving to be critical in the interpretation of results in single-gene mutants.

Artificial Selection

Artificial selection is highly effective at producing extreme-scoring individuals, when some proportion of the trait variability within a population is genetically influenced. In this case, mating of extreme-scoring individuals results in offspring that are more extreme-scoring as a group, relative to the original population mean. Rodent lines selectively bred to study ethanol reward and sensitivity have been described elsewhere and their use in the new genetic era discussed. These selection programs have been bidirectional, producing genetic animal models with extreme sensitivity opposite those with extreme insensitivity for a presumably single trait. Like standard inbred strains, a typical use of selected lines has been to search for correlated traits to selection–traits that are pleiotropically affected by some of the same genes. For example, short-term selected lines bred for gene mapping purposes were used to demonstrate a genetic relationship between ethanol consumption and genetic susceptibility to exhibit withdrawal. In the current era, a time when mapping of complex trait genetics has become approachable, selected line utilization has been adjusted.

The extreme nature of their selected traits makes them ideal for QTL mapping. For example, short-term selected lines were used to verify the locations of QTLs for home-cage ethanol preference drinking; P and NP rats, long-term selected lines bred for high (P and HAD) and low (NP and LAD) home-cage ethanol preference drinking, have also been used to narrow the gene search to chromosomal regions. Genotypic, rather than phenotypic, selection has also been used to verify a QTL for alcohol acceptance (consumption of ethanol in the absence of a water choice) in mice. In this case, animals were chosen for breeding based on their genotype at chromosomal regions thought to harbor QTLs for alcohol acceptance. The extreme traits of selected lines also make them suitable for examining the effects of genetic manipulations.

Once a gene has been identified as a candidate for regulation of the selection trait or a genetically correlated trait, a logical approach might be the use of antisense oligodeoxynucleotides against that particular gene. This approach has been used successfully for non-alcohol-related traits. For example, injection of the antisense oligodeoxynucleotide against angiotensinogen significantly reduced the elevated systolic blood pressure, which is the hallmark of SHR rats. Currently, problems with delivery of antisense molecules to the brain, with duration of action, and with difficulty in determining appropriate sites of injection, complicate the widespread utilization of this technology.

Locating Genes on a Particular Chromosome

We have already mentioned QTL mapping, a means by which genes for several ethanol reward traits have been localized to segments of particular chromosomes. A chapter that serves as a primer on QTL mapping has recently been published, and several other treatments of this topic also exist. We review here the findings to date with regard to nominated and confirmed QTL locations for traits thought to reflect sensitivity to ethanol reward and aversion. We mention when similar regions have been implicated among studies. However, we recognize that the same gene or a different linked gene could be the source of similarity. It is not surprising that ethanol consumption has been a focus of QTL mapping. Already mentioned are QTL studies using selected lines.

The first published QTL study for home-cage ethanol drinking offered ethanol vs water to recombinant inbred strain (RI) mice generated from C57BL/6J (B6) and DBA/2J (D2) progenitors. This panel of BXD RI strains was particularly appropriate for genetic analysis of this trait because of the well-known, extreme difference in ethanol preference and consumption between the B6 and D2 mouse strains. QTLs nominated by the RI data were followed by studies intended to confirm or eliminate them from further consideration. These studies utilized short-term selected lines and a genetically segregating B6 × D2 F2 cross, and in combination, confirmed significant linkage of home-cage ethanol consumption with QTLs on chromosomes 2 and 9. At about the same time, similar studies were being conducted by another group, also in BXD RI mice. This group ultimately confirmed QTLs for ethanol drinking on chromosomes 1, 4, and 9, using a B6 × D2 F2 intercross. A full genome scan for ethanol consumption QTLs in a (B6 × D2) × B6 backcross also identified a QTL on chromosome 2, which was male-specific, at a location somewhat more proximal to the location suggested by the studies.

A female-specific QTL mapped to chromosome 11 was not confirmed in a subsequent study, but two new linkages to markers on chromosomes 1 (female-specific) and 3 (male-specific) were suggested. The QTL on chromosome 1 was similar in location to that identified by, but it was not sex-specific in their F2 population. However, the QTL identified on chromosome 3 by was male-specific. Finally, another confirmation study used congenic strains, created using a classical backcross procedure, in which D2 alleles were introgressed onto a B6 background. Male B6 × D2 F1 mice were chosen for breeding based on low ethanol preference scores, and were crossed to female B6 mice. A full genome scan was accomplished for multiple congenic lines that ultimately confirmed linkages of ethanol consumption with markers on chromosomes 1 and 2. Locations were similar to QTL locations identified by others. Mice of genotypes other than those derived from B6 and D2 progenitors have also been used to map QTLs for home-cage ethanol drinking. A study in a set of RI strains derived from B6 and A/J progenitors (AXB/BXA RI) provided evidence for an ethanol drinking QTL on chromosome 2, but considerably distal to the location identified by others. Studies in B6.C Recombinant QTL Introgression strains, mice that carry approx 5% BALB/cJ (C) genome on a B6 background, have suggested linkage of ethanol consumption to markers on chromosomes 1, 2, and 15.

Screening for voluntary ethanol consumption in LSXSS RI mice, strains derived from selected lines bred for extreme sensitivity and insensitivity to ethanol's sedative effects, nominated regions on chromosomes 1, 2, 4, 5, 9, 15, and X. Several of these QTLs were in regions similar to those identified in other studies, particularly those on chromosomes 1, 2, 9, and 15. It should be clear from these descriptions that several chromosomal regions have been implicated by many independent studies to harbor genes that influence voluntary ethanol consumption. This question has been addressed directly. Based on the eight QTL mapping studies described above that used B6/D2-derived populations of mice, a meta-analysis was performed with the goal of combining the mapping information from these independent studies.

The QTLs that appeared consistently across studies were on chromosomes 2, 3, 4, and 9. The combined P values ranged from 1×10^{-7} to 1×10^{-15}, highly significant according to stringent criteria. Two other QTLs, one on chromosome 1 and one on 11, were reported as appearing less consistently, but still reaching overall stringent statistical criteria ($P < 0.0001$). These results indicate the remarkable convergence across studies that were not procedurally identical and pinpoint the chromosomal regions that should be most avidly pursued in the identification of genes that influence ethanol consumption in mice. Rats have also been used in the search for QTLs relevant to ethanol consumption. The first study that appeared in the literature utilized the F2 cross of the inbred P and NP rat lines, selectively bred for high and low levels of ethanol-preference drinking. A QTL was identified on chromosome 4 with a LOD score sufficient to meet suggested statistical criteria for significant linkage.

The gene for neuropeptide Y is located within the QTL region, which led to studies in knockout mice to examine the possible role of this peptide on ethanol consumption. However, QTL mapping in mice has not implicated the syntenic region. Subsequent work in P NP F2 rats nominated two additional QTLs on chromosomes 3 and 8, both in regions syntenic to those previously nominated in mice; the rat chromosome 3 region is syntenic with a region on mouse chromosome 2, and the rat chromosome 8 regions is syntenic with mouse chromosome 9. One other rat study used an F2 cross of noninbred LAD and HAD rats, similarly selectively bred for high and low ethanol drinking levels. Evidence was obtained for linkage of ethanol consumption to markers on chromosomes 5, 10, 12, and 16. Surprisingly, none of these QTLs overlap with those identified in the P × NP F2 cross. However, as the authors indicate, the genetic origins of the two sets of selected lines are different. Only a few other traits relevant to ethanol reward have been subjected to QTL analyses. One of these traits is ethanol-induced CPP.

Unlike rats, which tend to exhibit aversion to stimuli previously associated with ethanol, ethanol-naïve mice develop preference for ethanol-associated cues. A QTL study in BXD RI mice provided the strongest evidence for associations with ethanol-induced CPP on chromosomes 4, 8, 9, 18, and 19. The chromosome 4 region at about 56 cM and the chromosome 9 region at about 31 cM are within the confidence intervals of QTLs identified for ethanol consumption. Thus far, studies designed to confirm these nominated regions have not been published. Genetic correlations estimated from BXD RI means from this CPP study and the BXD drinking study showed a trend toward a positive genetic correlation ($r = 0.34$, $n = 17$, $.05 < p < 0.10$, one-tailed), offering weak support of a common genetic influence on these phenotypes. The other ethanol reward relevant trait subjected to QTL analysis is ethanol-induced CTA produced by pairing of a saccharin solution with injection of 2 or 4 g/kg ethanol. For 2 g/kg of ethanol, regions of chromosomes 1, 3, 4, 7, 11, and 17 were implicated. For 4 g/kg of ethanol, regions on chromosomes 1, 2, 4, 6, and 9 were nominated.

The nominated chromosome 2 QTL is in a region similar to that for ethanol consumption, whereas the nominated chromosome 9 QTL is more distal. Preliminary results suggest confirmation of the CTA QTL located on chromosome 1. Genetic correlations based on BXD RI strain means from this CTA study and strain means from our CPP study or our drinking study were not statistically significant, suggesting minimal genetic overlap for these traits. However, a recent study of ethanol-induced CTA in standard inbred strains reported a significant genetic correlation between CTA and home-cage ethanol preference.

More specifically, strains showing stronger ethanol-induced CTA tended to have a lower ethanol preference than strains showing weaker CTA ($r = 0.68$, $n = 13$, $p < 0.05$). Thus, in the standard inbred model, there is evidence of overlap in the genetic mechanisms influencing ethanol preference and ethanol-induced CTA. Gene by gene analysis will not fully define the array of genetic influences on a complex trait. A developing focus in QTL studies is identification of epistatic interactions that influence trait variability. Pairwise combinations of markers are analyzed to estimate the proportion of trait variability attributable to their interaction. Such an analysis has been applied to the ethanol consumption trait and identified a significant interaction between loci on chromosomes 2 and 3. In other words, the influence on ethanol consumption of the QTL on chromosome 2 was modified by genotype at a chromosome 3 region. Specifically, those mice that were homozygous for B6 alleles in both the chromosome 2 and 3 regions consumed more ethanol than those that were homozygous B6 for chromosome 2, but heterozygous or homozygous D2 on chromosome 3. These data, and others, illustrate the need to develop statistical models that routinely screen for interactive genetic influences.

Emerging Genetic Technologies

As in most areas of contemporary neurobiology, investigators interested in the genetic basis of ethanol reward have exploited emerging genetic technologies that allow inactivation of a specific target gene or the insertion of a foreign gene. In most instances, these approaches have involved permanent manipulation of the genome at the embryonic stage and the creation of a new mouse strain (e.g., a "*knockout*" or "*transgenic*" strain). However, it is also possible to temporarily inactivate a target gene (using antisense oligonucleotides) or to insert a novel gene (using an adenoviral vector) into specific brain areas of individual animals.

A detailed description of these technologies and their limitations is beyond the scope of this chapter. Interested readers can find overviews of these techniques and their application to the study of alcohol-related and other phenotypes elsewhere. This section focuses on use of gene-targeting manipulations to address the role of specific candidate genes in ethanol reward. Presentation of these studies is organized below according to the neurobiological system thought to be affected by the genetic manipulation.

Alcoholism and serotonin system

Given substantial evidence from animal studies linking the serotonin (5-HT) system to alcohol intake, several investigators have focused on the role of 5-HT receptor genes in ethanol reward. Interest in the 5-HT_{1B} receptor gene in particular was encouraged by an initial report that 5-HT_{1B}-/- mice drank twice as much ethanol as 129/Svter control mice across a range of ethanol concentrations in a two-bottle home-cage drinking procedure. This difference did not appear to be related to taste or calories because there were no differences in food intake or consumption of sucrose, saccharin, or quinine. A subsequent pair of conditioning studies extended these observations by showing that deletion of the 5-HT1B receptor reduced ethanol reward as indexed by CPP, but not ethanol aversion as indexed by CTA. However, later self-administration studies from the same laboratories and elsewhere have failed to show consistently greater ethanol intakes in knockout mice using either home-cage or operant procedures.

The reasons underlying these later failures are not well understood. One possibility is that the null mutation's effect on the drinking phenotype was lost due to changes in the background genotype caused by genetic drift or the presence of alleles from multiple 129 substrains. The role of the serotonin system in ethanol reward has also received attention in two recent candidate gene studies that involved insertion of transgenes. In the first study, transgenic mice that overexpressed the 5-HT_3 receptor gene in forebrain areas showed significantly lower intake of 10% ethanol in a two-bottle home-cage procedure than B6SJL/F1 control mice. Although conclusions are limited by use of only one ethanol concentration and lack of data on intake of other flavored solutions, this finding generally supports a modulatory influence of the 5-HT3 receptor on ethanol reward. The other transgenic study involved mice with a monoamine oxidase A (MAOA) deficiency produced by insertion of an interferon β (IFNβ) transgene into the MAOA gene. This enzyme deficiency produces a substantial elevation in whole-brain serotonin, as well as other amines.

Transgenic mice did not differ from C3H/HeJ control mice in intake of 10% ethanol or water in a home-cage two-bottle procedure, suggesting no effect of MAOA deficiency on ethanol reward. However, conclusions from this study are limited by use of only one ethanol concentration and relatively short test durations (only 2 or 24 h after 24 h of water deprivation). Moreover, given that serotonin enhancement in these transgenic mice is known to diminish with age, implications of this finding for understanding serotonin' s role in ethanol reward are uncertain in the absence of brain serotonin measurements in mice tested for ethanol drinking. Nevertheless, despite absence of an effect on drinking, transgenic mice of the same age showed reductions in other measures of ethanol sensitivity, including ethanol-induced sleep time and hypothermia.

Alcoholism and dopamine system

The dopamine (DA) system has been repeatedly implicated in the rewarding effects of most abused drugs, including ethanol. Thus, several candidate gene studies have targeted DA receptor genes, especially those encoding the D1 and D2 receptor subtypes. In an extensive study of home-cage ethanol drinking, mice lacking functional D1 receptors drank less ethanol and showed lower preference across a range of concentrations compared to heterozygous and wild-type littermate control mice in both limited and continuous-access two-bottle procedures. D1 knockout mice also drank less ethanol (12%) when it was the only fluid available for 24 h. A companion series of experiments examined effects of several pharmacological pretreatments on intake of 12% ethanol in a two-bottle choice procedure.

These experiments showed that treatment with selegiline (10 mg/kg), an MAOB inhibitor that increases synaptic DA levels, reduced ethanol intake in wild-type and heterozygous mice, but had no effect on D1 knockout mice. Similarly, treatment with a D1 receptor antagonist (SCH-23 390, 1 mg/kg) reduced ethanol intake in wild-type and heterozygous mice, but not in D1 knockout mice. In contrast, injection of a D2 receptor antagonist (sulpiride, 50 mg/kg) nearly eliminated the already low ethanol intakes of D1 knockout mice at a dose that had relatively modest effects on drinking in control mice. While acknowledging involvement of both DA receptor subtypes, the authors concluded that their data supported a greater role of the D1 receptor subtype in ethanol reward. However, given the substantial growth retardation (20-30%) noted in this genetic model, one must be cautious in interpreting these effects as specific to ethanol reward rather than a more general effect on feeding behavior, especially since intakes of other palatable substances were not reported. Several recent studies have used targeted mutation of the DA D2 receptor gene to provide information about its role in ethanol reward.

An initial study of ethanol drinking in a home-cage continuous-access choice procedure showed significant reductions in ethanol intake and preference across several concentrations in D2 receptor knockouts compared to either heterozygous or wild-type control mice, which did not differ. Alternative interpretations based on differences in taste sensitivity were eliminated by data showing no genotype effect on intake or preference for saccharin or quinine solutions presented without ethanol. The suggestion that reduced ethanol intake reflected a reduction in ethanol reward was supported by a subsequent conditioning study showing that D2 receptor knockout mice failed to develop ethanol-induced CPP. Consistent with findings from the drinking study, heterozygous and wild-type mice did not differ in place preference, suggesting that one functional D2 receptor allele is sufficient to maintain ethanol reward in this procedure.

The latter finding is of additional interest because place preference was maintained in heterozygotes despite a significant reduction in basal activity levels. Finally, an important role for the D2 receptor gene was indicated by a study showing that the null mutation significantly reduced responding for several ethanol concentrations in a continuous-access operant self-administration procedure. However, D2 knockout mice also responded less for water, food, and saccharin, suggesting a relatively broad, nonspecific impairment of motivated responding. The influence of the DA D2 receptor gene on ethanol drinking has received additional support in a very recent study that produced temporary overexpression of the receptor in nucleus accumbens of genetically heterogeneous rats by direct microinfusion of the gene via adenoviral vector.

More specifically, exposure to the gene-containing vector produced significant within-group reductions in 7% ethanol intake and preference in a continuous access home-cage choice procedure relative to intakes measured after exposure to a null vector, which had no effect on baseline drinking. Although interpretation of these findings would have been aided by examination of other palatable substances, these data raise the interesting possibility that an excess of DA D2 receptors reduces ethanol reward, a conclusion that appears at odds with earlier conclusions that ethanol reward is reduced by a lack of D2 receptors. Such data suggest that one may not be able to extrapolate behavioral effects of the receptor's absence to predict effects of variation in receptor levels. Although most candidate gene studies of DA system involvement in ethanol reward have focused on receptor gene manipulations, a recent series of experiments targeted the gene encoding DARPP-32, a phosphoprotein important for regulation of striatal dopaminergic systems.

DARPP-32 null mutants showed lower responding for ethanol and saccharin, but not for food or water in a continuous access operant self- administration procedure. Companion studies used conditioning procedures to support the conclusion of a deficit in ethanol reward as shown by absence of CPP, but no change in ethanol's aversive effect as measured by CTA. Thus, disruption of the DARPP-32 gene appeared to produce effects similar to those produced by targeted mutation of DA receptor genes.

Alcoholism and opioid system

Ethanol reward has also been linked to activity within the opioid system, encouraging examination of mice lacking μ-opiate receptors. In two operant self- administration studies, μ-receptor knockout mice consistently responded less for ethanol than wild-type control mice. μ-Knockout mice also consumed less ethanol in home-cage two-bottle tests, but only if they had prior ethanol self-administration experience in either the operant procedure or a home-cage single-bottle procedure. However, knockout mice also showed lower operant self-administration of water and sucrose, suggesting a broader involvement of the μ-receptor in ingestive behavior. A later report using a different μ-receptor knockout strain has generally confirmed and extended these observations by showing lower ethanol intakes in a home-cage drinking procedure and lack of ethanol-induced CPP in mice carrying the mutation.

Although these effects were statistically reliable only in females, findings from the previous study, which involved only males, argue against the suggestion that effects of t-receptor deletion are gender-specific. In light of the prominent role that endogenous opioids play in current theorizing about mechanisms of ethanol reward, it is not surprising that attention has also been given to gene manipulations that alter expression of the endogenous opioid peptide β-endorphin. In a home-cage two-bottle procedure, transgenic mice with β-endorphin deficiencies were found to drink more ethanol than sibling wild-type mice. The suggestion that β-endorphin normally interferes with sensitivity to ethanol reward was supported in another study that showed greater intravenous self-administration of ethanol by β-endorphin-deficient mice than by wild-type control mice.

Signaling pathways

Several recent studies of ethanol reward have targeted genes in protein kinase signaling pathways, especially those related to protein kinase C (PKC) and cAMP-dependent protein kinase (PKA). Mutant mice lacking the PKCε isozyme were found to self-administer less ethanol than wild-type sibling controls, both in an operant oral self-administration procedure and in a home-cage two-bottle choice procedure. However, PKCε-deficient mice did not differ in intake of food, water, saccharin, or quinine, suggesting a selective effect on ethanol reward. A selective effect on home-cage ethanol drinking has also been reported in mice with disruption of the gene encoding the regulatory IIβ (RIIβ) subunit of PKA. In this case, however, the null mutation increased ethanol consumption across several concentrations, while having no effect on intakes of food, water, sucrose, or quinine.

The effect on ethanol intake appeared to be specific to disruption of the RIIβ subunit because mutations that produced a deficiency in either of two other PKA subunits (RIβ or Cβ1) had no effect on ethanol consumption. The PKA signal transduction pathway has also been implicated in studies of mice with mutations in the gene encoding $G_{s\alpha}$, the α subunit of the stimulatory G protein of adenylyl cyclase (G_{nas}). These studies are especially noteworthy because effects of the mutation were examined on three different genetic backgrounds (C57BL/6J, 129/SvEv, and CD1 × 129/

SvEv). In all three cases, ethanol intake or preference in a home-cage two-bottle procedure was lower in mice with heterozygous inactivation of the G_{nas} gene (G_{nas}-/+) compared to wild-type littermates. Companion studies showed no effect of $G_{s\alpha}$ deficiency on intakes of food, sucrose, or quinine, suggesting a selective influence on ethanol reward. To provide additional information on the role of the PKA signaling pathway, these investigators also measured home-cage drinking in mice carrying one of two transgenes known to affect PKA activity.

In both cases, expression of the transgene was limited to the hippocampus and other forebrain areas by using the promoter from the gene encoding CaMKIIα. In the first case, transgenic mice with reduced PKA activity caused by expression of R(AB) (an inhibitory form of the regulatory subunit of PKA) drank less ethanol than wild-type littermates. However, transgenic mice that showed increased adenylyl cyclase activity caused by expression of a constitutively active mutant form of $G_{s\alpha}$ (*$G_{s\alpha}$ Q227L*) drank ethanol at levels identical to controls. The authors suggested that the failure to see an increase in ethanol drinking in the latter case may have reflected a ceiling effect related to testing the transgene on a genetic background (C57BL/6J) that already drinks ethanol at relatively high levels. Overall, this series of studies provides good support for a role of the PKA pathway in ethanol reward.

NPY's role

As noted earlier, QTL mapping studies in rats derived from the selectively bred alcohol-preferring (P) and alcohol-nonpreferring (NP) lines have identified neuropeptide Y (NPY) as a possible candidate gene for ethanol self-administration. More direct support for NPY's role has subsequently come from studies showing that targeted disruption of the NPY gene increases ethanol intake, whereas overexpression of the NPY gene reduces ethanol intake relative to wild-type littermate controls in a home-cage two-bottle procedure. The increase produced by NPY deficiency did not appear to be related to taste or calories as shown by the lack of effect on intake of food, sucrose, or quinine.

A later study extended these findings by examining effects of the NPY knockout on a different genetic background and by testing mice lacking the NPY Y5 receptor. These experiments showed that the enhanced drinking phenotype was retained when the NPY mutation was shifted from a mixed C57BL/6J × 129/SvEv background to an inbred 129/SvEv background, but only at the highest ethanol concentration (20% v/v). However, targeted disruption of the Y5 receptor (129/SvEv background) had no effect on ethanol consumption or preference.

Additional genes

Ethanol reward-related behaviors have also been examined in several other candidate gene studies that do not readily fit into the above categories. Gene targets in these studies include dopamine β-hydroxylase, which affects norepinephrine synthesis; the G-protein-coupled inwardly rectifying potassium channel GIRK2, the peptide-degrading enzyme neutral endopeptidase, angiotensinogen, bovine growth hormone, and catalase. To date, about 25 different mouse transgenic or knockout models have been tested using one or more of the ethanol reward phenotypes. The most commonly studied phenotype is two-bottle home-cage drinking, which was tested in nearly two-thirds of the published comparisons. In almost all of these drinking studies, a continuous-access procedure was used and intakes or preferences were based on measurements averaged over 24 h. Thus, most of the information currently available from candidate gene studies is based on a single procedure that, as noted earlier, has several limitations. Of particular interest is the finding that manipulation of candidate genes did not universally reduce ethanol intakes or fluid intakes more generally. For example, the outcomes from two-bottle studies were distributed such that about one-third of the studies showed an increase in ethanol intake or preference, about one-third showed a decrease, and about one-third showed no effect of the genetic manipulation.

Such findings increase confidence that effects of gene insertion or deletion are related to the targeted genes themselves rather than nonspecific effects of genetic engineering. Moreover, as noted above, many of the drinking and operant studies examined the effect of the gene manipulation on intakes of food or other palatable substances, thereby addressing whether effects were specific to ethanol or reflected a more general influence on ingestive or motivated behavior. Overall, however, there are several weaknesses in the candidate gene studies. For example, the influence of background genotype was addressed explicitly in only a few studies. In several cases, it was difficult to discern the exact nature of the background genotype. It was also sometimes difficult to determine whether wild-type and mutant mice were littermates derived from breeding of heterozygous parents or were bred separately from homozygous parents. Given the potentially important influence of background genotype on ability to detect effects of a targeted mutation, future studies should give greater attention to these issues.

Another weakness, which is not unique to the study of ethanol phenotypes, is that genetic alterations in these studies were generally present throughout development, raising the possibility that observed effects on target phenotypes reflect compensatory changes in other neurobiological systems. Moreover, although a few of the transgene studies restricted expression to the forebrain, most of these studies placed no anatomical limits on the gene manipulation. Both of these concerns may be addressed in the near future, with greater use of genetic technologies that allow for creation of tissue-specific mutations and "conditional" mutations that can be introduced or removed in individual mice under experimenter control. Finally, future studies involving genetic mutations should give greater attention to alternative interpretations that do not involve ethanol reward mechanisms. For example, only a few of the studies considered the possibility that the gene manipulation produced its effect by altering ethanol pharmacokinetics.

Another alternative interpretation that must be considered, especially in the case of conditioning procedures (i.e., operant, taste, or place), is whether the gene manipulation has affected sensory-motor or learning abilities that are critical to acquisition and performance of behavior in those tasks. At least ethanol drinking, is likely influenced by a relatively large number of genes acting through several different neurobiological systems. In some cases, the influence of these genes was already strongly predicted on the basis of previous neuro-pharmacological studies directed at the protein products of these genes (e.g., dopamine and opioid receptors). In other cases, however, these studies have revealed novel information on genetic influences that have not been, or in some cases could not be, identified using conventional approaches. The challenge for future investigators is to find a way to incorporate this complex pattern of information into a conceptual framework that will provide an integrated explanation of the genetic and neurobiological mechanisms underlying ethanol reward.

Regulatory Mechanism

A burgeoning focus in alcohol research is gene expression profiling for specific alcohol-related traits. Gene expression profiling is the study of changes in gene expression for multiple genes at one time. Although studying changes in gene expression is by no means new to alcohol research, advances in technology are beginning to alter drastically the way in which this research is conducted. If a candidate gene is the subject of study, expression may be compared by *in situ* hybridization, Northern analysis, Western analysis, or by other analogous methods. However, if the question is which genes are regulated by alcohol, the ability to screen thousands of genes or genetic sequences for expression changes, simultaneously, poses a major advantage. This approach has been termed microarray analysis because gene expression changes are assessed using an array of unique mRNA sequences, which have been adhered to a solid support, such as a small glass microscope slide. Hybridization with a double- labeled probe provides a signal that permits automated detection of gene up- or downregulation. This approach is being applied to human alcoholism, and is likely to make inroads into the question of genetic regulation of alcohol addiction.

Expression profiling has been used in cell lines treated with ethanol and in genetic animal models of altered ethanol sensitivity, but microarray analyses in animal models of ethanol reward are eagerly awaited. The information that can be derived from the literature about ethanol reward and gene expression comes largely from studies of candidate genes and immediate early genes (IEGs). In a putative study of "*craving*," rats were denied the beer they had become accustomed to consuming in a distinctive environment. The expression of c-fos was increased in corticolimbic and brainstem regions of craving rats, relative to rats given free access to beer or to control rats with no beer experience. Another study found an interaction between stress and ethanol consumption in B6 mice. Significant changes in IEG expression were found in some brain regions of stressed ethanol/sucrose-consuming B6 mice, relative to stressed sucrose controls. For example, lower c-Fos expression was found in the hippocampus and higher expression in the nucleus accumbens. However, no significant effects of ethanol were found on the gene expression of nonstressed mice.

In a follow-up study that achieved somewhat higher levels of ethanol consumption, c-Fos and FosB were increased in the nucleus accumbens, central nucleus of the amygdala, and Edinger-Westphal (EW) nucleus, in ethanol/sucrose-consuming mice, relative to mice offered sucrose alone or water. This study also found a significant reduction in c-Fos expression in the dentate gyrus from alcohol-experienced mice. Most recently, this research group has reported increased c-Fos expression in the EW nucleus of B6 mice when first given the opportunity to consume ethanol, which increased with additional ethanol experience. Furthermore, rats bred for high levels of voluntary alcohol drinking, Alko Alcohol (AA) rats, trained operantly to self-administer alcohol or alcohol in saccharin, have also exhibited increased c-Fos expression in the EW nucleus, relative to water or saccharin self-administering rats. The EW nucleus has previously been associated with oculomotor functions.

These results and others, for example, suggesting a role for the EW nucleus in response to stress, appeal for a reexamination of this nucleus and its connections as they relate to ethanol's motivational effects. Candidate gene expression studies include examination of dopamine D2 receptor gene expression in alcohol preferring AA vs alcohol-avoiding ANA rats. Using slot-blot to quantify mRNA, no significant differences were found between the rat lines in the dopaminergic nuclei examined. Another study used RT-PCR and found that ribosomal protein L18A mRNA and diacylglycerol kinase iota mRNA were differentially expressed between AA and ANA rats. In this study, the goal was to search for novel genetic substrates for ethanol preference, and did not target specific genes.

There is the possibility that these expression differences are unrelated to the ethanol- drinking phenotype, since selected lines would be expected to differ genetically for some trait-irrelevant genes due to genetic drift. One study examined the expression specifically of genes coding for receptors and enzymes involved in GABAergic and glutamatergic neurotransmission. Compared were rats given the choice of drinking ethanol or water for 2 mo or over 9 mo, rats given ethanol as the sole source of fluid for over 9 mo, and rats offered only water. mRNA content was evaluated in several brain regions chosen on the basis of known expression of the genes of interest. The most pronounced effects were seen in the parieto-occipital cortex. Here, reductions in mRNA levels for all eight of the $GABA_A$ receptor subunits examined were found in rats with free choice of ethanol for over 9 mo. mRNA levels for six of the $GABA_A$ subunits were also reduced in rats offered free choice of ethanol for 2 mo.

Only mRNA for the 3 subunit was reduced in the group of rats forced to consume ethanol. Similar results were found for some glutamate receptor subunits and enzymes associated with GABA and glutamate production, with respect to changes being associated primarily with the voluntary consumption groups. Data from this study must be carefully interpreted given differences among groups in daily ethanol consumption. However, it is interesting that larger changes were seen in the group of rats offered ethanol vs water for 2 mo than in those under the forced-consumption condition, since the later group consumed relatively more alcohol. Given the focus of alcoholism research on genes associated with ethanol metabolism, it is not surprising that the effects of ethanol consumption on the expression of *Adh-1*, *Ahd-2*, and *Cas-1*, three genes involved in ethanol metabolism, have been examined.

In this study, ethanol preference was characterized in B6, BALB/c, their F1, and six recombinant inbred strains derived from the B6 × BALB/c F2. Separate groups of mice of these strains were offered ethanol-containing liquid diet or isocaloric liquid diet as their only source of fluid and food. Using cDNA probes, liver mRNA for the genes of interest was quantified. Ethanol feeding resulted in increases in the level of mRNA for *Adh-1* and *Ahd-2*, across most genotypes. A smaller effect of ethanol was seen for *Cas-1*. A strong association between ethanol preference and change in *Ahd-2*-specific mRNA following ethanol feeding was found. Gene expression analysis in the study of ethanol's motivational effects will likely expand as chip technology and bioinformatics continue to improve.

One of the greatest advantages of the gene profiling method, relative to analysis of one gene at a time, is that coordinated changes in related systems can be detected. Detection of patterns of increased and decreased gene expression are likely to prove critical to our ultimate understanding of the genetic processes influencing alcohol addiction.

FUTURE PROGRESS

Tremendous strides have been made over the last 10 years in the search for genes that influence ethanol reward-related phenotypes. For example, as illustrated above, QTL mapping studies in both rats and mice have identified and confirmed several candidate chromosomal regions that contain genes thought to influence ethanol intake or preference in the two-bottle home-cage procedure. Although the QTL approach has not yet yielded evidence supporting the involvement of any specific gene in ethanol reward, these studies have pointed to several interesting candidates that have been pursued using other strategies such as targeted gene disruption. Efforts to narrow the chromosomal regions containing these QTLs are ongoing, and it seems likely that this approach will lead to the identification of specific genes in the very near future. Candidate gene studies using genetically engineered mouse strains (knockouts, transgenics) have also been quite successful in suggesting a role in ethanol reward for many genes across several different neurobiological systems, especially as indexed by the two-bottle home-cage drinking procedure.

The targets of these studies have included genes for receptors that were already strongly implicated in ethanol reward on the basis of neuropharmacological studies (e.g., D2 dopamine receptor, μ-opioid receptor), as well as signaling pathway genes and other genes whose influence on behavior has been more difficult to establish by conventional approaches (e.g., protein kinase Cε, Gsα). The application of gene-expression profiling technology to the study of ethanol reward-related behaviors holds great promise for the future, although present efforts in this area are still at a very early stage of development. This situation is expected to change quickly as access to this technology increases and costs are reduced. It may prove fruitful in future studies to compare basal gene expression levels and ethanol- induced expression changes in rodent lines that have been selectively bred for sensitivity to a reward-related phenotype such as ethanol intake or preference.

A selected line difference in the expression of a particular gene would strongly implicate that gene in producing the phenotypic difference, especially if expression differences were observed across multiple replications of the selection study. One might also adopt the strategy of studying basal and ethanol-induced gene expression in large panels of standard inbred strains that have been well characterized for ethanol-reward phenotypes. Genetic correlational analysis could then be used to identify the subset of genes whose patterns of expression levels or changes were most strongly associated with ethanol's rewarding or aversive effects. Although the field has advanced substantially through the application of all of these new genetic technologies, there are still several limitations in our current understanding. For example, most existing information is based on a single behavioral procedure: continuous-access, two-bottle home-cage drinking. Because behavior in this procedure is influenced by variables other than ethanol's postingestive pharmacological effects, it is critical that future genetic research give greater attention to examining ethanol's effects in other reward-relevant procedures, especially operant and Pavlovian conditioning.

By using multiple behavioral models, the field will eventually develop a better understanding of the relationships among these models and, it is hoped, will converge on a common set of genes that mediate ethanol reward and aversion. As correlational approaches such as QTL mapping and gene expression profiling yield an increasing number of candidate genes, gene manipulation strategies (i.e., targeted mutation, transgene insertion, antisense oligonucleotides, adenoviral vector insertion) will assume an even more important role in providing an experimental basis for eliminating "*false positives.*" An issue that has not yet been well addressed in the literature is the extent to which these techniques model natural genetic variation. For example, knowing that complete inactivation of a gene changes an ethanol-reward phenotype may not be very useful for understanding the normal function of that gene when the range of naturally occurring alleles does not include a "null" variant. Thus, it will be important for future studies and technologies to examine gene manipulations that more closely approximate natural genetic variation.

Future progress in this area of research will also be facilitated by continued appreciation of the fact that ethanol reward-related phenotypes are *polygenic.* That is, multiple genes determine these traits and each of these genes may have only a relatively small influence on total phenotypic variance. Moreover, as noted earlier, the impact of a given gene may depend critically on the form or level of expression of other genes (epistasis) and environmental conditions (gene environment interaction). When attempting to integrate and interpret findings, greater attention should also be given to the principle of *pleiotropy,* which refers to the fact that a given gene may affect multiple traits.

There is a strong tendency among alcohol researchers to place a very high value on genes whose influence appears alcohol-specific, even though it seems quite unlikely that natural selection pressure would have fostered retention of such genes. Rather, it seems likely that many of the genes that will eventually be found to influence ethanol reward and aversion were selected because of more general roles they play in determining an organism's sensitivity to motivational variables that affect survival (e.g., ability to distinguish between nutrients and toxins, ability to learn and remember based on previous experience). These are very exciting times for scientists interested in the genetic basis of complex behavioral traits. The postgenomics era has placed great emphasis on the development of tools and conceptual frameworks for understanding the functions and interrelationships among genes, environment and behavior. Advances during the last 10 years have nearly overshadowed those in the preceding 50 years, and progress over the next 10 years is likely to be even more astonishing.

8

Effect of Addiction on Nervous Activities

Addiction is characterized by compulsive drug seeking and taking despite adverse consequences. Addiction is a gradual process, which occurs in most individuals only after repeated drug exposure. Thus, it appears that individuals addicted to drugs undergo time-dependent alterations or neuroadaptations in the brain that occur during a course of repeated drug administration. Chronic drug intake produces these neuroadaptations in neuronal cell types located within specific brain regions. These neuroadaptations alter the function of these individual neurons and ultimately the neural circuitry that mediates the abnormalities in complex behaviors, which characterize aspects of addiction. Opiates, a class of drugs with high abuse potential, produce several types of behavioral abnormalities that are mediated by neuroadaptations within specific brain regions.

Chronic opiate administration produces dependence, a functional alteration that is described as the need for continued drug administration to avoid a withdrawal syndrome. The opiate withdrawal syndrome is characterized by physical and psychological symptoms; the latter include restlessness, extreme anxiety, and depressed mood. These negative emotional symptoms of withdrawal are thought to contribute to addiction by inducing relapse to drug use during the withdrawal phase. The physical dependence associated with opiates has been ascribed to numerous brain regions as well as spinal cord.

The psychological symptoms of withdrawal have been mapped partly to the mesolimbic dopamine system, which arises from dopaminergic neurons in the *ventral tegmental area* (VTA) of the midbrain and projects to the *nucleus accumbens* (NAc) and other forebrain limbic structures. Tolerance, another type of functional alteration seen after chronic opiate administration, describes the need for an increasing drug dose to achieve the same effect. Tolerance occurs to many opiate actions, and is thought to be mediated by neuroadaptations in the brain and spinal cord regions that mediate those actions.

Tolerance may contribute to addiction by causing escalation of drug doses. Opiates can also produce sensitization, or reverse tolerance, in which the same dose of the drug elicits progressively greater responses. Of particular importance is sensitization that occurs to the rewarding effects of opiates, which is believed to be mediated by neuroadaptations in the mesolimbic dopamine system. Sensitization may contribute to addiction by promoting relapse to drug use long after withdrawal periods subside. To date, numerous neuroadaptations have been described in the brain and spinal cord after chronic administration of opiates, and it has been increasingly possible to relate certain of these changes to the behavioral abnormalities associated with opiate addiction. This review focuses on one particular neuroadaptation, upregulation of the cAMP second messenger pathway, which is one of the best-established molecular and cellular mechanisms of opiate tolerance and dependence.

Upregulation of the cAMP pathway has been demonstrated in several regions of the nervous system, and related to different aspects of opiate action within those regions. Here, we focus on two such brain regions: the locus coeruleus (LC) and the nucleus accumbens (NAc).

Table 8.1 Upregulation of the cAMP pathway and opiate addiction

Site of upregulation	*Functional consequence*
Locus coeruleus	Physical dependence and withdrawal
Ventral tegmental area	Dysphoria during early withdrawal periods
Periaqueductal gray	Dysphoria during early withdrawal periods, and physical dependence and withdrawal
Nucleus accumbens	Dysphoria during early withdrawal periods
Amygdala	Conditioned aspects of addiction
Dorsal horn of spinal cord	Tolerance to opiate-induced analgesia
Myenteric plexus of gut	Tolerance to opiate-induced reductions in intestinal motility and increases in motility during withdrawal

LOCUS COERULEUS OF BRAIN

The LC, located on the floor of the fourth ventricle in the anterior pons, is the major noradrenergic nucleus in brain. The LC is a densely packed homogeneous nucleus that regulates attention, arousal, and autonomic tone. The LC's noradrenergic neurons have extremely widespread projections and innervate virtually all regions of the brain and spinal cord. Activation of the LC, which occurs upon precipitation of opiate withdrawal, increases noradrenergic transmission throughout the central nervous system and mediates some of the physical symptoms of withdrawal. Indeed, excitation of LC neurons is both necessary and sufficient to elicit certain behavioral signs of opiate withdrawal. Although physical dependence is not thought to be an important contributor to addiction *per se*, studies of opiate action in the LC have provided important insight into the types of neuroadaptations that opiates elicit in responsive neurons. Acutely, opiates inhibit the firing rate of LC neurons by binding to μ opioid receptors that are coupled to the $G_{i/o}$ family of G proteins. $G_{i/o}$ proteins couple directly to an inward rectifier K^+ channel to mediate an increase in outward K^+ current.

The coupling to $G_{i/o}$ also causes inhibition of adenylyl cyclase to decrease cAMP levels and reduce activity of cAMP-dependent protein kinase (PKA). This leads to a decrease in the conductance of a nonspecific cation channel, which has not yet been isolated at the molecular level. These two opiate actions, activation of an inward-rectifying K^+ channel and inhibition of a nonspecific cation channel, diminish LC neuronal excitability. In contrast, chronic opiate exposure increases the intrinsic excitability of LC neurons via upregu-lation of the cAMP pathway. In particular, chronic opiate exposure increases levels of expression of type I and type VIII adenylyl cyclase and of the catalytic and type II regulatory subunits of PKA.

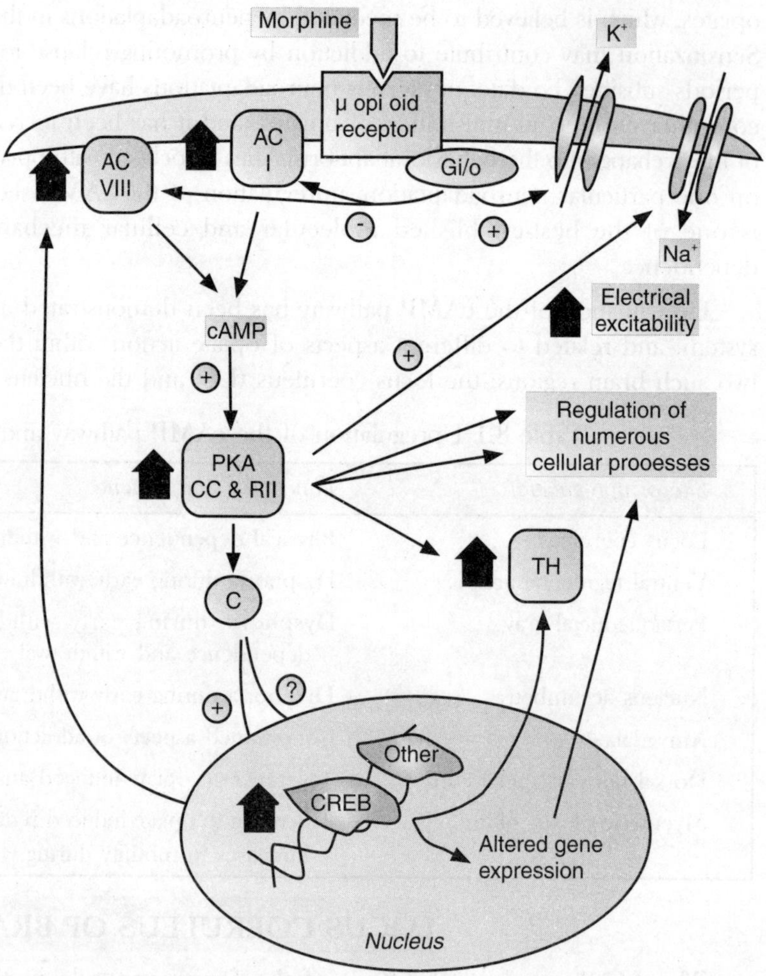

Fig. 8.1. Schematic illustration opiate actions in the locus coeruleus.

This upregulation has been shown to increase the excitability of LC neurons partly via activation of the nonspecific cation channel. Upregulation of the cAMP pathway can be viewed as a homeostatic response of LC neurons to opiate inhibition and, therefore, as a mechanism of opiate tolerance and dependence in these neurons. The continued presence of morphine and the up- regulated cAMP pathway results in LC neurons firing at near-normal levels. When opiates are removed, the up-regulated cAMP pathway is unopposed and leads to over-activation of these neurons, as seen during withdrawal. Interestingly, the time course for the recovery of the cAMP pathway to normal, over a course of hours to days, parallels that of the diminution of the withdrawal syndrome.

NEUROLOGICAL FUNCTION

Many long-lasting, activity-dependent changes in neuronal function are mediated by changes in gene transcription. This occurs via activity-dependent regulation of transcription factors, proteins that recognize and bind to specific DNA sites within the genes they control. These transcription factors then influence gene expression and contribute to the long-lasting phenotypic changes in neurons. One of the best-characterized transcription factors in the nervous system is CREB (cAMP response element-binding protein). Considerable work has shown that CREB is an important mediator of the upregulation of the cAMP pathway seen in the LC during chronic opiate administration. CREB, and its family members, ATF1 (activating transcription factor 1) and CREM (cAMP response-element modulator), are expressed ubiquitously and are highly regulated by extracellular stimuli.

CREB binds as a dimer to CRE (cAMP response element) consensus sites within the regulatory region of many genes. CREB is activated by its phosphorylation on a specific serine residue, Ser 133. Upon phosphorylation, CREB interacts with the adaptor proteins, CBP (CREB-binding protein) and p300, to stimulate (and in some cases inhibit) gene transcription. CREB can be phosphorylated on Ser 133 by many different protein kinases, including PKA (protein kinase A), Ca^{2+}-calmodulin kinase II and IV, Rsk2 (pp90 ribosomal S6 kinase), MAPKAP kinase 2 (MAPK-activated protein kinase 2), Akt/PKB (protein kinase B), p70 S6K (p70/S6 kinase), and MSK1 (mitogen and stress-activated protein kinase 1). These findings suggest that diverse types of extracellular stimuli and intracellular pathways converge at the level of CREB (and related proteins) to regulate the expression of CRE-containing genes. Addiction researchers focused attention on CREB based on the upregulation of the cAMP pathway that occurs in response to drugs of abuse. Thus, according to the scheme, the up-regulated cAMP pathway in the LC would be expected to cause increased activation of CREB and induction of several CRE-containing genes. Indeed, CREB is highly regulated by opiates in this brain region.

Acute morphine has no effect on CREB immunoreactivity, but, it decreases the level of CREB phosphorylation (activation) in the LC. After chronic opiate administration, levels of phosphoCREB return to control levels, while there is an increase in CREB expression. Levels of phosphoCREB increase dramatically upon the onset of opiate withdrawal. These data can be explained in the following way. Opiate inhibition of the cAMP pathway leads to decreased PKA activity and lower levels of phosphoCREB, which feeds back to increase transcription of CREB via a CRE site within the CREB gene. As the level of CREB rises it eventually is phosphorylated either by the now normal levels of PKA (which is itself induced by chronic opiates independently of CREB) or by other protein kinases.

Higher levels of phosphoCREB then increase expression of many genes that contain CRE sites, such as adenylyl cyclase type VIII and tyrosine hydroxylase. Further evidence for a role of CREB in physical opiate dependence comes from mice with a targeted mutation in the CREB gene, which show a reduction in opiate physical withdrawal. However, this mouse has several shortcomings for investigating detailed mechanisms of opiate dependence. First, the CREB mutation occurs throughout the mouse, not just in particular brain regions involved in opiate dependence such as the LC. Second, CREB expression is not completely lost in this mouse, since the mutation results in increased expression of an alternative isoform of CREB. Third, since the mutation occurs at the earliest stages of development, any phenotype seen could reflect abnormalities during development.

Future generations of genetically engineered mice that allow inducible CREB mutations targeted to specific brain regions of adult animals are necessary to define the role of CREB in opiate dependence more clearly. Although opiate regulation of CREB mediates many aspects of the upregulated cAMP pathway in the LC, other mechanisms also exist. Two examples are the upregulation of adenylyl cyclase type I and PKA subunits, genes that do not contain CRE sites. The ability of chronic morphine to increase these proteins is not affected by blockade of CREB in the LC.

Although the mechanism underlying induction of adenylyl cyclase type I is unknown, upregulation of PKA subunits appears to occur via a post-transcriptional mechanism. Sustained PKA activation in cultured cells leads to decreased levels of the catalytic and type II regulatory subunits, without affecting their mRNA levels. The ubiquitin-dependent protein degradation pathway apparently mediates the proteolysis of the PKA subunits after their activation by cAMP. Our hypothesis is that opiates, by inhibiting cAMP formation and PKA activation, lead to an accumulation of PKA subunits due to reduced degradation. Work is now needed to study this scheme within the LC in vivo.

NEURAL SUBSTRATE

As stated earlier, the mesolimbic dopamine system is a major neural substrate for the motivational and rewarding effects of opiates. This occurs via two mechanisms. Opiates increase dopaminergic transmission to the NAc by activating VTA dopamine neurons. This occurs indirectly through opiate inhibition of GABAergic interneurons within the VTA that inhibit the dopamine neurons. Opiates also act directly on opioid receptors expressed by NAc neurons. The rewarding effects of other drugs of abuse are mediated via similar actions in the VTA-NAc pathway, although each drug produces these effects via drug-specific mechanisms. In addition, the mesolimbic dopamine system appears to be play a similar role in mediating the actions of natural reinforcers, such as food, drink, sex, and social interactions. It is interesting, therefore, that chronic administration of opiates or several other drugs of abuse up-regulates the cAMP pathway in the NAc. The upregulation, similar to that observed in the LC, involves increased levels of adenylyl cyclase and PKA, although the specific isoforms of adenylyl cyclase that are induced have not yet been identified.

In addition, the drugs decrease levels of $G_{\alpha i/o}$, which would further augment activity of the cAMP pathway. As would be expected, chronic opiate or cocaine administration also causes a sustained increase in the functional activity of CREB in the NAc. Insight into the functional consequences of the up-regulated cAMP pathway in the NAc has come from several studies. Infusion of inhibitors of the cAMP pathway (e.g., PKA inhibitors) directly into the NAc makes animals less responsive to the rewarding effects of cocaine and, possibly, opiates. Conversely, infusion of activators of the cAMP pathway (e.g., adenylyl cyclase or PKA activators) have the opposite effects. Consistent with these findings are observations obtained from direct manipulation of CREB within the NAc by use of viral-mediated gene transfer.

Over-expressing CREB specifically within the NAc decreases the rewarding effects of cocaine and opiates, whereas expressing a dominant negative inhibitor of CREB (termed mCREB) increases the drugs' rewarding effects. This action of CREB appears to be mediated in part via regulation of dynorphin gene transcription. Dynorphin is an opioid peptide expressed in a subset of NAc neurons. Dynorphin acts on κ-opioid receptors to inhibit dopamine neuron cell bodies in the VTA and

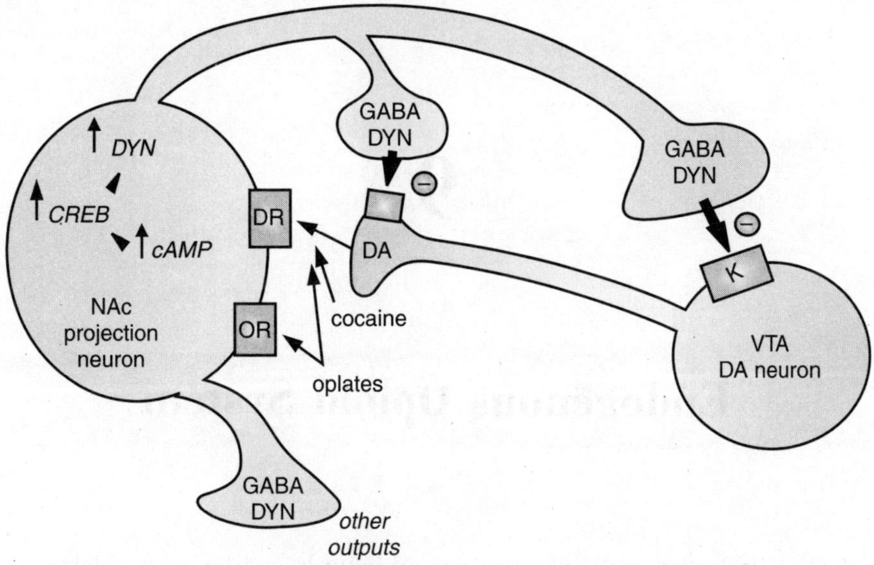

Fig. 8.2. Regulation of CREB by drugs of abuse.

their terminals in the NAc, and thereby inhibits drug reward. The dynorphin gene has been shown to be a target for CREB both in vitro and within the NAc in vivo.

Moreover, the effect of increased CREB activity in the NAc on drug reward can be completely blocked by a κ-receptor antagonist. Thus, it appears that CREB negatively regulates drug reward by increasing the gain of the dynorphin feedback loop. Recent work has provided a more complete understanding of the behavioral phenotype mediated by CREB activity in the NAc. Not only does CREB activity decrease sensitivity to drug reward, it also renders an animal less sensitive to a range of external stimuli, including both natural rewards (e.g., sucrose preference) and aversive stimuli (e.g., anxiogenic, aversive, and nociceptive stimuli). Based on these data, our hypothesis is that upregulation of the cAMP pathway induced in the NAc by chronic drug exposure numbs an animal's emotional responsiveness. This state presumably contributes to part of the negative emotional state that occurs during drug withdrawal and helps drive relapse to drug use.

CONCLUDING REMARKS

Great excitement surrounds the progress that is being made in understanding the alterations in specific neurons that occur following chronic opiate administration, and relating these adaptations to complex behavior. This review focused on one alteration, namely, upregulation of the cAMP pathway and CREB and the very different roles it subserves in two different brain regions. Upregulation of the cAMP pathway is, however, just one of a large number of changes at the molecular and cellular levels that have been documented after chronic drug administration. A current challenge is to understand how these many changes interact with one another to produce the complex abnormalities in behavior that characterize drug addiction.

9

Endogenous Opioid System

Derivatives of *Cannabis sativa,* such as marijuana and hashish, are the most widely consumed illicit drugs in humans. However, the potential ability of cannabis derivatives to produce dependence in humans is still a controversial issue. Several authors have reported that cannabis derivatives do not induce physical dependence in humans, whereas others describe some symptoms of abstinence in heavy users of strong cannabis preparations. The rarity of reports of these withdrawal reactions may reflect the fact that they are mild and seldom observed in cannabis users. However, cannabis derivatives produce clear subjective motivational responses in humans, leading to drug-seeking behavior and abuse. Many studies have used different animal models to clarify the consequences of chronic exposure to cannabinoid agonists and the abuse liability of these compounds. Tolerance and withdrawal syndrome are adaptive responses to the prolonged exposure of neurons to different drugs, but provide only a partial correlate of their addictive properties.

The main factor common to all drugs of abuse is their ability to induce drug-seeking behavior, which is due to the positive-reinforcing effects of the drugs. Indirect indices of reinforcement can be evaluated through the ability of a drug to module the reinforcing properties of other rewards (e.g., intracranial self-stimulation techniques) or to impart reinforcing properties on previously neutral stimuli or environments (e.g., place-conditioning paradigm). Drug reinforcement can also be evaluated directly by using operant self-administration paradigms. Biochemical and electrophysiological studies can also be helpful to clarify the potential addictive properties of drugs of abuse. Indeed, these kinds of studies have identified the mesolimbic dopaminergic system as the common neuronal substrate for the motivational and rewarding properties of drugs of abuse. The endogenous opioid system has been reported to be a common neurobiological substrate involved in the development of dependence to several drugs of abuse, including cannabinoids.

Neurons containing endogenous opioid peptides are largely distributed within the central nervous system, and three different subtypes of opioid receptors, μ-, δ-, and κ-opioid receptors, have been identified and cloned. Cannabinoid compounds induce their pharmacological effects by activating two different receptors that have been identified and cloned: the CB1 cannabinoid receptor, which is highly expressed in the central nervous system; and the CB2 cannabinoid receptor, which is localized in the peripheral tissues, mainly at the level of the immune system. Anatomical studies have found that CB1 cannabinoid receptor and μ-opioid receptor mRNA are collocalized in brain limbic areas associated with dependence.

CB1 cannabinoid and μ-opioid receptors are also found in a similar population of striatal GABAergic neurons. Furthermore, a recent study using electron microscopy has revealed that both CB1 cannabinoid and μ-opioid receptors are collocalized in somata and dendrites of the same striatal neurons, suggesting potential coupling to similar second messenger systems. Cannabinoid and opioid receptors are seven transmembrane domain receptors coupled to G proteins, and through these proteins, the activation of cannabinoid and opioid receptors produces a similar inhibition of adenylyl cyclase activity. The stimulation of cannabinoid and opioid receptors also modifies the activity of other second messenger systems, and both produce an increase in the activity of the mitogen-activated protein (MAP) kinase pathway.

Activation of cannabinoid and opioid receptors also has similar consequences on the permeability of several ion channels. Thus, both opioid and cannabinoid agonists increase the permeability of potassium channels and inhibit calcium influx. The pharmacological responses induced in vivo by opioid and cannabinoid agonists in different animal species also have some common features. Thus, both groups of compounds share several pharmacological properties, including antinociception, hypothermia, hypolocomotion, hypotension, sedation, and decrease of gastrointestinal motility. Based on these anatomical, biochemical, and pharmacological findings, several pharmacological and molecular studies have investigated the involvement of the endogenous opioid system in the different responses induced by acute and chronic administration of cannabinoid agonists.

The involvement of the endogenous opioid system in acute cannabinoid responses has been investigated mainly in the case of antinociceptive effects. Biochemical studies have also revealed that cannabinoids are able to modify the activity of the endogenous opioid system. Furthermore, the opioid system has been reported to be involved in several adaptive and motivational responses induced by repeated cannabinoid administration that are related to the development of addictive processes. This chapter focuses on the role played by the opioid system in these acute and chronic effects induced by cannabinoids. A first section briefly summarizes the extensive literature that has been devoted to the involvement of the opioid system in cannabinoid antinociception. Other sections are devoted to the participation of the opioid system in the various pharmacological responses of cannabinoids related to their addictive properties.

ANTINOCICEPTIVE RESPONSES

CB1 and CB2 cannabinoid agonists induce *antinociceptive responses* that have been revealed in several animal species, including mouse, rat, rabbit, cat, dog, and monkeys. These antinociceptive properties have been shown in several acute behavioral models of nociception: thermal models such as the radiant heat tail flick, tail immersion, and the hot plate, mechanical models evaluating motor or reflex responses, chemical tests such as the abdominal constrictions induced by phenylbezoquinone, acetic acid or formic acid and electrical stimulation of the paw, sciatic nerve, or tooth. Electrophysiological studies have largely confirmed these antinociceptive properties of cannabinoids. Cannabinoid agonists also induce antinociception in inflammatory models of pain, such as hyperalgesia induced by carrageenan, capsaicin, formalin, and Freund's adjuvant. Doses of cannabinoid agonists required to induce antinociception in inflammatory processes are usually lower than those required in other nociceptive models, in agreement with the anti-inflammatory properties of cannabinoid agonists.

Cannabinoid agonists are also effective in visceral models of pain, such as the bladder wall inflammation induced by turpentine administration, and neuropathic models, such as the painful mononeuropathy induced by loose ligation of sciatic nerve. In contrast with opioid responses, antinociceptive effects of cannabinoids in neuropathic pain are not decreased after repeated administration and seem to be independent of *N*-methyl-D-aspartic acid (NMDA) activity, which is closely involved in the development of tolerance. CB1 cannabinoid receptors are involved in the systemic antinociceptive responses of cannabinoid agonists, as revealed by the use of selective CB1 antagonists, oligodeoxy-antinucleotide directed against the CB1 receptor, and knockout mice lacking CB1 receptors. However, other cannabinoid receptors different from the CB1 receptor could also be involved in the central antinociceptive effects of cannabinoids.

At the peripheral level, both CB1 and CB2 cannabinoid receptors seem to participate in cannabinoid antinociception. Pharmacological studies have suggested that the endogenous opioid system could be involved in cannabinoid antinociception. Thus, opioid antagonists have been shown to attenuate antinociceptive responses induced by cannabinoids in some experimental conditions. Several studies have reported contradictory results concerning the ability of naloxone, a nonselective but preferential μ-opioid antagonist, to antagonize cannabinoid antinociception. Thus, naloxone (1 mg/kg) has been reported to block the antinociceptive effects induced by Δ^9-tetrahydrocannabinol (THC) on the mouse hot-plate test, but very high doses (10 and 20 mg/kg) were required to attenuate the responses of THC in this animal species in the tail flick and the abdominal constrictions induced by phenylbezoquinone. Other studies have also reported that THC-induced antinociception in the mouse tail flick was unaffected by naloxone administration. Similarly, the effects induced by the cannabinoid agonist WIN-55,212 on the electro-physiological responses of rat rostral ventromedial neurons to thermal pain were not modified by naloxone.

Naloxone was also unable to block the antinociceptive effects induced by intrathecal and intracerebroventricular administration of THC in the mouse tail flick test. Quadazocine, another preferential μ-opioid antagonist, has been reported not to modify the antinociceptive effects of THC in rhesus monkeys subjected to the tail-immersion test. However, other studies have reported that naloxone (1 mg/kg) attenuates the antinociceptive responses induced by THC in the mouse tail flick and hot plate tests and by 11-OH-Δ^8-THC in the mouse hot-plate test. Selective opioid antagonists have been used to study the specific involvement of the various opioid receptors. The selective δ-opioid antagonist ICI-174,864 failed to block the antinociceptive responses induced by THC in the mouse tail flick. Naltrindole, another δ-selective antagonist, did not modify THC-induced antinociception in the mouse tail-flick and hot-plate tests.

The administration of the κ-selective antagonist norbinaltorphimine, antisense oilgodeoxynucleotides directed against the κ-opioid receptor, or antiserum against endogenous dynorphins has been reported to attenuate cannabinoid-induced antinociception. κ-Opioid receptors seem to be involved in cannabinoid antinociception at the spinal level but do not seem to participate in cannabinoid supraspinal antinociception. Recent studies using knockout mice deficient in the various opioid receptors or opioid peptide precursors have provided new highlights to clarify the involvement of the endogenous opioid system in cannabinoid antinociception.

Thus, antinociceptive responses induced by a high dose of THC (20 mg/kg) in the tail-immersion and hot-plate tests were not modified in knockout animals deficient in μ-, δ-, or κ-opioid receptors. Therefore, the suppression of a single opioid receptor was not enough to alter this acute cannabinoid response. However, an attenuation of THC-induced antinociception in the tail-immersion test was observed in knockout mice deficient in the preproenkephalin gene, but derivatives from proenkephalin are not selective agonists of any opioid receptor. THC-induced antinociception in the tail-immersion test was also reduced in pro-dynorphin knockout mice, whereas the effects of THC in the hot-plate test remained unaffected in these animals. Most of the peptides derived from prodynorphin are preferential agonists of the κ-opioid receptor but are not selective compounds and can therefore activate other opioid receptors.

The different results obtained in studies using selective opioid antagonists and knockout mice could be due to the doses of opioid antagonists used. Indeed, most of these previous pharmacological studies used rather high doses, which could produce a crossreactivity with other opioid receptors. Synergistic effects of opioid and cannabinoid agonists have been reported on antinociception. This synergism has been shown after the administration of these compounds by intrathecal, intracerebroventricular, and systemic routes, suggesting that cannabinoid-opioid synergism occurs at both spinal and supraspinal levels.

μ-Opioid receptors are selectively involved in the antinociceptive responses induced by morphine. However, the facilitatory effects of THC on morphine-induced antinociception seem to be mediated by different opioid receptors, that is, δ and κ receptors, which could have important therapeutic implications. Another study has revealed that δ-opioid receptors are involved in THC-facilitated opioid antinociception, but it suggests that κ-opioid receptors would not be involved in this interaction. Finally, it has been also suggested that μ-opioid receptors could be involved in THC-potentiated morphine antinociception at the supraspinal level, whereas κ-opioid receptors would be involved at the spinal level. Further studies must be performed in order to clarify the exact mechanism involved in this cannabinoid-opioid synergism.

CANNABINOID AGONISTISM

Supraspinal, spinal, and peripheral mechanisms seem to be involved in the antinociceptive properties of cannabinoid agonists. Thus, intracerebreventricular, intrathecal, and local peripheral administration of cannabinoids are able to induce potent antinociceptive responses. Studies using local administration of cannabinoid agonists in various brain structures have identified the central areas involved specifically in cannabinoid antinociception. Thus, antinociceptive responses have been induced by local microinjection of cannabinoid agonists in the periaqueductal gray matter, rostral ventromedial medulla, submedius and lateral posterior nuclei of the thalamus, superior colliculus, central and basolateral nuclei of the amygdala, and A5 noradrenergic group in the brainstorm. Autoradiographic and immunocytochemistry studies have revealed that these brain areas contain CB1 cannabinoid receptors. These neuroanatomical structures are involved in pain transmission and perception as well as in the descending pain control system, and are also sensitive to local opioid microinjection-induced antinociception. Indeed, both cannabinoid and opioid compounds modulate the activity of the descending pathways controlling spinal nociceptive neurons. These modulatory effects are responsible, at least in part, for the antinociceptive responses of cannabinoids and opioids.

The rostral ventromedial medulla is a brain structure involved in these modulatory effects. Thus, endogenous cannabinoids and opioids seem to be tonically released and to control basal nociceptive thresholds through the modulation of neuronal activity in this brain structure. However, cannabinoids and opioids act in the rostral ventromedial medulla by independent mechanisms. The periaqueductal gray matter has also been reported to participate in the modulatory effects induced by both cannabinoids and opioids in the descending control of spinal nociceptive neurons. Noxious stimuli increase the release of endogenous anandamide in the dorso-lateral portion of the periaqueductal gray matter, which would inhibit pain transmission. The site of opioid action in the periaqueductal gray matter is different. Indeed, electrical stimulation of the ventral part of the periaqueductal gray matter, but not the dorso-lateral portion, produces antinociception mediated by the release of endogenous opioid peptides.

Cannabinoids as well as opioids activate this pain-control descending pathway in the rostral ventromedial medulla and the periaqueductal gray matter by blocking the inhibitory GABA inputs. Spinal cord is another important neuroanatomical site for cannabinoid and opioid antinociception. Intrathecal administration of cannabinoids blocks *c-fos* expression induced in the dorsal horn of the spinal cord by noxious stimuli. Similarly, opioid agonists are also able to inhibit nociception-induced *c-fos* expression in the dorsal horn of the spinal cord. CB1 cannabinoid and opioid receptors are abundant in the dorsal horn of the spinal cord responsible for pain transmission. CB1 cannabinoid receptors are also present in dorsal root. In contrast with opioid receptors, only a minority of cannabinoid receptors within the spinal cord seem to be localized on small-diameter primary afferent fibers responsible for pain transmission, and most primary afferent neurons that express CB1 receptor mRNA are those with larger-diameter fibers involved in the transmission of non-nociceptive sensitive inputs.

In agreement with this hypothesis, CB1 mRNA cannabinoid receptor is not collocalized extensively in rat dorsal root ganglia with neuropeptides involved in pain transmission, such as substance P, calcitonin gene-related peptide, and somatostatin. This localization is different from opioid receptors that are mainly collocalized in neurons containing substance P. In spite of the spinal localization of CB 1 cannabinoid receptors, cannabinoids are more effective in inhibiting the activity of nociceptive C fibers than the activity of non-nociceptive Aβ- or Aδ-fibers, and seem to inhibit the release of neurotransmitters responsible for pain transmission such as substance P or calcitonin gene-related peptide. CB1 cannabinoid receptors are also located on the peripheral terminals of the primary afferent neurons, and an anterograde axonal transport of cannabinoid receptors has been revealed from dorsal root ganglia toward the peripheral terminals of sensory nerves.

On these peripheral terminals, CB1 and CB2 cannabinoid receptors have been reported to inhibit nociceptive transmission, and both receptors seem to be activated by an endogenous cannabinoid tone. Similarly, the endogenous opioid system also participates in the control of pain at the peripheral level. Thus, the presence of peripheral opioid receptors has been reported in peripheral tissues, and endogenous opioid peptides seem to participate in the control of pain at this level. Therefore, cannabinoid and opioid agonists are able to inhibit pain transmission at supraspinal, spinal, and peripheral levels, and both groups of compounds modify the activity of the descending pathways controlling spinal nociceptive neurons, which also participates in their antinociceptive effects.

BIOCHEMICAL STUDIES

Biochemical studies have revealed that cannabinoid administration can increase the release of several endogenous peptides, which can contribute to the acute and chronic pharmacological responses induced by these compounds. Thus, acute intrathecal administration of THC and other cannabinoid agonists has been reported to enhance in vivo the extracellular levels of endogenous dynorphins in the spinal cord. These dynorphins seem to play an important role in the antinociception induced by cannabinoids at the spinal level. Different dynorphins seem to be released by the administration of several cannabinoid agonists. Thus, an increase of the extracellular levels of dynorphin A has been observed after THC administration, whereas CP-55,940, another cannabinoid agonist, has been reported to enhance the release of dynorphin B. However, the administration of the endogenous cannabinoid ligand anandamide did not modify the extracellular concentration of dynorphins. The antinociceptive responses induced by THC are potentiated by the central administration of inhibitors of enkephalin catabolism, suggesting that cannabinoids are able to enhance the release of endogenous enkephalins.

A recent in vivo microdialysis study has demonstrated this hypothesis. Thus, acute systemic administration of THC has been reported to increase the extracellular levels of endogenous enkephalins in the nucleus accumbens. However, acute intrathecal THC administration did not modify spinal concentrations of Leu-enkephalin. Cannabinoid administration has also been reported to enhance mRNA levels of endogenous opioid peptide precursors. Indeed, repeated THC administration increases preproenkephalin and prodynorphin mRNA levels in the spinal cord. Preproenkephalin mRNA levels were also enhanced in several brain structures, including the hypothalamus, after chronic THC treatment. An increase in the expression of preproenkephalin mRNA has also been reported in the striatum after acute THC administration. Levels of pro-opiomelanocortin mRNA were also enhanced in the hypothalamus after repeated THC treatment.

DEVELOPMENT OF CANNABINOID TOLERANCE

Chronic administration of cannabinoid agonists develops tolerance to most of their pharmacological responses. Indeed, several studies have shown tolerance to cannabinoid effects on antinociception locomotion, hypothermia, catalepsy, suppression of operant behavior, gastrointestinal transit, body weight, cardiovascular system, anticonvulsivant activity, ataxia, and corticosterone levels. This tolerance has been reported in rodents but also in pigeons, dogs, and monkeys. The development of cannabinoid tolerance is particularly rapid, and an important decrease of the acute response has been already observed after the second administration of a cannabinoid agonist. Tolerance reaches a maximal degree rather soon during chronic cannabinoid treatment. Different pharmacokinetic mechanisms have been suggested to be involved on cannabinoid tolerance, such as changes in drug absorption, distribution, bio-transformation, and excretion. However, the role of such pharmacokinetic mechanisms, if any, seems minor. In contrast, pharmacodynamic events play a crucial role in cannabinoid tolerance.

Indeed, a significant decrease in the total number of CB1-binding sites and the levels of mRNA coding for the CB1 cannabinoid receptor have been reported in several brain areas during chronic administration of cannabinoid agonists. A widespread decrease in mRNA levels of $G_{\alpha i}$ and $G_{\alpha s}$ proteins has also been reported in rats treated chronically with cannabinoids. Changes in G-protein expression have been postulated to be related to desensitization of CB1 cannabinoid receptors. Accordingly, a strong reduction of cannabinoid agonist-stimulated [^{35}S]GTPγS binding has been obtained in most brain regions of rats treated chronically with cannabinoids. Several studies have revealed the presence of a crosstolerance among different exogenous CB1 cannabinoid agonists for their main pharmacological responses, such as antinociception, hypolocomotion, catalepsy, and hypothermia.

The presence of crosstolerance between opioid and cannabinoid compounds has also been revealed. Thus, systemic THC administration can induce tolerance to the antinociceptive and cardiac rhythm effects produced by systemically administered morphine. However, tolerance to the antinociception induced by systemic morphine was not always observed in THC-tolerant animals. Morphine pretreatment can also induce tolerance to the antinociceptive effects of systemic THC. However, tolerance to antinociception induced by systemic and intrathecal THC administration was not always observed in morphine-tolerant rats, and even a potentiation of the antinociceptive effects of cannabinoids has been reported in animals pretreated with morphine. Crosstolerance between CB1 cannabinoid agonists and -opioid agonists on antinociceptive responses has also been reported. Thus, systemic repeated THC administration induces tolerance to antinociception produced by dynorphins (preferential κ agonists) and by the selective κ agonists U-50,488H and CI-977.

Systemic repeated administration of U-50,488H and CI-977 also renders mice tolerant to the antinociceptive effect of intrathecal THC administration. Furthermore, administration of antisense oligodeoxynucleotides that selectively blocked the expression of κ-opioid receptors increased the development of tolerance induced by chronic THC. These results are in agreement with the release of the endogenous κ-opioid agonist dynorphin induced by THC administration and reported in the previous section. However, no correlation between THC-induced dynorphin A release and development of tolerance to THC antinociception has been reported during chronic intrathecal THC administration. Development of tolerance to the antinociceptive effects of the endogenous cannabinoid anandamide seems to involve different mechanisms to those implicated in tolerance to other cannabinoids. Indeed, anandamide-tolerant animals did not show cross-tolerance to the antinociceptive responses induced by μ-, δ, or κ-opioid agonists, while THC-tolerant animals showed crosstolerance to κ-opioid agonists under similar experimental conditions.

Furthermore, controversial results have been reported on the possible crosstolerance between anandamide and other cannabinoid agonists. Recent studies using knockout mice deficient in the different components of the endogenous opioid system have provided new data about the mechanisms involved in cannabinoid tolerance. The development of tolerance to the hypothermic effects of THC was not modified in knockout mice lacking the preproenkephalin gene. However, these knockout mice showed a decrease in the development of tolerance to THC antinociceptive effects and a slight attenuation of tolerance to THC-induced hypo- locomotion. The development of tolerance to the different pharmacological responses of THC was not significantly modified in knockout mice deficient in the prodynorphin gene.

THC tolerance has been also investigated in knockout mice deficient in the various opioid receptors. Thus, the development of tolerance to THC-induced hypothermia, hypolocomotion and antinociception was not modified in knockout mice lacking μ- or δ-opioid receptor. κ-Opioid receptor knockout mice showed a slight decreased tolerance to THC hypolocomotor effects. However, the development of tolerance to THC-induced antinociception, and hypothermia was not significantly modified in κ knockout mice. These results indicate that the suppression of a single opioid receptor does not have significant consequences on the development of cannabinoid tolerance. Opioid peptide derivatives from the prodynorphin gene do not seem to participate in cannabinoid tolerance. However, peptide derivatives from the preproenkephalin gene partici-pate in the development of tolerance to antinociceptive effects.

MECHANISM OF OPIOID PARTICIPATION

Several studies have reported the absence of somatic signs of spontaneous withdrawal syndrome after chronic THC treatment in rodents, pigeons, dogs, and monkeys, even after the administration of extremely high doses of this compound. However, a recent study has reported somatic signs of spontaneous abstinence after the abrupt interruption of chronic treatment with the cannabinoid agonist WIN 55,212-2. Differences in the pharmacokinetic properties of THC and WIN 55,212-2 could explain these results. In contrast, the administration of the selective CB1 cannabinoid receptor antagonist SR 141716A in animals (mouse, rat, and dog) treated chronically with THC has been shown to precipitate somatic manifestations of cannabinoid withdrawal. In rodents, this cannabinoid withdrawal syndrome is characterized by the presence of a large number of somatic signs and the absence of vegetative manifestations.

The most characteristic somatic manifestations of cannabinoid withdrawal in rodents are wet-dog shakes, head shakes, facial rubbing, front paw tremor, ataxia, hunched posture, body tremor, ptosis, piloerection, hypolocomotion, mastication, licking, rubbing, and scratching. It is important to point out the dramatic motor impairment occurring during the cannabinoid withdrawal. Doses of THC required to induce dependence in rodents are very high, and SR 141716A-precipitated withdrawal is currently reported after chronic administration of doses from 10 to 100 mg/kg daily of THC. CB1 cannabinoid receptors are responsible for the somatic manifestations of cannabinoid withdrawal. Thus, SR 141716A administration in CB1 knockout mice receiving chronic THC treatment did not precipitate any manifestation of cannabinoid abstinence. Bidirectional interactions between cannabinoid and opioid dependence have been reported. Thus, administration of the

CB1 cannabinoid antagonist SR 141716A is able to precipitate several behavioral and biochemical manifestations of opioid withdrawal in morphine-dependent rats. Additionally, the opioid antagonist naloxone precipitated behavioral signs of withdrawal in rats treated chronically with cannabinoid agonists, but the severity of the abstinence was less intense than the withdrawal syndrome precipitated by the homologous antagonist of each system: SR 141716A in cannabinoid-dependent rats and naloxone in opioid-dependent ones.

Accordingly, SR 141716A administration in morphine-dependent mice did not precipitate jumping, the most important behavioral manifestation of opioid withdrawal in this animal species. In contrast with the previous results obtained in rats, a study has recently reported that SR 141716A administration did not precipitate any behavioral signs of withdrawal in morphine-dependent mice, and naloxone challenge did not produce any behavioral changes in cannabinoid-dependent mice. Molecular studies have reported a significant decrease in the severity of morphine withdrawal syndrome in knockout mice deficient in CB1 cannabinoid receptors. In addition, acute administration of several cannabinoid agonists, including anandamide, has been reported to attenuate the severity of naloxone-precipitated morphine withdrawal.

A similar effect was observed when THC was administered for a long period of time before starting opioid dependence. Thus, chronic pretreatment with THC (10 mg/kg once daily during 3 wk) before starting chronic opioid administration decreased the somatic manifestations of naloxone-precipitated morphine withdrawal and did not modify morphine rewarding effects. Therefore, long term preexposure to cannabinoids does not seem to modify motivational responses of opioids related to their addictive properties. Recent studies have investigated the involvement of the endogenous opioid system on the somatic expression of cannabinoid abstinence by using knockout mice. Thus, the severity of SR 141716A-precipitated cannabinoid withdrawal was decreased in THC-dependent knockout mice lacking the preproenkephalin gene, but was not modified in knockout mice deficient in the prodynorphin gene.

The behavioral expression of SR 141716A-precipitated THC withdrawal was also evaluated in knockout mice lacking the various opioid receptors. Cannabinoid withdrawal syndrome precipitated by SR 141716A administration was not modified in μ-, δ-, or κ-opioid receptor knockout mice treated chronically with 20 mg/kg (twice daily) of THC. Another recent study has reported a decrease in the severity of cannabinoid withdrawal syndrome in μ-knockout mice treated chronically once daily with 30 and 100 mg/kg of THC, but not those receiving 10 mg/kg of THC once daily. Therefore, endogenous opioid peptides derivative from preproenkephalin are important for the somatic expression of cannabinoid abstinence by acting on μ- and probably other opioid receptors.

The use of combinatorial opioid-receptor knockout mice lacking two or three opioid receptors will clarify these findings in the near future. Common features of the withdrawal syndrome to several drugs of abuse, such as opioids, psychostimulants, and ethanol, include an important elevation in extracellular levels of corticotropin-releasing factor in the mesolimbic system, and a marked inhibition of mesolimbic dopamine activity. Similar behavioral changes have been reported during cannabinoid withdrawal. Thus, an increased release of corticotropin-releasing factor and an enhancement of Fos immunoreactivity have been found in the central amygdala during SR 141716A-precipitated cannabinoid withdrawal. This alteration of corticotropin-releasing factor function in the limbic system may have a motivational role in mediating the stresslike symptoms and negative affect that accompany cannabinoid withdrawal. In agreement with this hypothesis, the spontaneous firing rate of ventral tegmental area dopamine neurons has been reported to be attenuated during cannabinoid abstinence. This decreased dopaminergic activity also seems to be related to the aversive/dysphoric consequences of cannabinoid withdrawal. However, changes in the activity of the dopaminergic system are not involved in the expression of the somatic signs of cannabinoid withdrawal. Cannabinoid and opioid withdrawal syndromes are associated with compensatory changes in the cyclic AMP pathway. Initially, acute activation of cannabinoid and opioid receptors leads to an inhibition of adenylyl cyclase activity.

In contrast, SR 141716A-precipitated THC withdrawal and naloxone-precipitated morphine withdrawal produce an increase in adenylyl cyclase activity in vivo in the central nervous system in rodents. In spite of these common biochemical mechanisms, different brain structures have been reported to be involved in the physical manifestations of opioid and cannabinoid withdrawal. Thus, locus coeruleus and other brainstem structures, such as the periaqueductal gray matter, are responsible for the somatic signs of opioid withdrawal. In the case of cannabinoid dependence, the cerebellum seems to play a crucial role for the somatic expression of THC withdrawal. Indeed, basal, forskolin- and calcium/calmodulin-stimulated adenylyl cyclase activities were selectively increased in the cerebellum, but not in other brain structures (frontal cortex, hippocampus, striatum, and periaqueductal gray matter). Furthermore, the behavioral manifestations of cannabinoid abstinence are markedly reduced by preventing the activation of protein kinase A (PKA) by microinfusion of the selective cyclic AMP inhibitor Rp-8Br-cAMPS onto the surface of the cerebellum. Similar increases in cyclic AMP levels and PKA activity have been reported during chronic THC treatment in the cerebellum, striatum, and cortex.

REWARDING EFFECTS

Intracranial self-stimulation is an experimental procedure that is widely used to explore brain circuits involved in reward as well as the rewarding properties of drugs of abuse. When an animal is equipped with an electrode placed in a brain rewarding circuit and given the opportunity to perform a behavioral response, such as pressing a lever, that is followed by a short-pulse train of electrical current via the electrode, the animal will initiate and maintain responding. A common property of most of drugs of abuse is to acutely facilitate electrical stimulation of brain reward loci, presumably due to their euphorigenic properties. Acute administration of THC has been reported to lower intracranial self-stimulation thresholds in rats, suggesting the activation of central hedonic systems. The opioid antagonist naloxone blocked the enhancing effects of THC on electrical intracranial self-stimulation reward, at doses of naloxone that themselves have no effect on brain reward, which reveals the involvement of the endogenous opioid system in the rewarding effects of THC. Interestingly, the facilitation of brain stimulation reward induced by most drugs of abuse is also reversed by naloxone administration.

Behavioural Model

Conditioned place preference is a behavioral model that is currently used to measure the rewarding properties induced by the administration of a drug. In this paradigm, the rewarding properties of a compound are associated with the particular characteristics of a given environment. A similar procedure can be used to explore the aversive properties of a drug. In this case, the animal will avoid staying in the compartment associated with a compound producing aversive/ dysphoric effects.

The administration of cannabinoid agonists currently produces aversive-like responses in the place-conditioning paradigm. Conditioned place aversion induced by cannabinoid agonists is abolished by the co-administration of SR 141716A, suggesting a selective involvement of CB1 cannabinoid receptors. Rewarding effects of cannabinoid agonists can also be revealed in this paradigm by using particular experimental conditions. Thus, THC produced conditioned place pref-erence in rats when administered at lower doses than those used to induce place aversion and when animals were exposed to a 24-h washout period between the two THC-conditioning sessions. THC also produces a clear conditioned place preference in mice using a long period of conditioning and avoiding the possible dysphoric consequences of the first drug exposure. Indeed, place preference was obtained with 1 mg/kg of THC when mice received a previous priming THC exposure in the home cage before the conditioning sessions.

Interestingly, doses of THC used in these studies were similar to those reported to facilitate intracranial self-stimulation in rats. Recent studies have used the place-conditioning paradigm in knockout mice to evaluate the involvement of the endogenous opioid system in the rewarding and aversive properties of cannabinoids. The motivational responses induced by THC have been investigated in knockout mice lacking μ-, δ-, or κ-opioid receptor. The rewarding effects induced by 1 mg/kg of THC, avoiding the possible dysphoric consequences of the first drug exposure, were abolished in μ-opioid receptor knockout mice but were not modified in mice lacking δ- or κ-opioid receptor. The dysphoric effects induced by a high dose of THC (5 mg/kg) were not modified in δ-opioid receptor knockout mice, slightly attenuated in μ-knockout animals, and completely abolished in mice lacking κ- opioid receptor.

These κ-opioid receptor knockout mice also showed place preference to THC (1 mg/kg) in animals that had not received any priming exposure, revealing the crucial role of this opioid receptor in THC-induced dysphoria. Results obtained with knockout mice deficient in the prodynorphin gene are in agreement with this hypothesis. Thus, the conditioned place aversion induced by a high dose of THC (5 mg/kg) was completely abolished in these prodynorphin knockout mice. Interestingly, this interaction between the cannabinoid and opioid systems also seems to be bidirectional. Indeed, the rewarding effects induced by morphine in the conditioned place preference paradigm were blocked in knockout mice deficient in CB1 cannabinoid receptors. Furthermore, the selective CB1 cannabinoid antagonist SR 141716A blocks the acquisition of morphine conditioned place preference, as well as the rewarding effects of other drugs of abuse, such as cocaine, and natural stimuli such as food.

Intravenously Self-administration Drugs

The procedure by which animals are permitted to intravenously self-administered drugs has provided a reliable method to evaluate directly the reinforcing properties of a psychoactive compound, and is probably the clearest indication in animals of its addictive potential in humans. Numerous studies have shown that THC is unable to induce a self-administration behavior in any of the animal species studied. Animals that have already learned to self-administer other drugs of abuse did not self-infuse THC.

THC pharmacokinetic properties seem to be crucial for the behavioral responses on the self-administration paradigm. Indeed, WIN 55,212-2, a synthetic cannabinoid agonist that has a shorter half-life than THC, was intravenously self-administered in mice in a concentration-dependent manner, according to a two-phase "bell-shaped" curve. Another synthetic cannabinoid agonist, CP 55,940 has been reported to sustain intracerebroventricular self-administration in rats. A recent study has also revealed intravenous THC self-administration behavior, for doses much lower than those used previously, by monkeys that have a previous history of cocaine self-administration. SR 141716A completely blocked self-administration induced by WIN 55,212-2, CP 55,940, and THC indicating a selective involvement of CB1 cannabinoid receptors in the reinforcing properties of these cannabinoids. Bidirectional interactions between cannabinoid and opioid systems have also been reported by using self-administration techniques.

Indeed, morphine-induced intravenous self-administration was abolished in knockout mice lacking the CB1 cannabinoid receptors. In agreement with this result, blockade of CB1 cannabinoid receptors by SR 141716A administration partially reversed heroin-induced intracerebroventricular self-administration. These results indicate that the endogenous cannabinoid system is involved in opioid reinforcement. Interestingly, the endogenous opioid system also participates in the reinforcing properties induced by cannabinoids. Indeed, the opioid antagonist naloxone partially blocked self-administration induced by CP 55,940. THC self-administration behavior was also attenuated by the administration of the opioid antagonist naltrexone.

Bioelectrophysiological Investigations

The mesolimbic dopaminergic system has been proposed as a common neuronal substrate for the rewarding properties of drugs of abuse. Indeed, prototypical drugs of abuse, including opioids, psychostimulants, alcohol, and nicotine, increase the discharge rate of mesolimbic dopamine neurons. This activation is associated with increased dopamine output in innervated projection structures, such as the nucleus accumbens. In vivo microdialysis studies have revealed that acute administration of THC and other cannabinoid agonists also increases extracellular efflux of dopamine and its metabolites in nucleus accumbens of freely moving rats. In agreement with these biochemical results, THC and other cannabinoid agonists produced a dose-dependent increase in the spontaneous firing rate of ventral tegmental area dopamine neurons.

The biochemical and electrophysiological effects of cannabinoids on in vivo dopamine transmission in the nucleus accumbens were prevented by the administration of SR 141716A, indicating the selective involvement of CB1 cannabinoid receptors.The endogenous opioid system has also been reported to be involved in the effects of cannabinoids on mesolimbic dopamine activity. Thus, administration of opioid antagonists was reported to block the increase in extra-cellular efflux of dopamine in the nucleus accumbens induced by the acute administration of cannabinoid agonists. However, other studies have reported that naloxone did not modify THC-induced increase of the firing rate of dopaminergic neurons projecting to the nucleus accumbens and neostriatum.

CONCLUDING REMARKS

The involvement of the endogenous opioid system in cannabinoid-induced antinociception, tolerance, and dependence has been evaluated by using pharmacological approaches and more recently by the use of new molecular tools. Pharmacological studies have suggested that the endogenous opioid system could be involved in cannabinoid antinociception. κ-Opioid antagonists were particularly effective to attenuate these cannabinoid antinociceptive responses. However, molecular studies have revealed that the suppression of a single opioid receptor did not modify THC antinociception.

The different results obtained when using these two experimental approaches could be due to a crossreactivity with several opioid receptors when using opioid antagonists. In agreement with this hypothesis, molecular studies have demonstrated that opioid peptides derivatives from preproenkephalin and prodynorphin, which are not selective agonists of any opioid receptor, participate in cannabinoid antinociception. Furthermore, biochemical studies have revealed that cannabinoid administration increases the release of endogenous enkephalins and dynorphins in different brain areas. The endogenous opioid system has been also reported to be involved in several adaptive and motivational responses induced by repeated cannabinoid administration that are related to their addictive properties.

Pharmacological studies have suggested the involvement of the opioid system, and particularly κ-opioid receptors, in the development of tolerance to cannabinoid antinociception. Molecular studies using knockout mice have revealed that the suppression of a single opioid receptor has not significant consequences on the development of cannabinoid tolerance, whereas peptide derivatives from preproenkephalin gene participate in the development of tolerance to antinociceptive effects. Bidirectional interactions between cannabinoid and opioid dependence were demonstrated by using pharmacological approaches and have been now clarified by using knockout mice deficient in opioid receptors and peptide precursors. Thus, endogenous opioid peptides derivative from preproenkephalin, but not from prodynorphin, are important for the expression of cannabinoid abstinence by acting on μ- and probably other opioid receptors. The suppression of a single opioid receptor did not modify cannabinoid withdrawal, although the severity of cannabinoid abstinence was reduced in μ knockout mice when using very high doses of THC.

Cannabinoid and opioid withdrawal share several biochemical common features, such as upregulation of the cyclic AMP pathway, elevation in extracellular levels of corticotropin-releasing factor in the mesolimbic system, and inhibition of mesolimbic dopamine activity. However, different brain structures have been reported to be involved in the expression of the withdrawal syndrome to these two drugs of abuse. The cerebellum seems to play an important role in the somatic manifestations of cannabinoid withdrawal, whereas other brainstem structures, such as the locus coeruleus, are responsible for the somatic signs of opioid abstinence. Several behavioral models, such as intracranial self-stimulation, conditioned place preference, and intravenous self-administration, have revealed the rewarding properties of cannabinoid agonists.

Cannabinoid rewarding properties, similarly to other prototypical drugs of abuse, seem to be related to the activation of the mesolimbic dopaminergic system. The endogenous opioid system has been also reported to play a crucial role in cannabinoid rewarding effects. Thus, opioid antagonists are able to block cannabinoid responses on intracranial self-stimulation, conditioned place preference, and intravenous self-administration, as well as the biochemical effects of THC on mesolimbic dopamine transmission. Furthermore, the rewarding properties of THC in the place-conditioning paradigm were abolished in knockout mice deficient in μ-opioid receptors, and the dysphoric effects induced by high doses of THC in this paradigm were suppressed in knockout mice deficient in κ-opioid receptors and prodynorphin gene. In conclusion, pharmacological and molecular studies have demonstrated that the endogenous opioid system plays a key role in cannabinoid-induced antinociception, tolerance, physical dependence, and rewarding effects. The exact involvement of each component of the endogenous opioid system in some of these cannabinoid responses remains to be further elucidated.

10

Physiological and Subjective State of Alcoholism

Alcohol dependence and addiction, here equated with alcoholism, is a clinically well- defined disorder in which normal behavioral control is lost by the individual, leading to particularly severe consequences. Several key aspects of alcoholism have been modeled in experimental animals. These animal models mimic various behavioral aspects seen in human alcoholics, such as loss of control over drinking, tolerance, physical dependence, craving, and relapse. There are opposing views in the field of drug abuse research regarding the term "*craving*"–whether it describes a physiological, subjective, or behavioral state, and if it is necessary at all to explain addictive behavior or it is an epiphenomenon that is not necessary for the production of continued drug use in addicts.

An Expert Committee gathered by the United Nations International Drug Control Programme (UNDCP) and the World Health Organization (WHO) agreed on the definition of craving as "the desire to experience the effect(s) of a previously experienced psychoactive substance." Markou et al. conceptualized craving within the framework of incentive motivational theories of behavior and modified the definition of craving as "incentive motivation to self-administer a psychoactive substance." Such an operational definition of craving has the advantage of making the phenomenon of craving accessible to experimental investigation and making it measurable. On the basis of this definition, animal models of alcohol craving and relapse have been developed.

LONGER INTAKE OF ALCOHOL

Long-Term Alcohol Self-Administration with Repeated Deprivation Phases

A promising animal model to study craving and relapse behavior in heterogenous Wistar rats or in alcohol-preferring P and HAD-rat lines is a long-term alcohol self-administration procedure with repeated deprivation phases. Alcohol-experienced animals show a transient increase in alcohol consumption and alcohol preference after a period of forced abstinence (alcohol deprivation), which is termed the *alcohol deprivation effect.* The alcohol deprivation effect reflects an incentive motivation to take alcohol–that is, alcohol craving–and can also be seen as relapse-like drinking behavior. It can be observed in long-term alcohol-drinking rats that have developed a behavioral alcohol dependence, as well as in nondependent rats, both under home-cage drinking and under operant self-administration conditions; and in monkeys and humans. Interestingly, the alcohol deprivation effect is prolonged and enhanced in alcohol-preferring P and HAD-rat lines after repeated deprivation phases and changes its characteristics with repeated deprivation phases.

Thus, the alcohol deprivation effect in long-term alcohol self-administering rats that had experienced repeated deprivation phases has interesting characteristics: during an alcohol deprivation effect these animals consume large amounts of highly concentrated alcohol solutions, even at unusual times. Pronounced changes in the diurnal rhythm of drinking activity were observed in long-term alcohol-drinking rats that had repeated deprivation phases. Tested in a fully automated electronic drinkometer device, age-matched control animals showed normal drinking activity: drinking activity during the active night phase was high, whereas drinking activity during the inactive light phase was very low, reaching zero for some hours. In contrast, in long-term alcohol-drinking rats during the alcohol deprivation effect, the pattern of drinking activity changed completely.

In particular, during the inactive phase, most of the animals still showed high drinking activity. Moreover, some animals were found that even demonstrated level drinking–that is, drinking activity during the dark and light phases no longer differed. Such drinking activity is far beyond normal controlled behavioral mechanisms seen in the appropriate control animals. Furthermore, alcohol drinking behavior during an alcohol deprivation effect cannot be modified by taste adulteration with quinine or the additional choice of a highly palatable sucrose solution. These findings suggest that the observed alcohol drinking behavior is pharmacologically and not nutritionally motivated. In conclusion, alcohol drinking during the alcohol deprivation effect, especially following repeated deprivation phases, seems to consist of an uncontrolled incentive motivation to self- administer the drug. This statement is fully compatible with the operational definition of craving by Markou et al. An incentive motivation to drink a highly concentrated alcohol solution following deprivation is further supported by the introduction of various progressive ratio tasks.

Under those operant conditions, animals have to work more and more in order to receive a reinforcer. Here the breaking point (the number of consecutive lever responses in order to receive one reinforcer) for alcohol responding is significantly higher following deprivation compared to baseline responding. However, it is important to note that the measurement of an alcohol deprivation effect assesses only a behavioral outcome and cannot tell us anything about a subjective state associated with an incentive motivation to drink alcohol. Nevertheless, the fact that clinically effective antirelapse drugs also reduce the alcohol deprivation effect lends predictive value to this animal model for the development of new and better drugs for the treatment of alcoholism. In addition, alcohol self-administration with repeated deprivation phases can be used as an animal model to further study the neurobiological and molecular basis of craving and relapse.

Measurement of Craving

The reinstatement model is also used for the measurement of craving and relapse behavior. In this paradigm, the animal is trained to self-administer a drug and is then subjected to extinction–that is, the animal is tested under conditions of non-reinforcement until operant responding appears to be extinguished. When the animal reaches some criterion of unresponsiveness, various stimuli are presented. A stimulus is said to reinstate the drug-seeking behavior if it causes renewed responding—lever pressing—without any further response-contingent drug reward. At least five conditions can reinstate responding:

1. Drug priming, that is, the injection of a small dose of the drug;
2. Stress;
3. Conditioned stimuli;
4. Withdrawal; and
5. Electrical brain stimulation.

Although reinstatement of intravenous self-administration of psychostimulants and opioids has been established for many years, only a few attempts have been undertaken to transfer this paradigm into the alcohol field. In 1995 the first alcohol reinstatement study in rats was reported by Chiamulera and co-workers. In this study, rats acquirred operant responding for alcohol over several months. After stable lever pressing was obtained between subsequent sessions, the rats were tested in extinction, meaning that animals received water instead of alcohol following lever pressing. After 8–10 extinction sessions, reexposure to a small quantity of ethanol was able to reinstate previously extinguished alcohol-seeking behavior. These results are consistent with the widely reported description of the "*first drink*" phenomenon: ingestion of a small quantity of alcohol may induce in abstinent alcoholics a strong subjective state of craving and then relapse to drug-taking behavior.

The "*priming effect*" due to alcohol preload may be evident even after years of abstinence from the drug. Only very recently has the alcohol reinstatement paradigm been followed by other research groups. It could be demonstrated that intermittent foot shock stress can also reinstate previously extinguished responding for alcohol. Furthermore, it has been shown that alcohol-associated olfactory cues and other cues can reinstate extinguished alcohol-seeking behavior. In conclusion, reinstatement of alcohol-seeking behavior has similar characteristics in comparison to other drugs of abuse and can be used to study the neurobiological and molecular bases of craving and relapse.

Behavioral Sensitization of Psychoactive Drugs

Repeated administration of psychoactive drugs can have neuroadaptive consequences that lead either to a decrease (tolerance or desensitization) or an increase (sensitization) of their behavioral effects. Sensitization processes have been implicated in the development of compulsive drug use. One of the most prominent theories on the significance of drug-induced sensitization, proposed by Robinson and Berridge, posits that compulsive drug-seeking behavior is a result of a progressive hypersensitivity of neural systems that mediate "*incentive salience*," resulting in a transformation of ordinary "*wanting*" into excessive craving. Although this theory is extremely difficult to test, drug-induced sensitization processes may give some indirect insights into behavioral and neurobiological processes of craving. In contrast to most other drugs of abuse, ethanol-induced behavioral sensitization is difficult to demonstrate.

Sensitization to the locomotor stimulant effects of low doses of ethanol has been documented only in some mouse genotypes, such as the inbred strain DBA/2J. While DBA/2J mice are known to develop sensitization to ethanol, they have also been characterized as alcohol avoiders based on alcohol self-administration studies, raising questions regarding the relevance of examining sensitization in these mice. However, recent evidence indicates that DBA/2J mice are sensitive to the motivational effects of ethanol in that they show ethanol-induced place preference and intravenous self-administration of ethanol, suggesting a causal link between ethanol-induced sensitization and reinforcement processes. This suggestion is further supported by the finding that voluntary ethanol consumption induces sensitization to ethanol´s locomotor activating effects in a high-alcohol-preferring mouse line. On the other hand, sensitized DBA/2J mice did not show enhanced alcohol intake in a subsequent drinking experiment, and prior voluntary intake did not alter ethanol-induced behavioral sensitisation, showing that ethanol-induced behavioral sensitization may not be linked directly to ethanol-induced reinforcement processes in DBA/2J mice.

In summary, in some mouse strains repeated intermittent injections of ethanol can induce behavioral sensitization to this drug, and there is some evidence that ethanol-induced behavioral sensitization is linked to reinforcement processes. In rats it is even more difficult to demonstrate ethanol-induced sensitization; however, a recent study showed that in Sprague-Dawley rats that were divided into either high or low responders to novelty, high responders to novelty exhibit a sensitized locomotor response to a very low challenge dose (0.25 g/kg). Thus, depending on the mouse or rat strain, ethanol-induced behavioral sensitization can be induced by repeated intermittent injections. This provides the possibility of studying neurobiological and molecular mechanisms of ethanol-induced sensitization processes. However, findings from those studies have only limited significance to extend our understanding of neurobiological and molecular mechanisms underlying alcohol craving.

CLINICAL ASPECTS OF CRAVING

The neurobiological and molecular bases of alcohol craving and relapse are still not well understood; however, preclinical as well as clinical data strongly imply that craving and relapse for alcohol (and other drugs of abuse) can be induced through different mechanisms. A first pathway may induce alcohol craving and relapse due to the mood-enhancing, positive reinforcing effects of alcohol consumption. This pathway seems to involve opioidergic and dopaminergic systems in the ventral striatum. The role of the dopaminergic system may lie in the direction of attention toward reward-indicating stimuli, whereas the induction of euphoria and positive mood states may be mediated by opioidergic systems.

Associative learning may, in turn, transform positive mood states and previously neutral environmental stimuli into alcohol-associated cues that acquire positive motivational salience and induce reward craving. A second and potentially independent pathway may induce alcohol craving and relapse by negative motivational states, including conditioned withdrawal and stress. This pathway seems to involve the glutamatergic system and the corticotropin-releasing hormone (CRH)-system. Chronic alcohol intake leads to compensatory changes within these systems. During withdrawal and abstinence, increased glutamatergic excitatory neurotransmission as well as increased CRH release lead to a state of hyperexcitability that becomes manifest as craving, anxiety, seizures, and autonomic dysregulation. Moreover, cues associated with prior alcohol intake that are not followed by actual drug consumption may induce conditioned withdrawal.

Alcohol Deprivation Effects

Involvement of opioidergic systems in the alcohol deprivation effect

A first pathway that may induce alcohol craving and relapse due to the mood- enhancing, positive-reinforcing effects of alcohol consumption and that seems to involve opioidergic systems was studied in long-term voluntary alcohol-drinking rats that had repeated deprivation phases. In particular, the role of opioid receptors on the alcohol deprivation effect was studied. Three different types of opioid receptors have been identified: μ-, δ-, and κ-opioid receptors. Opioid receptor blockade, either by the unspecific antagonists naloxone and naltrexone or by selective μ-opioid receptor antagonists, reduces ethanol consumption. Furthermore, μ- opioid receptor knockout mice do not self-administer alcohol either under operant or under home-cage drinking conditions. These findings lead to the conclusion that at least a part of the rewarding effect of ethanol is mediated by the activation of μ-opioid receptors, which in turn reinforces ethanol intake. Holter and Spanagel determined under which treatment conditions naltrexone is effective in reducing the alcohol deprivation effect in long-term alcohol-experienced rats.

The effects of chronic naltrexone treatment via osmotic minipumps and repeated intermittent naltrexone injections were studied. Chronic naltrexone treatment did not reduce the alcohol deprivation effect but enhanced alcohol preference. In contrast to chronic treatment, intermittent injections of naltrexone attenuated the alcohol deprivation effect. These opposing effects can be explained on a pharmacological level: chronic opiate receptor blockade leads to an up-regulation of opioid receptors, rendering the endogenous opioid system more sensitive to the effects of alcohol, whereas intermittent naltrexone injections at moderate doses do not induce functionally relevant opioid receptor changes. These findings emphasize the importance of the treatment regimen for obtaining a suppressant effect on alcohol drinking and relapse behavior.

Thus, a treatment regimen with a low dose and frequency of administration that prevents drug accumulation is presumably necessary to maintain the reductive effect of opioid receptor blockade on alcohol relapse drinking. In view of these findings it seems questionable whether the use of naltrexone depots (some clinical trials have been already initiated) would be an appropriate treatment regimen. The role of δ-opioid receptors in ethanol reinforcement is less clear. Some of the published reports that used selective δ-opioid receptor antagonists showed an attenuation of voluntary alcohol drinking in laboratory animals, whereas others show no effect of this receptor subtype in ethanol consumption. In contrast, δ-opioid receptor knockout mice showed increased ethanol consumption.

The selective δ-opioid receptor antagonist naltrindole had no effect on the alcohol deprivation effect in voluntary long-term alcohol-drinking rat, demonstrating that it is unlikely that δ-opioid receptors play a critical role in alcohol craving and relapse behavior. The role of the dynorphin/κ-opioid receptor system in ethanol reinforcement is even less clear. κ-Opioid receptor stimulation, in contrast to μ- and δ-opioid receptor stimulation, has aversive motivational consequences in animals. To further elucidate the role of the dynorphin/κ-opioid receptor system in ethanol reinforcement and relapse behavior, Holter et al. investigated the effects of the highly selective κ-opioid receptor agonist CI-977 and of the long-acting selective κ-opioid receptor antagonist nor-BNI on the alcohol deprivation effect in long-term ethanol-experienced rats. The results of this study show that chronic treatment with the selective κ-opioid receptor agonist CI-977 strongly increased the alcohol deprivation effect, whereas the long-acting selective κ-opioid receptor antagonist nor-BNI had no effect on the alcohol deprivation effect in long-term ethanol-experienced rats. In conclusion, the κ-opioid receptor system can play a role in ethanol reinforcement in cases of increased receptor activation, as it was experimentally induced in this study by chronic treatment with a selective κ-opioid receptor agonist.

The subsequent increase in ethanol intake might be an attempt to counteract the aversive motivational consequences of this treatment. Previous findings indicate that increased dynorphin activity might occur during the early phases of alcohol withdrawal, but disappear later on. Hence, an increased endogenous κ-opioid receptor activation might enhance the probability of relapse during the early phases of alcohol withdrawal. However, since the κ-opioid receptor antagonist nor-BNI had no effect on the alcohol deprivation effect, endogenous κ-opioid receptor stimulation does not seem to be involved in relapse-like drinking after protracted abstinence.

Evidences of the mesolimbic dopaminergic system

Several lines of evidence indicate that ethanol activates the mesolimbic dopaminergic system. Alcohol injected intravenously increased firing of dopamine neurons in the ventral tegmental area and acute administration of alcohol results in preferential release of dopamine from the nucleus accumbens shell region. Dopamine D1 and D2 antagonists, administered either systemically or locally into the nucleus accumbens, decrease home-cage drinking and operant responding for alcohol. Furthermore, alcohol intake and preference are markedly reduced in mice lacking dopamine D2 receptors, and these mice show reduced ethanol-induced conditioned place preference.

Examination of dopaminergic neuron function in the nucleus accumbens associated with the alcohol deprivation effect revealed that relapse behavior was accompanied by enhanced dopamine release, which is in line with preliminary findings in voluntary long-term drinking rats that had repeated deprivation phases. These animals showed augmented dopamine release during the onset of an alcohol deprivation effect compared to age-matched control animals that were alcohol naive. The dopamine D2 receptor agonist lisuride was tested in voluntary long-term alcohol-drinking rats. Lisuride treatment significantly increased alcohol intake, indicating a "*pro-craving/pro-relapse*" effect of this compound.

The dopamine D2 receptor antagonist flupenthixol was also tested. A "*pro-craving/pro-relapse*" effect of flupenthixol treatment similar to that of lisuride treatment was observed in voluntary long-term drinking rats. These studies show that dopamine D1 and D2 receptors are important in reinforcement processes involved in the acquisition of alcohol drinking, but it seems that these receptors play no critical role in alcohol craving and relapse and if at all can induce a "*pro-craving/pro-relapse*" effect due to pharmacological stimulation. In another study, voluntary long-term ethanol consumption led to specific changes in dopamine D3 receptor gene expression, whereas ethanol drinking did not alter the mRNA expression of dopamine D1, D2, D4, or D5 receptors.

However, chronic bilateral infusion of endcapped phosphorothioate dopamine D3 receptor antisense oligonucleotides into the nucleus accumbens shell region had no effect on the alcohol deprivation effect, although this treatment led to a selective reduction of this dopamine receptor subtype that was accompanied by suppressed food-reinforced behavior. In summary, mesolimbic dopamine release and the activation of its receptors play an important role in the acquisition of alcohol-seeking behavior; however, whether dopamine receptor activation during craving and relapse plays a critical role is uncertain.

Evidences of the glutamatergic system

Numerous microdialysis studies revealed that, depending on the brain region and the rat strain, low doses of ethanol can increase extracellular glutamate levels whereas high intoxicating doses can decrease glutamate levels. The mechanism behind the inhibitory effect of high intoxicating doses of ethanol on glutamate release is not clear. Although multiple mechanisms have been implicated in this action, considering the general inhibitory influence of γ-aminobutyric acid (GABA) on glutamatergic neurotransmission, it may be suggested that the inhibitory effect of ethanol on glutamate is due to an initial increase of GABA release, which in turn inhibits the release of glutamate. Whatever the mechanism might be, the inhibitory effect of high intoxicating doses of ethanol leads to several adaptive responses within the glutamatergic system following its chronic administration.

Indeed, extracellular glutamate levels are enhanced during withdrawal, and these changes underly the occurrence of some physical signs of withdrawal such as seizures. However, a recent study by Dahchour and De Witte demonstrated that, following repeated alcohol withdrawal episodes, no enhancement of glutamate levels in the hippocampus can be observed despite the fact that the severity of ethanol-induced withdrawal seizures, as well as the duration of seizures, is exacerbated in rats that had a prior history of repeated alcohol withdrawal episodes. These data suggest that the phenomenon of "*kindling*" observed following repeated alcohol withdrawal phases may not be due simply to a further increase in glutamate, but that other mechanisms, that is, changes in glutamate receptors, may be implicated in the occurrence of seizures.

It also seems to be unlikely that enhanced glutamatergic neurotransmission in the nucleus accumbens and other brain sites following long-term abstinence is associated with craving and relapse. However, it has recently been demonstrated that ethanol-conditioned stimuli can induce an increase in extracellular glutamate levels in the amygdala. Therefore, one might speculate that conditioned responses to extracellular glutamate may participate in environmental cues-induced conditioned craving for ethanol and relapse behavior. As pointed out later, however, changes in glutamate receptors seem to be more likely involved in craving and relapse behavior.

Electroneurophysiological studies

Electrophysiological and neurochemical studies show that ethanol at behaviorally relevant concentrations (>5 m*M*) inhibits *N*-methyl D-aspartic acid (NMDA) receptors expressed in neurons and in recombinant expression systems. The fact that ethanol-induced inhibition of NMDA receptors differs across brain sites and that NMDA receptor subunits are differentially distributed throughout the central nervous system suggest that differences in the subunit composition are important determinants of ethanol sensitivity. Despite intensive research, the site of action of ethanol on the NMDA receptor remains unknown, but, there is good hope that an "alcohol receptor" on the extracellular domaine of the receptor complex will be identified soon. Following long-term alcohol intake, adaptive responses such as changes in the number and affinity of synaptic glutamate receptors and glutamate transporters will occur to counterbalance for the acute inhibitory effect of ethanol on NMDA receptor function and glutamate release. Initial binding studies detected increases in NMDA receptor density after chronic ethanol treatment, whereas others have not.

A very extensive study was recently performed by Rudolph and collaborators in order to determine if the variation observed in past NMDA receptor-binding studies might be related to genetic differences between strains and/or the chronic ethanol protocol used. The outcome of this comprehensive study and other unpublished reports indicate that robust increases in NMDA receptor binding do not occur, and suggest that NMDA receptor supersensitivity observed following chronic ethanol administration is not due simply to changes in the density of NMDA receptors. In fact, because of the complicated pharmacology, allosterism, and subunit heterogeneity of the NMDA receptor complex, it seems insufficient to use classical binding parameters. Therefore, following chronic alcohol treatment more information can be obtained by *in situ* hybridization studies or competitive RT-PCR studies aiming for different splice variants of the NR1 subunit and/or different NR2 subunits. In addition, highly sensitive and specific antibodies are available to analyze relative amounts of NR subunits.

Using these tools it became clear that chronic alcohol administration leads to specific changes in NMDA receptor subunit composition in various brain sites, which results in enhanced NMDA receptor function. Furthermore, following long-term alcohol self-administration, enhanced NMDA receptor function and long-lasting changes in NMDA receptor composition have been found. All these studies support the hypothesis that changes in NMDA receptor subunit composition represent neuronal adaptive responses to alcohol-induced inhibition of the NMDA receptor and that these changes are long-lasting and can contribute to withdrawal responses as well as to states of hyperexcitability that can occur long after cessation of alcohol intake and can eventually contribute to alcohol craving and relapse. The role of NMDA receptors in modulating alcohol self-administration has only recently been examined.

The functional NMDA receptor antagonist acamprosate was effective in a series of preclinical studies in reducing alcohol consumption and relapse. Competitive NMDA receptor antagonists attenuated operant responding for ethanol without affecting baseline levels of water self-administration; however, the selectivity of such an effect was questioned by demonstrating that the competitive NMDA receptor antagonist CPPene decreased both ethanol and saccharin self-administration. Noncompetitive NMDA receptor antagonists such as PCP and memantine were also tested, and both compounds reduced alcohol-reinforced behavior. However, both compounds have also modified other types of operant behavior in control experiments. In a more recent study, chronic treatment with a novel low-affinity, noncompetitive NMDA receptor antagonist, MRZ 2/579 (neramexane)–the cyclohexan MRZ 2/579 has very similar characteristics to memantine–has been evaluated in nonoperant and operant models of alcohol drinking behavior.

Nonoperant two-bottle-choice ethanol drinking was not affected by a 6-d sc infusion of MRZ 2/579 (9.6 mg/d). Only intermittent injections of the drug (5 mg/kg) but not constant infusion led to significant decreases in operant responding for ethanol. These promising results with low-affinity noncompetitive NMDA receptor antagonists led to further studies on alcohol craving and relapse behavior. In order to study alcohol craving and relapse, a new model of long-term alcohol self-administration with repeated deprivation phases was used.

Low-affinity noncompetitive NMDA receptor antagonists such as memantine or the cyclohexan MRZ 2/579 were tested in this model. Chronic administration of memantine or MRZ 2/579 via osmotic minipumps selectively reduced relapse behavior. In alcohol discrimination experiments, memantine as well as MRZ 2/579 dose-dependently generalized to the ethanol cue suggesting that low-affinity noncompetitive NMDA receptor antagonists might exert their reducing effect on alcohol intake and relapse behavior by generalizing for some of the stimulus properties of ethanol.

Chronic effects of ethanol

Acute and chronic effects of ethanol on other ionotropic glutamate receptors such as AMPA (alpha-amino-3-hydroxy-5-methyl-4-isoxazole propionate) and kainate (KA) receptors have also been studied. It has been found that ethanol at physiologically relevant concentrations (≤ 50 m*M*) can suppress both AMPA and KA receptor-associated currents. However, little is known about adaptations in AMPA/KA receptors after chronic ethanol treatment. In a single study on the role of AMPA/KA receptors on the reinforcing properties of ethanol, Stephens and Brown have found that neither a competitive AMPA/KA antagonist, NBQX, nor the preferential noncompetitive AMPA antagonist, GYKI 52466, exerted any selective effects on ethanol-reinforced behavior in rats trained to self-administer ethanol, sucrose, or saccharin on a progressive ratio schedule of reinforcement.

GYKI 52466 had no effect on operant responding, but increased spontaneous locomotor activity. In contrast, NBQX significantly decreased operant responding for ethanol, sucrose, and saccharin, but only at the doses that suppressed spontaneous locomotor activity. The authors concluded that non-NMDA ionotropic glutamate receptors may not play any specific role in the positive-reinforcing properties of ethanol. This conclusion is supported by a recent finding in AMPA receptor knockout mice: these mice did not differ in alcohol intake and preference from wild-type animals. The AMPA antagonist GYKI 52466 was also studied in voluntary long-term alcohol drinking rats following repeated deprivation phases. This compound did not affect the alcohol deprivation effect, showing that AMPA receptors do not play an important role in alcohol craving and relapse behavior.

Evidences of the CRH system

Stress is consistently viewed as an important factor in alcohol drinking in laboratory animals. Alcohol is postulated to reduce physiological effects induced by stress. As a result, alcohol ingestion itself or alcohol-induced relief from stress becomes reinforcing. CRH regulates endocrine responses to stress and mediates stress-related behavioral responses via extrahypothalamic sites. The CRH signal is transmitted by two types of CRH receptors, termed CRH1 receptor (CRHR1) and CRH2 receptor (CRHR2), which differ in their pharmacology and expression pattern in the brain. Dysregulation in the CRH system has been attributed to a variety of stress- related psychiatric disorders including alcoholism.

The detection of altered levels of endogenous CRH in distinct brain areas of selectively bred, ethanol-preferring animals is evidence for a possible relationship between the activity of the CRH system and voluntary alcohol consumption. Since CRH acting at CRHR1 plays a key role in mediating the central stress response, CRHR1 activation might also affect alcohol-drinking behavior. Therefore, the effects of the novel nonpeptide CRHR1-antagonist R121919 on the alcohol deprivation effect in long-term alcohol-experienced rats was studied. This compound did not influence relapse-like drinking.

The conclusion from this study is that CRHR1 activation is not involved in the expression of an alcohol deprivation effect; however, it is known that following repeated alcohol deprivation phases a subsequent alcohol deprivation effect is not associated with behavioral or endocrinological stress responses, offering an explanation as to why selective CRHR1 blockade does not affect the alcohol deprivation effect. Interestingly, mice lacking a functional CRHR1 markedly increased their alcohol intake in response to repeated stress and deprivation phases, although acquisition of alcohol drinking under resting conditions was not influenced by the dys-functional CRH system, as alcohol consumption of knockout mice did not differ from wild-type mice during this phase.

Reinstatement Alcohol-seeking Behavior

As already mentioned, one pathway that may induce alcohol craving and relapse due to the mood-enhancing, positive-reinforcing effects of alcohol consumption involves the opioidergic systems. The motivational effects of positive mood states and external alcohol-associated cues on alcohol craving and relapse (= alcohol-associated positive mood states) may be blocked by opioid receptor antagonists. Reinstatement studies of alcohol-seeking behavior have shown that naltrexone attenuates the reinstatement of responding elicited by conditioned cues. In this study reinstatement was tested with conditioned auditory stimuli, olfactory stimuli, and a combination of both stimuli. Auditory stimuli alone did not lead to reinstatement of responding, whereas olfactory stimuli and in particular the combination of both stimuli led to a strong reinstatement of responding.

Naltrexone (0.2 g/kg) selectively decreased responding elicited by the ethanol-associated olfactory cues; specifically, naltrexone reversed the increases in responding induced by the ethanol-associated cues but did not alter responding in the presence of water-associated cues. This finding supports a role for opioid receptor activation in the motivational effects of ethanol-related environmental stimuli and further suggests that craving and relapse elicited by conditioned stimuli can be prevented by opioid receptor blockade. In another study by Le et al. reinstatement of alcohol-seeking behavior was elicited by either priming injections of ethanol (0.25-1.0 g/kg) or exposure to intermittent foot shock. Naltrexone pretreatment (0.2-0.4 g/kg) blocked alcohol-induced, but not stress-induced, reinstatement.

In another study, naltexone (1-3 mg/kg) potently and dose-dependently inhibited reinstatement of alcohol-seeking behavior induced by noncontingent deliveries of the liquid dipper filled with 8% ethanol solution, which is comparable to an ethanol priming injection. Together these studies show that naltrexone suppresses reinstatement of ethanol-responding induced by either ethanol priming or conditioned cues. In contrast, naltrexone is ineffective when reinstatment is initiated by stress. These finding are in agreement with the hypothesis that activation of opioid receptors underly the induction of "*alcohol-associated positive mood states*" and of cue-induced "*alcohol-associated positive mood states.*"

NMDA receptors in reinstatement

Acamprosate acts as a functional NMDA receptor antagonist. Together with other noncompetitive, low-affinity NMDA receptor antagonists, the therapeutic effectiveness of these compounds may be exerted by decreasing stress- and cue-induced conditioned withdrawal and withdrawal relief craving (= alcohol-associated negative mood states). It remains to be established whether acamprosate or noncompetitive NMDA receptor antagonists might alter reinstatement of ethanol seeking after extinction induced by stress and cue-induced conditioned withdrawal, but, preliminary findings may indicate that the noncompetitive NMDA receptor antagonist MRZ 2/579 nonselectively inhibits reinstatement of ethanol- and water-seeking behavior induced by conditioned stimuli in the rat.

In a very recent study, Vosler et al. used a discriminative two-lever test to examine the ability of dizocilpine to reinstate ethanol-seeking behavior. In this study, one group of rats was trained to lever- press for ethanol and another group to lever-press for sucrose. After extinction, rats were injected with ethanol (0.5 g/kg). The ethanol group showed reinstatement of lever responding, whereas the sucrose group showed minimal responding following ethanol priming. In contrast, dizocilpine increased responding in both groups, suggesting a loss of discriminative control.

Role of CRHR1 blockade

Most recently, the role of CRHR1 blockade in reinstatement of alcohol seeking induced by intermittent foot-shock stress was studied in rats. Rats were given alcohol in a two-bottle free-choice procedure (water vs alcohol) for 30 d and were then trained for 1 h/d to press a lever for alcohol for 1 mo in operant conditioning boxes. After stable drug intake was obtained, lever pressing for alcohol was extinguished by terminating alcohol delivery. Subsequently, reinstatement of alcohol seeking was determined after exposure to intermittent foot shock in different groups of rats that were pretreated with selective CRHR1 antagonists. The CRHR1 antagonists D-Phe-CRF or CP-154,526 attenuated foot-shock-induced reinstatement of alcohol seeking. On the other hand, the removal of circulating corticosterone by adrenalectomy had no effect on foot-shock stress-induced reinstatement of alcohol-seeking behavior. These data suggest that CRHR1 activation contributes to foot-shock stress-induced reinstatement of alcohol seeking via its actions on extrahypothalamic sites.

Systematic Studies

Very little systematic work has been performed so far to elucidate the neurobiological and molecular mechanisms underlying ethanol-induced behavioral sensitization. Several studies have shown that the noncompetitive NMDA receptor antagonist dizocilpine prevents the development of drug-induced sensitization processes, and recently, Broadbent and Weitemier showed that dizocilpine also prevents the development of sensitization to ethanol in DBA/2J mice. The inhibtory effect of dizocilpine on the development of behavioral sensitizatzion to ethanol could be explained in terms of state-dependent effects. Thus a sensitized response to ethanol can occur only under the influence of dizocilpine. However, this possibility was ruled out by further experiments demonstrating that dizocilpine treatment inhibits the development of sensitization, independent of a state-dependent effect.

CONCLUDING REMARKS

Studies in voluntary long-term alcohol-drinking rats with repeated deprivation phases and reinstatement experiments clearly show that there are at least two different neurobiological pathways that induce alcohol craving and relapse. The first pathway involves the opioidergic system (in particular, μ-opioid receptors) and probably the mesolimbic dopaminergic system and may induce alcohol craving and relapse due to the mood-enhancing, positive-reinforcing effects of alcohol consumption. Associative learning may, in turn, transform positive mood states during alcohol drinking and previously neutral stimuli into alcohol-associated cues that acquire positive motivational salience and induce craving and relapse. These conditioned cues also depend on the activation of the opioidergic system and probably on the dopaminergic system. A second pathway involves the glutamatergic system (in particular, NMDA receptors) and the CRH system and may induce alcohol craving and relapse by negative motivational states including withdrawal and stress.

Associative learning may, in turn, transform negative mood states during alcohol withdrawal and previously neutral stimuli into alcohol withdrawal-associated cues that produce an aversive state and induce craving and relapse. These conditioned cues might also depend on the activation of the glutamatergic system, since it has recently been shown that the NMDA receptor modulator acamprosate inhibits conditioned abstinence behavior in rats. Clearly, more systematic work is needed to define fully these different neurobiological/molecular pathways involved in alcohol craving and relapse behavior. Nevertheless, the findings presented here already have important implications for relapse prevention in humans.

Since the neurobiological pathways of craving and relapse inducing mood states are different, pharmacological treatments that are normally aimed at a selective biological substrate should be specific to certain motivational states and stimuli, that is, effective in reducing relapse induced by a given mood state and stimuli but unable to counteract relapse induced by another mood state and stimuli. Consequently, relapse treatments, to be successful, should be adapted to the factor that has a greater chance to induce relapse in a given subject. Developing a new approach that takes into account relapse-specific mood states and stimuli can certainly be considered a breakthrough in the management and prevention of relapse in alcoholism.

11

Tobacco Addiction

Tobacco produces dependence or addiction in humans. Dependence can be defined as a maladaptive pattern of substance use or drug taking over an extended time period. The American Psychiatric Association, in its DSM-IV manual for psychiatric diagnosis, suggest an operative diagnosis of tobacco dependence when three or more of the following seven symptoms or signs are identified in a subject:

1. Persistent desire and unsuccessful attempts to quit
2. Use of large amounts of drug and for a longer period than intended
3. Continued use in the face of medical, familial, and social problems
4. Important social, familial, and recreational activities given up or reduced because of drug craving
5. Expenditure of a great deal of time and activity in relation to drugs
6. Tolerance
7. Physical dependence (withdrawal)

Interestingly, most of the symptoms can be related either to the loss of control over the drug use habit, or to the physiological adaptation to the drug. Since only three symptoms are required for the diagnosis, dependence can be present without the signs of tolerance and withdrawal, once considered necessary prerequisites of dependence and addiction. In fact, in the 1964 Report of the U.S. Surgeon General, chronic exposure to nicotine was recognized as not producing physical dependence, and was regarded as "*habit forming*" rather than addicting.

However, in the 1988 Report of the U.S. Surgeon General the definition was changed, recognizing nicotine as addictive, indicating that tolerance and withdrawal syndrome are neither necessary nor sufficient criteria for diagnosis. This definition was included in the DSM-IIIR, and then in DSM-IV. Psychological dependence, that is, liability detectable by behavioral changes, became a key qualifier of addiction. In this review the terms *dependence* and *addiction* are used interchangeably, following other authors who equate the two terms.

Constituents of Tobacco

Among tobacco's 3500 different constituents, nicotine has been considered the most likely responsible for tobacco's addictive effects. When smoked, tobacco releases nicotine, which readily enters the blood circulatory system and rapidly accumulates in the brain, exerting its neuroactive action. To date, the neurobiology of nicotine dependence is only partially understood; the knowledge about the underlying molecular mechanisms so far accumulated is based mostly on findings obtained in animal models of nicotine dependence and, to a lesser extent, on observations in humans.

According to a well-accepted psychopharmacological model, substance use or drug-seeking behavior is controlled by four main processes: the positive reinforcing effects, the aversive effects, the discriminative effects, and the stimulus-conditioned effects of the drug. In more explicit terms, an individual consumes nicotine because it:

1. Produces pleasure (positive emotions)
2. Reduces the aversion due to abstinence (negative emotions)
3. Signals discrimination about its presence (assuming motivational value)
4. Promotes, as a conditioned stimulus, the unconditioned learned component of the drug-taking response

In the real world all these processes are interacting in determining the behavior of individuals smoking tobacco, and have been modeled according to the different weight given to certain processed. The common factor to all these models is the presence of compulsive drug-taking.

In Vivo Drug Taking Paradigms

This review is focused mainly on in vivo drug-taking paradigms as a way to study the neurobiology of dependence. In naturalistic conditions, smokers titrate their optimal nicotine dose by taking different numbers of puffs per unit of time using inhalation devices (cigarettes, cigars, etc.). They consume several cigarettes per day, often clustering them around specific events, for example, at home after work, or after lunch. Similarly, in an experimental setting, smokers can be trained to self-administer nicotine intravenously and to titrate their optimal dose per session or daily. The determinants leading to the way nicotine is abused, that is, smoking, chewing, or intravenous nicotine self-administration are described elsewhere. The most commonly accepted reason for taking an addictive drug is because of its positive reinforcing effects.

A drug serves as positive reinforcer when its use increases the probability of eliciting a response on which it is contingent (i.e., smoke inhalation maintaining cigarette use in smokers; nicotine infusion maintaining lever pressing in rats trained in a Skinner box). In this review we selected studies where molecular and neural mechanisms were investigated in subjects in which the effects of nicotine were measurable as reinforcing properties, that is, the disposition of an individual to take the drug compulsively and experience the drug effects again, is monitored, quantified, and expressed as increased probability of responding. This definition is consistent with research paradigms used for addictive drugs in general.

In our attempt to reduce the confounding factors intrinsic to several animal models of dependence, that is, the inclusion of equivocal or indirect measurements of such disposition, only operant behavior paradigms in which nicotine serves as positive reinforcer were selected. These paradigms, in particular intravenous (iv) drug self-administration, offer both acceptable theoretical support for the modeling of the positive reinforcing properties of addictive drugs and technical robustness in view of molecular investigations.

Neural Mechanism of the Reinforcement

Most of our knowledge about the neural mechanism of reinforcement stems from the experiments of Olds and Milner using intracranial self-stimulation (ICSS) procedures. ICSS can be triggered when the electrodes are targeting several discrete brain regions, spanning from the ventral tegmental area (VTA) to the lateral hypothalamus, via the medial forebrain bundle, to the nucleus accumbens and prefrontal cortex. The ensemble of these structures has been loosely defined as reward system. At first, nicotine (0.2-0.4 mg/kg) was shown to enhance the rate of responding for ICSS in the rat VTA or medial prefrontal cortex without changing the current threshold. Years later, Huston-Lyon et al., using an auto-titration procedure, showed that nicotine indeed lowered the current threshold for ICSS in a dose-dependent fashion, demonstrating the reinforcing efficacy of low-dose nicotine. More recently, nicotine was shown to produce a leftward shift of the ICSS curve relating the frequency of currents and the response rate in rats.

These data indicate that nicotine is acting directly on the reward system, affecting its efficiency in positive terms. Other lines of investigation, using neurochemical-electrolytic brain lesions, local microinjections of antagonist drugs, or local measurements of neurotransmitter levels with microdialysis probes, have shed some light on the relevance of certain structures in mediating the rewarding effects of addictive drugs. According to recent neurobiological models, it is surmised that: (a) the complex processing involving cognitive and associative mechanisms associated with drug-taking, such as subjective attribution, cue assessment, and craving, is assumed to depend on the prefrontal and associative cortex, as well as the amygdala and hippocampus; (b) the mesocorticolimbic dopamine (DA) system, projecting from the ventral tegmental area to the nucleus accumbens and medial prefrontal cortex, is implicated in the control of instrumental behaviors and their outcomes; (c) the dorsal striatum, which receives projections from the DA neurons of the substantia nigra, is suggested to participate in habit formation.

Finally, chronic exposure to pharmacological doses of addictive drugs, by acting on some components of the reward system, in particular the mesocorticolimbic DA system, are believed to produce adaptive modifications of neural functioning. These changes result in several behavioral changes, some not related to changes in reward-dependent behavior–for example, locomotor sensitization, place preference, latent inhibition, increased DA release to addictive drug challenge, increased sensitivity to stress exposure, reduced palatable food consumption, and maintenance of drug self-administration.

It is interesting to note that most of these behaviors are observed neither in rats chronically self-administering nicotine nor in human smokers. In addition, it is not automatic to suggest that exposure to nicotine, independent of the paradigm of administration, produces dependence in individuals. In fact, continuous vs pulsed administration, or passive vs active self-administration, may differ importantly in defining the total dose effects, the neurobiological changes in the substrates, and the tolerance or sensitization profile of the drug.

POSITIVE REINFORCING EFFICACY

The authors are aware that if a drug sustains drug-taking behavior, that is, nicotine self-administration or smoking, this does not tell how it does so. Many factors have been proposed to contribute to the overall positive reinforcing efficacy of the drug, that is, nicotine reward, and these have been reviewed extensively. Accordingly, a drug may:

1. Reinforce the stimulus–response habit directly
2. Modulate internal affective states and producing euphoria or reducing anxiety
3. Modulate other reinforcer effects for example, enhancing social or sexual reinforcers
4. Modulate attentional or perceptual mechanisms

It has been suggested that the first and second are distinctive mechanisms by which the drug acts directly on the neural substrate and mediate the reinforcing properties, whereas for the third and fourth the drug exerts its effects indirectly. In drug self-administration the subject starts by learning about the response-outcome contingency (incentive, stimulus-reward learning, goal-oriented behavior), and then the action gradually becomes a habit (stimulus-response learning). Certain addictive drugs may influence self-administration behavior by acting on stimulus-reward learning, leading to the acquisition of a drug-reinforced habit over time. In this case environmental stimuli can become associated with the effects of the drugs, acquiring secondary reinforcing properties (conditioned reinforcer).

Interestingly, using so called second-order schedules, conditioned reinforcers may take control over operant responding, indicating persistence of drug-taking induced by conditioned cues in the absence of the drug. It has been argued that, at least for human studies, substance use (i.e., smoking) is habitual, and relapse is most often triggered by exposure to eliciting cues, including environmental conditioned stimuli described above, or a priming stimulus with a single dose of the same drug. In the latter case, the drug acts as a discriminative stimulus, producing interoceptive effects that contribute to the initiation of drug-taking behavior. Interestingly, low doses of nicotine can induce relapse of nicotine self-administration in rats following extinction.

White maintains that both forms of stimulus-reward and stimulus-response leaning coexist in parallel in the brain of subjects who self-administer addictive drugs, and this appears particularly true for nicotine. Eliciting cues, both environmental and interoceptive-discriminant, play an important role in determining the subjective feeling of craving for the drug. These conditioned processes have to be taken into account when addressing experiments on the molecular correlates of self-administration procedures. However, it is believed that they play important roles in sustaining drug self-administration only if the drug is able to maintain its activating properties on the neural substrate of the reward system over time.

Relevance of Nicotine

Difficulties in giving up the habit of smoking are often reported to be also related to the nicotine withdrawal syndrome, that is, increase nervousness, frustration, anger, and desire to smoke. Nicotine withdrawal syndrome can also be seen in rats passively infused with high doses of nicotine and then acutely treated with mecamylamine. Stress, depression, and other negative affective states, and peripheral signs induced by nicotine withdrawal, have been suggested to be the basis of nicotine craving, and may constitute an important component of the maintenance of drug-taking behavior. In smokers, the time to the first cigarette in the morning after awakening is considered a main feature in the Fargenstrom Questionaire for Smoking Dependence. The value of such an indicator is based on the overnight abstinence from smoking due to sleep, which should induce a state of physiological unbalance (i.e., withdrawal) that will drive a motivation for drug-seeking behavior. This phenomenon has been identified as a sort of psychological withdrawal syndrome.

Such a syndrome can be measured objectively in rats by increases of ICSS threshold after the interruption of passive chronic infusions with nicotine, suggesting a disruption of the reward system. However, relief from withdrawal symptoms has been regarded as possibly related to negative reinforcement, and it may belong to a separate category when compared with the positive reinforcement properties of a drug. Interestingly, in certain paradigms nicotine exerts aversive effects, and it was suggested that nicotine might serve as a negative reinforcer. For example, Henningfield and Goldberg showed that three smokers who failed to self-administer nicotine intravenously learned an operant paradigm to avoid the scheduled injection of nicotine, suggesting aversion. However, to date, no evidence of a role of negative reinforcement has been observed in animals trained to nicotine self-administration. In addition, a recent elegant experiment indicates that rats trained to stable levels of intravenous nicotine self-administration do not show signs of psychological withdrawal in a Social Interaction Test at 24 and 72 h after the last daily self-administration session.

The Social Interaction Test, a very sensitive procedure to measure emotional negative states, showed that all rats were significantly more anxious than saline-exposed control rats at any time intervals following the self-administration sessions up to 72 h, suggesting a mild, generalized condition of negative emotional state. The lack of worsening of this condition over time following the last self-administration session suggests that psychological withdrawal is not involved in motivating the animal to start the following session 24 h afterward. Therefore, a prevalence of the positive reward component over the psychological withdrawal component is suggested in determining drug-seeking behavior in nicotine self-administering rats. In this regard, nicotine probably differs from other addictive drugs.

Intravenous Nicotine in Experimental Animals

Initially, nicotine self-administration was successfully developed in subhuman primates. Some early works also described different protocols in rats. It was only during late 1980s that Corrigalll and Coen (1989) developed reliable schedules of iv nicotine self-administration in rats, which were replicated and used by other groups. Under these specific methodological conditions, rats acquire stable self-administration. Critical schedule parameters are:

1. The short infusion time (1 s vs 5-6 s usually set for other drugs)
2. The limitation of aversive effects due to nicotine overdosing/overload (e.g., limited daily access to self-administration session, a "timeout" period after each infusion, checking nicotine solutions pH)
3. The fixed-ratio schedule between responding and nicotine infusion delivery

Overdosing, high intake levels, and any other exposure to potential aversive effects of nicotine need to be controlled into the schedule of reinforcement and in the protocol. Most of the published studies described schedule ranging between 1 and 3 h duration, with methodological conditions essentially similar to those described by Corrigall and Coen in 1989. However, Valentine et al. (1997) showed that rats could also be exposed to unlimited access to nicotine self-administration (24 h/d), with responding stable for weeks. Male Long-Evans, Sprague Dawley, Wistar, Holtzman, but not Lewis or Fisher-344 rat strains have been shown to acquire nicotine self-administration. Female Sprague-Dawley also acquire nicotine self-administration, but with rates of responding different from male subjects, suggesting a gender-dependent variability to the reinforcing effects of nicotine.

Training to lever-press for food reinforcement, cocaine infusion, or previous nicotine pretreatment facilitated the acquisition of iv nicotine self-administration, even if training histories did not appear necessary. Diet restriction is a preferred procedure, since it allows better control of rat body weight, an important factor to keep constant infusion volumes (infusion unit, mL/kg body weight) throughout experimental sessions. Nicotine is iv self-injected by rats in a dose-related fashion but within a narrow dose range, nicotine 0.03 mg/kg/infusion unit dose being the most widely used dose for training and baseline performance. In contrast to other drug self-administration protocols, such as for cocaine, it is not possible to identify a dose-response curve "*descending limb*" for nicotine self-administration, where one increase of drug unit dose corresponds to a proportional decrease of the rate of responding.

Only a steep dose-response curve, with a given threshold and a plateau of response at various active doses, has been reported for nicotine self-administration. Further increases in nicotine doses are associated with motor depression and a decrease of responding. Environmental factors influencing smoking behavior have also been shown to modulate nicotine self-administration in rodents. Preexposure to a single injection of low-dose nicotine or to environmental stimuli previously paired with nicotine self-administration is able to reinstate nicotine-seeking behavior, even after weeks of abstinence. Recently, iv nicotine self-administration has also been developed in mice, and effects obtained with different strains of mice or with nicotinic ligands were described. The development and characterization of these protocols offer the opportunity to use transgenic mice to study the effects of gene mutation on the reinforcing properties of nicotine.

PHARMACOLOGICAL STUDIES OF NICOTINE REWARD

When nicotine enters the brain, it binds to neuronal nicotinic acetylcholine receptors (nAChR), heterogenou sly expressed in central nervous system neurons. Much like their cousins in the neuromuscular junction, brain nAChR are probably pentameric complexes arranged around a central pore that is permeable to K^+, Na^+, and Ca^{2+}. In the mammalian brain, eight α subunits (α2–α7 and α9–α10) and three β subunits (β2–β4) are differently combined to define two principal subfamilies of receptors: the α-bungarotoxin-sensitive subfamily, consisting of homopentametic α7 nAChR subunits, and the hetero-oligomeric subfamily, made of different combinations of the other subunits and the β(2–4) subunits. Experiments performed in mutant mice lacking the α7-subunit nAChR receptors indicate that practically all the high-affinity α-bungarotoxin binding found in the mouse brain is due to α7-subunit-containing nAChR. Similar experiments performed in mutant mice lacking the β2 and α4 subunits indicate that almost all of the high-affinity ^{3}H-nicotine-binding sites of the mouse brain are represented by an α4β2 nAChR. However, persistence of high-affinity binding sites for other nicotinic ligands with high affinity for α4β2 receptors, such as ^{3}H-cytisine or ^{3}H-epibatidine, were still found in restricted brain regions of both mutant mice, in particular the medial habenula and the interpeduncular nucleus, structures known to selectively express α2β4 and α3β4 nAChR.

Chronic Nicotine Administration Produces AchR Upregulation

Recent reports consistently indicate that ^{3}H-nicotine binding, but not ^{125}I-α- bungarotoxin binding, is upregulated in human postmortem prefrontal cortex, hippocampus, entorhinal cortex, and, to a lesser extent, striatum of smokers when compared with age-matched controls. Interestingly, nicotine binding is normalized in ex-smokers, confirming the reversibility of the upregulation of ^{3}H-nicotine-binding sites observed in rodent brain following chronic passive nicotine administration. These results indicate that smoking and nicotine exposure produces upregulation of nAChR, most likely α4β2, but does not directly support receptor hypersensitivity, that is, increased effects of nicotine on postsynaptic target cells.

Recent unpublished results obtained in rats self-administering nicotine (0.03 mg/kg/injection) for at least 3 wk indicate that ^{3}H-nicotine-binding sites are upregulated in the parieto-frontal cortex, but not the VTA, suggesting a potential relevant role of the cerebral cortex in maintenance of self-administration. This interpretation is highly speculative, since it is not clear yet if this receptor upregulation stands for functional hypersensitivity (similar to denervation hypersensitivity) or reduced membrane turnover of inactivated receptors. Pharmacological and electrophysiological experiments show that prolonged exposures to nicotine produces desensitization of nAChR into a high-affinity binding state, suggesting the latter case. Finally, recent experiments in a cell line carrying heterologous α3β2 showed that a fraction of the upregulated receptor is still active, suggesting that a similar phenomenon may apply to *in vivo* receptor upregulations.

Reinforcing Effects of Nicotine

The involvement of nAChR in the reinforcing properties of nicotine is also supported by evidence that the nonselective antagonist mecamylamine reduces nicotine self-administration in rats. Experiments performed on mutant mice lacking the β2 subunit and previously trained for cocaine self-administration showed a marked attenuation of nicotine self-administration when cocaine was substituted with nicotine. The β2 mutant mice were also able to lever-press for food, and the selective attenuation of lever pressing contingent to nicotine infusions indicates the key role of β2-contaning nAChR in sustaining nicotine self-administration. Again, this effect was clearly mediated by the mesolimbic DA system.

Neuroadaptation to Nicotine Exposure

The mesocorticolimbic DA system has been the object of intense research in the field of drug addiction because:

1. It is a relevant component of the reward system.

2. It is involved in stimulus-reward learning and in the incentive motivational effects of drugs.
3. When exposed to chronic treatment with addictive drugs, it is the site of neuroadaptive changes, such as sensitization or dysregulation associated with negative emotional states.

Midbrain DA neurons contain mRNA for all the principal nAChR subunits. However, it is not clear if all of them form functional receptors. High-affinity binding for ^{125}I-α-bungarotoxin and ^{3}H-nicotine suggests that α7 and β2-containing nAChR proteins, respectively, are actually expressed. Antibodies against α4 have been recently used to confirm the presence of this subunit as translated protein, strongly supporting the presence of a functioning α4β2 nAChR. Pharmacological doses of nicotine are known to stimulate DA neurons of the VTA preferentially vs those of the substantia nigra, and most likely by a direct action.

Pioneering studies by Fuxe and collaborators using the Falk-Hillarp and related techniques showed that exposure to nicotine or smoked tobacco increases the rate of disappearance of fluorescent DA from mesolimbic DA terminals in rats. These data suggest that nicotine enhances the impulse flow and release of DA in nucleus accumbens. Neurochemical experiments using *in vivo* microdialysis show that acute nicotine induces DA release in the terminal fields of midbrain DA neurons, in particular in the shell of the nucleus accumbens. Nicotine effects in the nucleus accumbens are antagonized by mecamylamine microinjected into the VTA, but not into the nucleus accumbens. This result suggests that, in vivo, nAChRs located in the VTA region are of relevance in controlling DA overflow.

Interestingly, acute nicotine at doses of 50-100 μg/kg also increases the spontaneous burst firing of extracellularly recorded DA VTA neurons, most likely via NMDA receptor-dependent activation, and burst firing of VTA neurons has been associated with DA release. The mechanisms seem to be dependent on nAChR containing α4 or α3, but not α7. Repeated daily exposures to systemic nicotine result in sensitization of its stimulatory effects on DA overflow in the nucleus accumbens and prefrontal cortex in a manner that recalls the effects of chronic exposures to cocaine or amphetamine. On the other hand, sensitized responses to acute systemic challenges with nicotine are antagonized by mecamylamine, and can be induced by local microinjection into the VTA. These results point to a key role of nAChR located in the VTA region to induce sensitization to nicotine effects following chronic nicotine treatments. Interestingly, when nicotine is constantly infused using subcutaneous osmotic minipumps, and not injected, it results in desensitization and tolerance of nicotine effects on DA release.

Recent data indicates that repeated nicotine administration (0.4-0.5 mg/kg) can result in sensitization of DA release in the core of the nucleus accumbens and DA release tolerance in the shell, but not in the medial prefrontal cortex, suggesting a complexity due to differences in neuroanatomical compartments. All in all, these data suggest that repeated passive exposure to nicotine can enhance DA neurotransmission in the mesocorticolimbic DA system. However, to date, there is no evidence that this effect is also produced in rats trained to nicotine self-administration, casting doubt about the relevance of DA release sensitization for the reinforcing effects of nicotine. In contrast to the central role of VTA neurons in the sensitization of DA release to nicotine, recent microdialysis experiments performed by locally perfusing nicotine into the terminal fields of midbrain DA neurons showed a dose-dependent increase of mecamylamine-sensitive DA release produced by acute nicotine. Chronic nicotine treatment further enhances DA release in response to nicotine when locally infused via microdialysis probes into the nucleus accumbens, striatum, and prefrontal cortex in rat.

Experiments performed in vitro on synaptosomes confirm that DA is released via α4β2 and α3β2 nAChR-dependent Na^{+-} and Ca^{2+}-sensitive channel-operated mechanisms. Interestingly, when DA release was studied in striatal slices *ex mortem* in rats injected daily with nicotine, enhancement of nicotine-induced DA overflow response was measured, whereas desensitization was seen in synaptosomes obtained from rats continuously infused with nicotine. Therefore, nAChR expressed at the synaptic terminals of midbrain DA neurons is likely to contribute to the development of mesocorticolimbic DA system sensitization to nicotine effects only under specific conditions.

Attenuation of systemic nicotine

Attenuation of systemic nicotine and food-induced DA overflow in the nucleus accumbens was also produced by microinjections of the α7-subunit-selective antagonist methyllycaconetine (MLA) into the VTA. These α7-dependent effects are most likely mediated by presynaptic nAChR located on glutamatergic neurons projecting to the VTA.

Electrophysiological experiments on brain slices containing VTA DA neurons of rats previously continuously infused with nicotine for several days show desensitization of α7-dependent currents, suggesting that chronic nicotine attenuates α7-mediated synaptic transmission. However, in vivo studies indicate that MLA was unable to block the discriminative properties of nicotine in rats nor was nicotine self-administration, suggesting that α7 is not involved in the discriminative effects of nicotine, most likely mediated by hippocampal or prefrontal cortex rather than the DA systems.

Role of chronic nicotine on thyrosine hydroxylase (TH)

In vivo acute nicotine stimulates the activity of tyrosine hydroxylase (TH), the rate- limiting enzyme in the synthetic pathway of DA, in DAergic terminals of the nucleus accumbens, whereas tolerance to this effect is observed following repeated nicotine administration. Conversely, acute nicotine does not change the levels of TH immunoreactivity in the VTA and substantia nigra, whereas increases are observed following repeated nicotine administration. No data are available on the effects of chronic nicotine self-administration on TH levels and activity.

Mesocorticolimbic DA System

The involvement of mesocorticolimbic DA system in the reinforcing effects of nicotine is supported by several experimental data.

1. Selective 6-hydroxydopamine (OHDA) lesions of the nucleus accumbens reduce nicotine self-administration.
2. Low doses of systemic administration of neuroleptics attenuate nicotine self-administration in rats.
3. In humans, neuroleptics increase nicotine plasma levels in smokers, suggesting that the individuals smoke more to counteract the effects of the drug and experience rewarding effects.
4. Postmortem DA and DA metabolite levels were found to beincreased in the striatum of smokers when compared with nonsmoker controls.

5. The indirect DA agonist bupropion has been used successfully to reduce the relapse rate in smokers who have manifested the intention to quit smoking.

In rats, nicotine self-administration is attenuated by VTA microinfusions of the α4β2-preferential competitive antagonist dihydro-β-erythroidine, but not by the muscarinic antagonist atropine. This procedure perturbates the nicotinic-dependent cholinergic input from the pedunculopontine tegmental nucleus (PPT) into the VTA, probably by reducing nAChR-dependent facilitation of DA neuronal firing. Thus, the rewarding effects of nicotine self-administration are most likely mediated by nAChR, and not by muscarinic transmission. The lack of enhancing effects of nicotine on firing in VTA neurons and on DA release in nucleus accumbens and striatum observed in the β2 mutant mice strongly suggests a role for β2-containing nAChR in mediating the reinforcing properties of nicotine through the activation of the mesolimbic DA system. In support of a direct effect of nAChR on DA release, no changes of D1-D4 DA receptor levels and affinity nor of DA transporter were reported in mutant mice.

Transcriptional Regulation

Chronic exposure to addictive drugs indirectly stimulates transcription of specific genes by increasing intracellular cyclic adenosine monophosphate (cAMP), which in turn results in activation of multifunctional protein kinases and phosphorylation of several cellular proteins, including transcription factors, in target neurons of the mesocorticolimbic DA system. A limited but growing number of scientific reports indicate that nicotine also produces some adaptive changes in the target neurons of DA terminal fields, namely, the nucleus accumbens, striatum, and prefrontal cortex, most likely via DA release and D1 receptor mediation. One of the most well studied effects of D1 receptor activation in target neurons is the transcriptional regulation of the *immediate early gene* (IEG) *c-fos,* and medium-late genes of the Fos-related antigen family. Expression of *c-fos* has been used largely to provide transcriptional activation brain maps induced by various stimuli, ranging from natural to pharmacological.

Acute nicotine passive administration induces expression of *c-fos* that is not restricted to the mesocorticolimbic DA system terminal fields, but includes several other rat brain regions, in particular the superior colliculus, the interpeduncular nucleus, the raphe, but not the midbrain DA neurons. This activation is inhibited by mecamylamine. In the striatum, a dose-responsiveness was observed, with the lower dose (0.4 mg/kg) inducing *c-fos* in the medial and central portions. In the nucleus accumbens, the response was smaller than in the striatum, with prevalence in the shell. The acute administration of epibatidine, an nAChR agonist with preferential affinity for the α4β2 form, induced c-fos IR in the prefrontal cortex, medial striatum, nucleus accumbens, amygdala, and superior colliculus. In the nucleus accumbens and striatum, but not the prefrontal cortex, this effect was inhibited by pretreatment with D1-D4 receptor antagonists, but not D2 antagonist. This result is in agreement with the nicotine-induced DA overflow measured in the terminal fields of the mesocorticolimbic DA system, and the D1 mediation of reinforcing properties of DA released in the nucleus accumbens by addictive drugs. Finally, repeated nicotine administration in rats results in preferential Fos increases in visuomotor and limbic structures, including basal ganglia and prefrontal cortex.

Nicotine self-administration

Nicotine self-administration increases Fos immunoreactivity in most of the brain regions that are activated by passive nicotine treatment, in particular the striatum, the shell and core of the nucleus accumbens, the lateral septum, the prefrontal cortex, and the cingulate cortex. Also, the retrosplenial cortex, the piriform cortex, the periventricular thalamic nucleus, and the superior colliculus showed increase of Fos immunoreactivity. In contrast to acute nicotine, nicotine self-administration did not increase Fos immunoreactivity in the hypothalamus, locus coeruleus, amygdala, and dentate gyrus. Passive chronic nicotine administration produced patterns similar to those obtained with acute nicotine, showing little tolerance to the challenging dose of nicotine, and a more general activation of the core of the nucleus accumbens when compared to the selective increase in the ventral shell observed in nicotine self-administration.

These differences are important, since brain maps of immediate early gene expression simultaneously represent the direct pharmacological effects of nicotine on neural networks and the activation associated with the internal dispositional state that controls nicotine taking, which includes the reinforcing effects of nicotine. A more extensive study of the transcriptional regulation of the c-fos-related antigen family (FRA) expressed in the terminal fields of the mesocorticolimbic DA pathway was performed in rats trained for nicotine and cocaine self-administration. FRA heterodimers constitute the AP1 complex that transcriptionally regulate a large number of plasticity-related genes. FRAs, and in particular the 35-kDa component recently identified as Δ-Fos, are medium-late onset genes and their products persist for several days in the nucleus of the target neurons. Therefore, once induced by nicotine or cocaine, FRA-containing AP1 complexes may regulate plasticity-related genes for a long period. In one experiment, rats were trained to self-administering cocaine or nicotine.

Increased expression levels of FRA immunoreactivity were found in the anterior cingulate cortex, prefrontal cortex, nucleus accumbens, medial striatum, but not amygdala, of rats from both groups. In a second experiment the binding of AP1 complex to neuronal DNA was measured in brain tissues dissected out immediately before the last self-administration session. The results showed a persistent increase of AP1 complexes in the nucleus accumbens in cocaine and nicotine self-administering rats, but not in the prefrontal cortex or striatum. This differential effect points to a primary involvement of target neurons in the nucleus accumbens as cellular substrate for the maintenance of drug self-administration and/or the stimulus-reward learning produced by these two drugs. Interestingly, the lack of increase of AP1 complex in the prefrontal cortex suggests that long-term changes in target neurons of this DA terminal field are of modest relevance in drug self-administration maintenance.

cDNA microarrays

cDNA microarray is one among several large-scale biology techniques proposed during the last few years to address the possibility of parallel massive measurement of changes in gene expression products of living cells or tissue following pharmacological manipulations. Only preliminary reports and one full article on nicotine treatments in rodents have been published so far. In the latter study, Konu et al. used cDNA microarrays containing 1117 genes and expressed sequence tags (ESTs) to assess the transcriptional response to chronic nicotine treatment in rats, by comparing 4 brain regions, the prefrontal cortex, nucleus accumbens, VTA, and amygdala. The results indicate 94 genes whose expression was altered above threshold. Results from principal component analysis and pairwise correlations suggested that brain regions studied were similar in terms of their absolute expression levels (with amygdala showing the lowest activity), but differed in the composition of transcriptional profiles in response to chronic nicotine.

Accordingly, prefrontal cortex and nucleus accumbens were significantly more similar to each other than to either VTA or amygdala, supporting the value of the concept of terminal field of the mesocorticolimbic DA system as relatively homogeneous neurobiological substrates for nicotine pharmacological properties. Several genes involved in cellular signaling, structure/cytoskelectal maintenance, metabolism, cell cycle, and transcriptional regulation were found to be affected by chronic nicotine. In particular, the MAP-kinase, the phosphatidylinositol, and EGFR signaling pathways showed quite consistent changes and were proposed as possible targets in response to nicotine administration. These results, as exciting as they appear, need a cautionary note for an appropriate interpretation.

In fact, technical limitations can importantly question the reproducibility of obtaining the same up- or down-regulation for the same gene using different large- scale biological approaches. In addition, given the relatively small percentage of the genome featured into the cDNA microarray (approx 2%), these conclusions can be considered preliminary in all senses. When quasi-exhaustive cDNA microarrays containing >90% of the genome are used, it may well be that other pathways will be most affected, shifting the overall interpretation of comparative relevance. Independent of any criticisms, these types of experiments are opening the way to a novel "global" approach to understanding the neurobiological substrate of nicotine dependence, where all the cellular biochemical pathways are assessed in parallel, and large-scale computational models will be used to make inferences about results.

Neuroadaptation to Nicotine Exposure

Are all the positive reinforcing effects of nicotine mediated only by mesocorticolimbic DA? Some discrepancies in the experiments cited above, and evidence of neuroadaptation in other neurotransmitter systems, may open some alternatives. Two examples follow.

1. A cDNA microarray experiment reveals effects on biochemical pathways whose changes have not been related to DA effects (e.g., EGRF).
2. In rats, cocaine self-administration completely downregulates *c-fos* expression in the nucleus accumbens, whereas nicotine does not, suggesting a D1 receptor-independent mechanism for activating transcription of *c-fos.*

In fact, here, as in the anterior cingulate cortex, nicotine may act directly either on postsynaptic nAChR located in cortical or accumbens interneurons, or in pre-synaptic nAChR located on nordadrenaline, glutamate, or GABAergic terminals, exerting important modulatory roles.

Noradrenaline (NE)

Nicotine stimulates the in vitro release of NE from slices or synaptosomes from hippocampus and cerebral cortex via $\alpha3$- or $\alpha4$-containing nAChR. In rats, acute nicotine administration increases

TH mRNA in the locus coeruleus, whereas chronic nicotine administration increases TH levels in the locus coeruleus and telencephalic terminal fields (cerebral cortex, hippocampus, etc.). Acute nicotine also increases NE release in the hippocampus as measured by microdialysis. This release is antagonized by injections of mecamylamine into the locus coeruleus, but not into the hippocampus, while intrahippocampal α-BG or MLA antagonizes the NE release.

These data suggest somatic mediation of NE release by α4β2 nAChR located near the locus coeruleus, and distal mediation of NE release by presynaptic α7 nAChR in hippocampus. Finally, chronic nicotine exposure enhances the nicotine- induced NE release. Hippocampus has been involved in mediating some discriminative effects of nicotine, supporting a potential role of NE in this component of drug self-administration. In addition, NE substrates can be involved in mediating the antismoking effects of bupropion, a mild DA- and NE-uptake blocker. However, no NE antagonist or selective NE neurochemical lesions are known to affect nicotine self-administration in rats.

Serotonin (5-HT)

Serotonin has been suggested to participate to the control of motivated consummaroty behavior, and evidence indicates some involvement in cocaine self-administration, in particular via 5-HT_{1B} receptors. However, a role for serotonin in mediating the reinforcing effects of nicotine seems to be excluded by the lack of effects of several serotonergic drugs in affecting nicotine self-administration. These results are at variance with other preclinical observations showing motivational effects of 5-HT_3 antagonists, for example in the rat place preference paradigm. These results were not confirmed in human smokers.

In addition, when smokers were treated with antidepressant seroronin-uptake inhibitors such as fluoxetine, no therapeutic effects were shown. There is a general agreement that antidepressants are effective only in the subpopulation in which depression and smoking are co- morbidities. Moreover, data on the effects of chronic nicotine on the molecular expression patterns of serotonin-related enzymes or receptor in CNS are missing, leaving the picture rather incomplete.

Glutamate

Glutamate is probably the most common excitatory neurotransmitter of the brain, and glutamatergic neurons are integral part of the reward system. Several electrophysiological effects of nicotine are mediated by a direct effect on glutamatergic neurons. For example, in anesthetized rats, nicotine increases the firing of locus coeruleus dose-dependently via a glutamate-dependent mechanism. More interesting, nicotine stimulates presynaptic α7 nicotinic receptors within the VTA localized on glutamatergic afferents from the medial prefrontal cortex, producing an increase in glutamate concentrations that stimulates the NMDA receptors expressed by DA-containing neurons in the VTA. According to Svensson and collaborators, the resulting enhanced burst firing of VTA neurons would enhance DA release in the nerve terminal regions.

In this view, glutamate-containing neurons are instrumental for the building of sensitized DA functions that underlies the condition of nicotine dependence. Glutamate has been involved as one of the mediators of sensitization to nicotine. Co-administration of the NMDA receptor antagonist MK801 or D-CPPene during the nicotine pretreatment phase attenuates the development of locomotor and intra-accumbens DA release sensitisation, suggesting a potential role in determining nicotine dependence. The effects of glutamatergic drugs on nicotine self-administration have been rarely addressed. So far, there is no evidence of direct effects of systemically administered glutamate antagonists.

Opioid peptides

Chronic nicotine increases increases β-endorphin in the hypothalamus, and Met-enkephaline in the striatum and nucleus accumbens only 24 h after last nicotine administration, and returns to basal level 7 d afterward. This time-dependency partially explains the lack of effects reported in other studies. Increase of Pro-dynorphin following chronic nicotine was reported in nucleus accumbens. The role of opiates in nicotine dependence is not fully understood. Naloxone does not affect intravenous nicotine self-administration in rats, nor does it modify DA release in nucleus accumbens in rats chronically infused with nicotine. However, naloxone precipitates a withdrawal syndrome in rats chronically infused with nicotine, and reduces cigarette consumption in smokers, suggesting a motivational role as part of the central stress system.

Initial reports of craving reduction in smokers with naloxone were not confirmed in other studies with naloxone or naltrexone. In rats, Corrigal and collaborators reported that the δ-receptor agonist DAMGO, injected into the VTA, shows modest effects on intravenous nicotine self-administration. Overall, these data indicate a minor involvement of opiate peptides in the positive reinforcing aspects of nicotine dependence.

Gamma aminobutyric acid (GABA)

GABAergic neurotransmission is a recognized component of the reward pathway. Administration of a single dose of nicotine increases GABA release from the nerve terminals and synaptosomes of several brain structures, including the VTA-substantia nigra and hippocampus, via β2-containing nAChR. In VTA-substantia nigra, nicotine effects depend on viable presynaptic D1 neurotransmission, suggesting a presynaptic modulatory effect on the feedback loop from striatum-nucleus accumbens to VTA. Corrigal and collaborators showed that intra-VTA microinjections of baclofen, a $GABA_B$ receptor agonist, attenuates intravenous nicotine self-administration in rats, supporting a direct role of GABAergic neurotrasmission in the reinforcing properties of nicotine. The antimotivational role of $GABA_B$ receptor agonism is also supported by the baclofen-induced complete blockade of γ-hydroxybutyric acid self-administration in mice.

However, clinical data in 24-h-abstinent smokers administered 20 mg of baclofen showed no effects on number of cigarettes smoked or craving score during the following 3-h fee-smoking period. Mild sedative-like effects (increased "*relaxing*"), and changes in the sensory perception of smoked cigarettes (increased "*harsh*" and decrease "*liking*") were also reported, suggesting a role for facilitation of smoking cessation.

Recent evidence suggests that this role is shared by another pro-GABAergic drug, γ vinyl-GABA (Vigabatrin). Administration of 75-150 mg/kg of γ-vinyl-GABA to rats completely antagonizes the nicotine-induced DA release in nucleus accumbens and the development of nicotine- induced conditioned place preference. Presently, nicotine–GABA interactions are a topic for intense research, and a likely target for future antismoking therapy.

CONCLUDING REMARKS

The available data are insufficient to propose an acceptable model of the molecular and cellular substrates of the positive reinforcing properties of nicotine. Possible important information can be derived by a better knowledge of the genetic risk of smoking, in particular if the genetic differences are affecting the reinforcing properties of nicotine. The evidence so far accumulated does not support this hypothesis directly. For instance, significant association between two markers of the tryptophan hydroxylase gene (TPH C218A and C779A) studied in 780 genotyped subjects and the age of smoking initiation, but not nicotine dependence progression, was recently found. In another study, association between polymorphism of the dopamine β-hydroxylase gene and smoking cessation was also found. Both studies are suggestive of potential genetically determined differences in the reward system, but only indirectly, since they address clinical readouts that are not directly measuring reinforcing effects.

Hypothetically, a genetically determined mesolimbic DA reward system that can be easily sensitized by nicotine may predispose to smoking because of its enhanced reward-related signaling associated with drug-taking. This genetically determined defective mesolimbic functioning has been sometimes indicated as reward deficiency syndrome. Genetic differences are also known to be important in determining addictive drug effects in rodents, in particular nicotine, including strain differences in the capacity to acquire and maintain operant responding, in particular intravenous drug self-administration. In spite of these limitations, a working hypothesis is presented here with the aim of conservatively summarizing the essential features of the data discussed so far. Nicotine exerts its positive reinforcing effects by acting via the $\alpha4\beta2$ nAChR located on the somatodendritic membranes of the DA cells and possibly its sensitising effects via $\alpha7$ nAChR located on glutamate terminals.

Direct neuroadaptive effects of nicotine on a supposedly highly sensitive reward system, in particular the mesolimbic DA system, are believed to influence the stimulus-reward learning and to be the major player in sustaining nicotine self-administration. However, environmental cues previously positively associated with nicotine taking and, to a lesser extent, changes of the internal affective state toward anxiety-depression, are thought to participate in triggering drug-taking behavior. These dispositional states are suggested to be produced by some of the neural networks related to the reward system, as indicated by the diffuse changes of gene expression patterns in the terminal fields of the mesocorticolimbic DA system, in particular the nucleus accumbens and prefrontal cortex. Further research matching molecular investigation to behavioral paradigms, selectively addressing each phenomenological component of the behavioral processes underling nicotine dependence and its genotypic characterization, will provide the information necessary to describe the relevant molecular players in these important disorders.

12

Vaccination Programme

The misuse of drugs, both legal and illegal, presents society with a number of significant problems. Health impact on addicts include chronic conditions such as cirrhosis, lung cancer and HIV, and acute or short-term effects such as overdose. Drug misuse is also linked to risk taking and, in many cases, to criminal and antisocial behaviour. The misuse of potentially addictive drugs and the resulting harms have a complex impact on the relationship between the individual and society, raising issues of the addict's behaviour and their capacities to act responsibly. However, contemporary debates have focused on the biological and neurological alteration of the brain associated with addiction, as this is the area where we are gaining most new knowledge, rather than on the social sciences. While there are undoubtedly good prospects for significant benefits to individuals and to society from advances in understanding and technologies in this area, there is also some uncertainty about the moral implications of these developments.

Many of the ethical questions concerning neuroscience and addiction research have been given renewed emphasis by the negative connotations, including stigmatisation and discrimination, associated with drug misuse. Here we do not comment on research ethics per se, since the literature is well established in this area. In neuroscience and addiction research, special concern should be devoted to confidentiality and privacy, because of the nature of illicit drug misuse and the circumstances and behaviour sometimes associated with addiction, the capacity for informed consent in addicts and their voluntary participation in therapeutic and non-therapeutic research, the constitution of the research team, especially with regard to members' competency to deal with comorbid conditions and criminal activity, and incentives to participate in research (e.g., socio-economic constraints); and possible coercion to participate (e.g., from attending medical professionals). In this report, we concentrate on some major avenues of research that are of interest because of their potentially imminent and significant application to human subjects.

We are concerned with unpacking such developments within an ethical framework broadly conceived around '*human rights*', and which essentially protects one's autonomy – the capacity to lead one's life according to one's own values – including emergent principles such as competency, consent, privacy and confidentiality. Autonomy is not to be understood as mere '*independence*' from others, nor as the 'freedom to do whatever one wants'. Rather, it is the capacity to make decisions for oneself, without undue influence, interference or coercion from others, and without harming or imposing excessively on others. Respect for the autonomy of a given person entails the requirement to respect the autonomy of others; all human rights jurisprudence is consistent on this issue.

In the body of this chapter, we discuss various situations where the respect for autonomy of a given person and the respect for the autonomy of others requires balancing. While our emphasis is on the autonomy-undermining effects of addiction, and therefore may suggest an overly paternalistic attitude to protecting people from drug misuse, ie doing something in someone's best interests without concerning oneself too much with whether the person wants this or not, certain psychoactive substances may have therapeutic uses and other benefits. For example, drugs that improve alertness or maximise concentration may allow people to make more choices concerning their potential aspirations. Therefore, while we focus on the addictive component of drug misuse, there may develop socially approved reasons why drugs may legitimately be used if the risk of addiction is reduced (so the benefits of the drug outweigh the risk of addiction) or removed entirely. We do not look at individual aspects of why one tries or starts misusing potentially addictive drugs, the reasons why an addict embarks on a particular type of behaviour, nor how society should deal with addicts (implicating the so-called '*moral*' and '*medical*' models of addiction). We are presuming that the goals of the new developments are to either treat addiction (i.e., addiction is a disease); to limit the effects of a drug on the individual (and therefore potentially '*cure*' addiction); or to prevent or reduce the risks of developing an addiction.

MISUSE OF DRUGS

At present, treatments aimed at stopping the craving felt by addicts or to prevent relapse can be modestly effective. Their success can improve significantly when used alongside behavioural and counselling services. These treatments are designed to reduce the intake of the drug, prevent craving in periods of abstinence, and to reduce relapse. Research has concentrated on developing drugs with fewer adverse side-effects, and genotyping may better match patients to existing pharmacological treatments for addiction. Recent attention has been directed to developing a means of inactivating the drug before it can gain access to the central nervous system.

These vaccines are designed to block or significantly reduce the effects of certain drugs on the brain by inducing antibody formation, producing an effect similar to vaccination against infectious disease. Removing the 'high' may reduce the craving for the drug. Furthermore, vaccination may stop a person from becoming addicted in the first place by removing the incentive to take the drug. Vaccination poses a range of ethical issues in its own right, and is also a good model for thinking about the ethics of treating and preventing addiction by other means. There have been recent advances in the potential vaccination of individuals from common drugs of misuse such as cocaine, alcohol, tobacco and nicotine, marijuana, and methamphetamine. Although it is presently not known whether these treatments will have any side-effects, they may have better rates of compliance than existing oral drugs. They would be administered less often than conventional drugs, perhaps every six months, and could not be *'switched off'* during their relatively prolonged effectiveness.

This creates the problem that should anything go wrong with the treatment, it may be difficult to reverse the vaccine and also that such treatments would only affect the neurobiological effects of the drug. They would leave untouched any underlying behavioural pathology or dependency. So vaccination would have to be accompanied by counselling and support. A vaccine would have to be efficient and safe. It should also be selectively discriminatory, so that it would not unintentionally immunise against beneficial drug uses such as pain relief. If these criteria were met, the ethically significant questions would concern the balance between individual freedom and public welfare. First, however, issues of privacy and confidentiality must be raised.

Consenting Addict

There are few concerns about vaccinating a genuinely *consenting addict* to alter the progress of their addiction or prevent relapse after discontinued drug use. But questions arise about confidentiality and privacy. For example, if the cocaine vaccine or its antibodies were identifiable in tests, a positive result would betray a status open to stigmatisation. The risk is that non-addicts, vaccinated because of a predisposition to addiction, as well as ex-addicts and actual addicts, might all be grouped together because of the presence of the vaccine antibodies. The fear of stigmatisation should be weighed against the potential benefits from treatment. As we shall see, this may be a question of a choice to have treatment or forced treatment to protect others. Confidentiality regarding the presence of antibodies that may rightly or wrongly associate the status of a recovered addict with that of an actual addict is essential, as is self-determination in allowing the individual to decide who can have access to this information.

Need of Vaccination

A key distinction would be between those who are voluntarily vaccinated and those who are required to be vaccinated. Vaccinations have traditionally been directed at a whole population, but there may often be reasons to target individual communities that have a high propensity to some disease. In the case of psychoactive substance use, vaccination could be used to prevent and treat the health effects of drug misuse, and also to control the use of illegal drugs, since the costs of drug misuse for society are high. Vaccination could be used:

1. As an aid to coping with drug withdrawal and possible relapse,
2. To vaccinate persistent drug users,
3. To vaccinate potential users.

Vaccination programme at nation level

Population-wide vaccination programmes have been largely successful in decreasing the incidence of certain diseases, and have removed some almost entirely. Could a nationwide drug vaccination programme be implemented to reduce the incidence of drug addiction? There are a number of arguments against such a measure. First, the vast majority of people do not misuse drugs, nor will they ever be exposed to them. Even among those who take a potentially addictive drug, addiction is not certain. It is therefore questionable whether mass vaccination would be justified in light of the relative risks, and compared with the cost and potential harms of repeated vaccination. It is also possible that certain beneficial aspects of medicinal drug use, such as pain control, which uses the active components of some drugs of misuse, would become ineffective in those vaccinated. Some potentially harmful drugs, such as alcohol, also serve a socially beneficial role, and although regulated (to promote responsible use), their (mis)use is not prohibited in adults, though some resulting behaviour may be criminalised.

It would be considered paternalistic and intrusive to vaccinate against their use. Would other drugs be viewed differently? In determining an appropriate balance between individual rights and social stability, a difficult choice may need to be made between restricting or removing the right to self-determination from the vast number of individuals not at risk in order to prevent the misuse of drugs by a minority. We all have the right to live our lives relatively free from the interference, however well-meaning, of others and of the state. But equally, we do not have the right to inflict harm on others. The difficult balance in legislation and public policy between allowing and encouraging individual freedom and restricting that freedom to protect others will continue to be controversial – it is the very stuff of modern politics. In this area, the issues are complex because our interests in public safety and the prevention and detection and punishment of crimes against the person and against property seem to conflict with our interests in privacy and self-determination. This is a particularly acute problem since the interventions under discussion here are medical, and social trends since the 1950s have encouraged a very rights-based approach to medical treatment.

Many doctors, for instance, will be loath to use their skills in the service of what they might see as a non-medical issue. Equally, many doctors will see the preventive approach as appropriate, particularly given their experience of treating the consequences of drug overdose or alcohol- related violence and injury. And this debate is likely to be repeated throughout society, not just within the medical profession. One outcome of this sort of debate may be a swing back away from the privileging of individual autonomy or personal rights in medical ethics as much as in criminal justice. However, it is impossible to discern how likely this is – or how far it could go. Insofar as vaccination forms part of medical practice, vaccination will be constrained, in competent persons at least, by the requirement that the vaccinee consent to vaccination. If vaccination is court-mandated, with the doctor carrying out the vaccination at the request of the court, such interventions will still normally require informed consent (and would normally be optional, with other more traditional punishments or rehabilitative measures as the alternatives).

A challenging policy issue would concern the vaccination of mentally non-competent individuals; in such cases much more would lie on the judgement of the individual doctor that this vaccination was in the medical best interests of the vaccinee as well as being court-approved. If it is not justifiable to vaccinate the entire nation, it could be feasible, and more affordable, to target certain groups within it. While there may be legitimate reasons to screen a certain population for a realistic link to addiction (based on ethnic, social or comorbid grounds), groups could be selected on more dubious arguments. What groups are candidates for vaccination? It is evident that areas of social deprivation or of high 'visible minorities' have a higher prevalence of drug use. Likewise, their inhabitants are disproportionately represented within the criminal justice system in connection with drug-related crime. There may be certain groups that would benefit from directed vaccinations based on consensual and willing participation. But we must be very wary of the use of (mis)information to further alienate, stigmatise or discriminate against vulnerable groups, and of the risk of causing further social problems by such actions.

Voluntary capacity

Even with a competent adult volunteering to participate in a vaccination programme, we must consider issues of confidentiality and privacy. Entering a drugs programme in a voluntary capacity could leave the participant open to stigmatisation and discrimination. Nor are the benefits and disbenefits of volunteering for treatment as clear cut as one may expect. The individual has to consider taking up a '*patient's role*' and being seen to be part of the world of drug use. The vaccination will probably need periodic renewal and it may perpetuate withdrawal, adding to the relative unattractiveness of vaccination compared with taking the drug itself. Moreover, addicts are often addicted to, or at least using, more than one drug, and there is often comorbidity between drug misuse and other mental conditions, such as depression.

Vaccination would not be a treatment for addiction itself, and would need to be used alongside other interventions, for example, counselling, for prolonged abstinence from drug misuse. This is demonstrated by the relatively low success rate of existing drugs used in addiction therapy, which at most have only modest outcomes in longterm abstinence. It is also possible that restrictions on the health service provision of such treatments would apply for cost reasons. They could include measuring compliance and willingness, and possibly requiring further rehabilitative measures. Finally, making vaccination available to episodic drug users raises the concern that they could use the drug and experience the 'high' in the knowledge that there may be a fast track out of addiction, at least for those able to afford it. Using a vaccine that palliates the addictive effects of drugs may allow people to use such substances for pleasure and not risk the harms of addiction, or may create socially acceptable uses for previously addictive drugs.

In the case of pleasure drugs, the possible social changes that this development may promote are unknown. Speculatively, it may lead to increased use of drugs for pleasure on the perceived availability of a '*quick fix*'. This raises the question of whether policies should continue to counter such use when the benefits outweigh the harms. The outcome may either be that more people come to perceive present paternalistic measures as unwarranted, or that spiralling drug use per se would lead to stricter legislative measures. On a positive note, the use of drugs that are no longer addictive may lead to the expansion of or changes in the range of so ially approved purposes. The potential use of such drugs in employment, for example, in maintaining alertness in driving, or commercial flight, may be significant for their social acceptance.

Medical intervention

Perhaps the most controversial questions raised by the vaccination of individual populations relate to the use of force. Since this is a medical intervention, the spectre of people being treated against their wishes is raised. Does addiction to an illegal drug give the state the right to impose vaccination? It could be justified as providing an alternative to sanctions such as prison. The underlying assumptions are that treating an offender's drug dependence will reduce the likelihood of their reoffending, and so reduce the associated criminal activity; and that coerced treatments, based on the disease model of addiction, are in the best interests of the addict. The former assumption relates to the public interest in controlling and reducing crime; the latter relates to paternalistically overriding the choices of the addict in their best interests (whatever the addict him or herself may think about them!). There are two types of addict to mention here: those who, in addition to breaking the law by taking an illegal drug, are involved in activities that are antisocial or illegal; and those who take the illegal drug but are not breaking any further laws – the crime committed is the taking of the drug. Should these two groups be treated differently should vaccination ever become possible?

It is not unknown for methadone or other remedial measures, such as attendance at a support group, to be attached to a court judgement or housing-support programme. Might similar constraints be put on offenders who are also suffering from mental health problems, or who are claiming social security payments for unemployment or are homeless and seeking housing? Such choices may help or push offenders towards a more acceptable way of life, perhaps involving stable employment or housing for 'clean' individuals. However, there is evidence that coerced or compulsory drug treatment doesn't always work.

If a safe vaccine were available, it could either be offered in addition to a prison sentence or fines, or as an alternative to more severe punishment, for example, as a requirement of probation as opposed to imprisonment. But both approaches have the problem that they fail to meet the requirements of informed consent. A choice between imprisonment or a medical intervention is clearly coercive. Would the addict make the same choice if he or she were not in this situation? And what are the ethical concerns of the medical profession's involvement in such measures?

On the other hand, from the point of view of criminal justice, coercion is an essential element of punishment, and the coercive element of this offer, while questionable from the point of view of medical ethics is perhaps unproblematic from the point of view of penal ethics. And indeed many convicted offenders and many victims of crime may prefer the vaccination approach to imprisonment as a way of dealing with the causes of crime and the personal problems frequently involved at the root of criminal behaviour. If a criminal justice system is seeking to deter or prevent serious crime, it is unlikely that any addict would be 'offered' the vaccine as an alternative to the sentencing option, because knowing that there is an alternative to prison would not deter drug misuse or drug-related crime.

But a retributive theory of punishment, claiming that any punishment is justified because the offender has committed a wrong act and that the degree of punishment should be in proportion to the crime, may point to a different answer. In this case, we may offer the addict who commits no further crimes the option of vaccination, along with other sanctions such as community service, since the wrong act is taking the drug, and this may therefore be the most appropriate level of punishment. This would also be consistent with a 'reformist' theory of criminal justice, which sees its role as the rehabilitation of offenders. Additionally, imprisonment may reinforce addictive behaviour, cause the addict to switch to more dangerous drug habits, and expose them to further harms such as HIV and hepatitis. Control of drugs within prisons is a recognised problem worldwide.

Uncontrollable use of legal drugs

In the case of the misuse of drugs that are regulated but not prohibited, such as alcohol and tobacco, one must be careful that the status of addict does not become a crime. Of course, the uncontrollable use of legal drugs leads to criminal activities – such as antisocial behaviour, theft, violence and domestic violence – but this doesn't necessary follow. In the case of legal drugs and where no additional crimes have been committed, the option of vaccination must be voluntary, since the addict is not breaking any law. Such a vaccination would normally be provided as part of a wider treatment plan. This may not be the case where criminal convictions follow from problem drug use short of addiction (for example, driving while under the influence of a psychoactive drug). In such situations, we can foresee vaccination being offered as part of the rehabilitation or punishment of the offender.

During pregnancy

The main issue here is whether a case could ever be made to compel a pregnant mother to be vaccinated in her own interests and those of the child. The concern is that drug misuse may harm the foetus or continuation of the pregnancy, or that the child will be born with an addiction or addiction-related condition. Vaccination if offered as a treatment, and taken by choice, by an addicted pregnant mother could protect her child; but what would be the possible outcome if treatment was forced, and on what grounds may this be ethical, if at all? One side of the argument holds that, because the woman has decided that she wants to carry the pregnancy to term, she has responsibilities for the health of the foetus. This may be matched by the state's perceived duty to protect all human life – including the foetus – from intentional harms.

The other side of the argument holds, in line with the present understanding in UK law, that the foetus has no legal status. This suggests that neither the state nor the pregnant woman has any moral responsibilities towards the foetus except, in the case of the woman, those that she wishes to take on. Somewhere between these competing claims is the argument that compelling the mother could only lead to further difficulties in her pregnancy, such as adverse effects on the foetus, the course of the pregnancy, and for the mother's health. Coercive programmes may persuade pregnant addicts to avoid or not present themselves for prenatal care due to unwanted testing and treatment. The key issue is whether society or the state have any accountability to promote responsible pregnancies that protect the mother, the foetus, or both, and, if it does, how far these measures can go. Some would argue that as society has to deal with the consequences of births of babies affected by maternal drug and alcohol use, it has some right to intervene to limit harm to the unborn caused in this way.

Society already does intervene in a variety of ways to promote child health in the womb – through encouraging the taking of folic acid supplements and the offer of antenatal classes and antenatal screening, for instance. On the other hand, society has tended to reject approaches that would penalise mothers for actions during pregnancy which might harm their unborn child – from the decriminalisation of abortion to the rejection of so-called 'wrongful life' suits by the courts. The spectre of eugenics and the prospect of children suing their parents for negligence are considered by many to be stern warnings against taking such a coercive or punitive approach to maternal behaviour during pregnancy.

Drug protection for children

A further significant question is whether a strategy for immunising individuals before they try a drug should include children. As Nutt and Lingford-Hughes (2004) point out (drawing on the paradigm of child vaccination), 'cocaine dependence is more damaging to many individuals than measles'. The term '*vaccine*' may also be unfortunate because it could raise unrealistic expectations among parents of complete drug protection ('*immunity*') for their children. The question is not simple. It needs to consider both enrolment into a vaccination programme as a targeted population and as an individual. In the case of the former, one might look at targeting all minors as an 'at-risk' group, or use socioeconomic indicators that suggest that certain groups of minors are susceptible to drug addiction.

In the case of individual vaccinations, the significant question could be whether it would be permissible for parents to have their children vaccinated in their children's best interests. In all these cases, the vaccination would be pre-emptive and parental concerns that their children could be exposed to drugs in the future could fuel the programme. Can a case be made to immunise children against a potential future of drug use? If the vaccine requires repeat injections (as seems likely), one risks the possibility of conditioning children to injecting substances. Furthermore, one has to ask when such a programme would begin, what it would cost, and the competence of the person being treated. There may be a point when the child wishes to discontinue the treatment. Would it still be worthwhile to force the adolescent to accept the injections, with consequences to medical and parental relationships? Perhaps the greatest concern would be the intrusion into the life of a child – not just at the time of the vaccination but also into his or her future behaviour.

Hasman and Holm (2004) argue that drug misuse is not an infection and therefore the discomfort and risk of harm to the child is not justified. The vaccination is a violation of the child's ability to assess the benefits and harms for himself or herself. This denial of important decision-making capacities for the child, they argue, limits the child's right to an open future. They conclude that whether such a decision does limit the future options of the child will be contestable. However, as Hall (2002) has pointed out, objections to the 'open future' argument depends on a gross overestimation of the efficiency of childhood vaccination against addiction. For example, an adult who has been vaccinated as a child but who was determined to become addicted still could. It would simply be more expensive to do so.

Mind Altering Effect

Vaccinations could potentially affect the individual in two ways. First, they may alter the individual's 'mind' itself. Drug vaccinations, by their very nature, are '*mind-altering*' in that they change the functioning of the brain. This is not the same as a vaccine that produces antibodies to ward off infection. Instead, it has been described as the use of pharmacotherapy to expand the '... drug war battlefield ... to a new terrain directly *inside* the bodies and brains of drug users'. This, according to the CCLE, is a 'threat to cognitive liberty', since one may be able to eliminate the desire to use illegal drugs, and thus the conscious actions of addicts or potential addicts. One may, however, question these claims of '*mind-altering*' effects.

The vaccines are meant to change biological function transiently and affect the brain indirectly by attenuating the effects of psychoactive drugs. Indeed, it is intended that such vaccines will operate outside the brain, targeting and mopping up the drug before it enters the central nervous system, therefore not affecting the brain directly in the way that neuropharmaceutical drugs do. The concern for the CCLE is that vaccines may be used to control the conscious actions of individuals even before they themselves act on them. Furthermore, vaccinations for drug use could be applied further, by vaccinations against other lifestyle choices (enforcing social conformity) and addictive behaviour such as gambling. One must look at the potential responses of the actual, reformed or potential addict to the presence of the vaccine. Will those on vaccine programmes try to overcome the vaccination, thus risking overdose or harm from increased doses of the drug and its contaminants? Or will they switch to other drugs?

There are also risks, demonstrated in the completion of methadone maintenance treatment programmes, that tolerance can be reduced. So if the patient relapses, they overdose because they have no tolerance to the heroin at the levels they used to take, and can't judge the amount they need. However, it is also important to consider that, when taken by choice, vaccines could provide a valuable addition to the available treatments for those dependent or addicted to any given drug which is causing medical and social harm.

PHARMACOGENETIC STUDIES

Pharmacogenetics, the study of the biological response and effects of different drugs on the individual, and pharmacogenomics, the examination of the genome and gene-gene interactions with regard to drugs, offer the possibility of '*genomic medicine*', tailoring drugs to the individual and foreseeing the likely effects that they may have - including a potential susceptibility to addiction. Pharmacogenetics and pharmaco-genomics have the potential to allow us to foresee and avoid side-effects and tailor treatment regimes more accurately to patients' needs. '*Predictive genomic medicine*' may give us insights into an individual's likelihood of addiction or of developing a drug-related disease, allowing individual choices and societal policies based on genetic predispositions. It is also possible that advances in genomics will drive down the current spending of the health service on treatment, cutting down on expensive treatment in favour of detection and avoidance programmes.

Specific to drug addiction, pharmacogenomics may identify genetic profiles with greater or lesser susceptibility to the action of the drug, while behavioural genomics could identify genetic profiles with greater or lesser susceptibility to addictive behaviour. Therefore, someone could be highly susceptible to becoming drunk on alcohol, but not likely to become addicted to alcohol; or less than usually susceptible to the effects of alcohol, but likely to develop an addiction to it. All of these possibilities have advantages and disadvantages from an ethical point of view. It is important that the ethical issues of genetic research on addictive disorders are considered, because:

1. The predictive nature of genetic information has the potential to affect people's lives adversely;
2. Genetic information carries implications not just for individuals, but also for families, communities and populations;
3. Genetic information has the potential to be used to stigmatise and victimise, but also to be used for treatment and improving individuals' safety and welfare.

Questions that have arisen in the context of addiction and genetic information are:

1. Should an individual be responsible for their actions if they know of a genetic predisposition to addiction or drug-related disease?
2. How would a family (or community) respond to a member with a genetic disposition to addiction (raising issues of the misattribution of blame and guilt, and the subsequent effects, to relationships)?
3. Should parents be responsible in using measures to detect 'predispositions' to addiction and to prevent it by whatever means in their children?

At present we know very little about the answers to these questions, and part of the reason for this is that society at large is only now beginning to handle genetic information about risk, disease and behaviour on a regular basis. What we can infer on the basis of existing knowledge about the social response to genetic information may be relatively unreliable as society changes in response to the wealth of new genetic knowledge we are garnering. However, it is plausible that the available repertoire of responses to addiction and lay theories of what it is and how it is acquired or treated, being based on long-standing lay theories of free-will and responsibility, will shape how genetic explanations are understood, rather than being radically overhauled by such new kinds of explanation.

Advanced Hereditary Information

Advances in understanding the genetic basis of disease through the establishment of genetic databases and the mapping of the human genome make genomic medicine (ie medicine organised around the systematic use of genetic information in diagnosis, prognosis and treatment) a realistic but probably distant proposition. In the case of single-gene conditions, where most progress in gene detection has occurred, few effective interventions are available and, as a whole, they are relatively rare in the population. The more common multifactoral diseases are dependent on complex gene-environment relationships and are not open to the same simple analysis. Regardless of these biological hurdles, genomic medicine could potentially be used to inform or regulate an individual's use of certain drugs. This possibility presents us with questions of privacy, confidentiality, discrimination, and the misuse and usefulness of probabilities.

For instance, how will knowing genetic susceptibility to addiction or drug-related disease change the behaviour of individuals? How will drug and drug- related disease screening programmes affect individuals and groups? And how should genetic information be given to 'at risk' individuals? There is strong evidence to suggest that genetic factors play a role in the effects of drugs on the person, such as their absorption, metabolism and excretion, and these will sometimes correspond to a susceptibility to a drug-related disease. Advances in pharmacogenetics may allow researchers to identify genes and gene products that are implicated in addiction, tolerance, sensitisation, dependence, craving and relapse. From this information, pharmacogenomic research may identify genes that represent a higher probability of addiction and indicate how large a dose, or doses, of a drug may cause addiction in the first place. The genetic basis of addiction reflects genetic traits that influence one's behaviour with regard to drug taking and any resultant addiction.

The effect will depend on the genotype and environmental cues and stresses, as well as the specific effects of any drug on an individual. These two factors are linked in addiction. So, for example, an addiction may need to be 'triggered' in any individual, but the sensitivity of the trigger may depend on the type of drug taken, how much and how often, as well as the user's genetic basis. A certain gene or group of genes may reveal a disposition to addiction to a given substance or behaviour, but perhaps not to others, and its effect may be different for different individuals.

It has been postulated that the mapping of the human genome and the use of genetic technology, such as gene chips – or DNA array technology – to allow the rapid identification of thousands of gene products will facilitate the identification of potential addicts. This emphasis on '*gene chasing*', while potentially offering more effective and lower risk therapies, also raises concerns, such as a potential underemphasis on the social mechanisms of exposure to drugs. One of the dangers is in the '*folk*' understanding of genetic traits, which emphasises a deterministic understanding of genotype and phenotype. We have to acknowledge the complex interaction of genetics, behaviour (and therefore life choices) and environment on an individual's phenotype and genotype. The already complex interaction between drug addiction, genes and the environment may be further complicated by the genetic factors involved in diseases themselves. Certain diseases may be caused, catalysed or compounded by the interaction of certain drugs with '*disease*' genes. Furthermore, the likelihood of taking a drug, the amount taken and the duration of use may depend on other genetic factors, including those related to addiction.

It may become possible to screen the population to identify individuals who may be predisposed to developing specific diseases, some of which may be related to drug taking. Such information may be used to benefit the individual by developing safer, better-targeted and more efficient treatments for illness and other medical conditions. It may also be used to inform people of their genetic susceptibility to disease, which may motivate them to change their behaviour. But this evidence must be considered alongside the use of a specific drug, how and where it is taken, how it affects that individual, and a myriad of other environmental cues. Finally, the identification of 'genes for . . .' has been partly responsible for the creation of medical conditions out of what had previously been considered to be the normal difficulties an individual might face. So-called '*medicalisation*' has been argued to have taken place in the characterisation of, for example, *general anxiety disorder* (GAD), social anxiety disorder and the gendered condition of 'female sexual dysfunction' (in response to female sexual anxieties and the move to promote impotence drugs). Behaviour currently viewed as normal such as shyness (or GAD) may, in the future, be reclassified as pathological and be investigated for medical intervention.

The concern here is that while people who are concerned about these traits or behaviours will be encouraged to see them as medical problems rather than as something to do with their own personal attitudes and behaviours, or to see something which may from some points of view be valuable or desirable as problematic and requiring change. That these shifts in perception may be driven principally by market-seeking commercial behaviour, rather than principally by patients' or consumers' needs, has been a consistent worry in modern societies. It is surprisingly difficult to separate out what is 'genuine' and what is 'invented' in discussing these kinds of need or desire, however. The outcome may be that the perspective that '*genes*' cause disease will lead to pressure to purchase remedies for these conditions and that it becomes the individual's responsibility to take measures to avoid them, rendering it unnecessary for the community to address them. Yet, on the other hand, the relationship between biological theories, commercial interests and social wants is complex and difficult to predict or disentangle.

There is a legitimate public and commercial interest in trying to understand and provide remedies for conditions that are debilitating or undermining for those who undergo them. We might question whether shyness is really a medical condition; but that is not to dispute that shyness is not something which shy people find deeply upsetting or disempowering. One consequence of the '*medicalisation*' of shyness may be the growth of a market parallel to that of medicine which frankly and openly accepts '*lifestyle*' conditions as its central concern. What would be controversial would be the boundary between lifestyle and healthcare interventions, especially given public funding of the latter but not the former. Genetic factors in addiction and disease prevalence raise the opportunity for research to be misused or misunderstood. People may pay attention to genetic risk factors when environmental factors are far more significant, or far easier to manipulate, or both. For instance, if there was a genetic risk factor for smoking-related lung cancer (or addiction to nicotine), some might argue that this means that people without these risk factors can smoke safely (or nonaddictively).

Yet the baseline risk to people of regular tobacco smoking is very high, and these genetic risk factors merely make it worse. A further risk inherent in genetics and addiction studies is that the interpretation of research may compound influences one way or another. The identification of susceptibility genes for addiction or addiction-related diseases would tempt individuals who lacked these genes to misuse drugs 'safely', while those with the genes might be dissuaded from trying drugs. But there are a number of problems with such a simple interpretation. It fails to recognise that the effects of drugs vary from person to person. Individuals may vary in their sensitivity to the active ingredients' effects, in their tolerance of them and in the severity of their withdrawal. There are benefits from genetic studies, however, not least that individuals at risk of becoming dependent or who have a higher susceptibility to a certain disease can be identified and warned and appropriate healthcare measures taken, including screening for the disease. Side-effects can be avoided, preventive steps by individuals and society can be better undertaken and understood, and potentially a more sophisticated and perhaps more tolerant approach taken to the care and support of people at risk or in the grip of destructive behaviours.

Individual's Behaviour

How will genetic information alter an individual's behaviour? With Huntington's disease, the late onset of the condition can be preceded by prolonged periods of waiting, putting potential carriers of the gene in a situation of uncertainty about their own future. Among those who chose to be tested and were positive for the Huntington's gene, there was a reported higher rate of suicide as a result of a fatalistic response to the onset of the disease. Screening alongside genetic counselling has been shown to reduce this. The ethical concern is whether those affected individuals had a right to know, or not know, future probabilities, so that they could make, or not make, important decisions on the basis of the best available knowledge (which is relevant in this case with regard to passing the gene on to their children); or whether they have a responsibility to know the results of such tests so that they can make conscientious decisions. Yet these risks may not materialise; and we should also consider that in many cases people will choose to avoid passing on genetic '*conditions*' such as shyness, addictions, or other behavioural traits to potential children. This may be achieved by choosing not to have genetically related children or screening gametes to be used in IVF, or through screening or terminating existing embryos or foetuses.

This may not be because the conditions necessarily create problems for society, but because they create problems for individuals, either because of uncertainty in their own actions or of how they feel society sees these actions. This leads us onto the potential risk of measures being imposed on individuals. A positive result to a test for a potential predisposition could validate forced measures by the state. We may also see the '*treatment*' of previously normal behaviour through medical interventions, increasing the anxieties of individuals who fail to measure up to this new '*normalisation*', and diverting funds and support from improvements in environmental and socioeconomic circumstances. State intervention is not necessarily bad - it is bad only where the intervention is ineffective, or bad value for money, or unjust, or unaccountable.

The role of the state in using coercive treatment or prevention measures needs to be controlled by a robust and functioning democracy, and by adherence to human rights norms. Sometimes this might involve the imposition of coercive measures in the public interest. More often than not, coercion will not be necessary, or will be outweighed by other factors. With regard to drug use and addiction, learning of a genetic risk factor for addiction may create a fatalistic attitude to drug use, but, unlike in the previous example, these predispositions are multifactoral and, at most, provide only an assessment of how likely a person is to develop an addiction or how difficult it will be to overcome it. Nonetheless, for some people, the idea that someone has a genetic risk factor for addiction may be interpreted as meaning that this person is completely unable to control their behaviour and so should have his or her rights strictly limited on the grounds of danger to self or others. This would be a dramatic overreaction to information which is at most likely to give information about relative risks.

Yet this does underscore how difficult it is to think about the relationship between physical determinism (everything I do is caused by my physical make-up) and free will. Genetics and neuroscience may add little to this debate other than a new idiom for talking about it. In the context of drug addiction, this debate may have two outcomes: individuals will not try to stop taking the drug, or they may turn to other forms of help. What needs to be demonstrated and understood is that genetic information may allow an individual to decide that certain activities may put them at risk, and that changing their behaviour may reduce this risk - and not that there is no point in changing behaviour. Such use of genetic information is not new. Family histories have been used for some time to highlight risks, although there is evidence that genetic knowledge of a predisposition does not necessarily change behaviour.

While it is commonly asserted that a 'gene for ...' actually means a tip in the balance of probabilities, which may be very small if one considers environmental causes and life choices, a swing in the other direction - that one does not have the '*gene*' - may lead to increased reassurance that drug misuse will not harm one. The key message should be that drug misuse puts anyone who does it at risk. Genetic factors may merely alter the likelihood of side-effects, overdose, addictiveness or the direct psychoactive effects of the drug. For most drugs, however, the basic pattern of response will not vary much between specific individuals with specific variations in genetic make-up. Key to this debate will be how genetic information is presented and explained to individuals (including the task of effectively portraying data on '*risk*') and how this motivates individuals to change their lifestyles (including the effects of anxiety, demoralisation and fatalistic attitudes).

Genetic Drift

One use of genomic information is to confirm genetic differences, in the context of this report those relating to susceptibility to addiction, the processes of addiction, and diseases associated with addiction in different populations. The benefit of such an approach is to identify individuals or populations who may be more susceptible, and then to take measures to treat, monitor, or pre-emptively avoid the onset of such disorders. However, whether such information is useful is debatable. The ethical question is how do policy makers evaluate and decide on the weight placed on such data? As we have already emphasised, predispositions have both a genetic and an environmental aspect, and this corresponds to a risk. How society deals with this risk is dependent on incomplete data, or data which are highly complicated to interpret, and on cultural values. This is important because of the sensitivity of discussions about inequalities on the basis of perceived genetic '*evidence*' of ethnic differences.

One must avoid naïve inferences about genetic causation without evidence. Furthermore, the importance to be placed on such information must be carefully considered, as should the means of dissemination of such findings. Here, it is important to emphasise the qualitative variables that present risk factors for diseases, treatment responses and adverse effects of pharmacological agents, allowing tailored treatments. The highlighting of such beneficial applications of genomics will be important in the study of addiction.

Neuropharmacological Studies in Man

It may become possible to alter the genes of an individual in order to alter their addictive behaviour. Neuropharmacological interventions already exist for this purpose and are more easily accomplished than genetic interventions. Gene therapy interventions that alter genes in situ to change the mechanisms of addiction may be difficult to achieve, like other gene therapies, whereas it is comparatively easy to inject a pharmacotherapeutic drug to counter the effects of addiction, once the basis of addiction is known. One controversial means of achieving changes in genetic susceptibility to addiction that may be possible in the future would be at the preimplantation stage of embryogenesis. It would involve 'knocking out' or selecting some genes before the embryo developed. Apart from the complex arguments concerning embryo selection and termination, this raises ethical concerns similar to those mentioned in vaccinating children – that of a 'right to have an open future'.

Given that the use of this technology to avoid even serious genetic disease is socially controversial at present, it may well not be brought far into development. Yet it is perfectly possible that society would take a more actively welcoming approach to the use of *preimplantation genetic diagnosis* (PGD) to prevent the birth of babies with '*socially harmful*' genes. The history of eugenics suggests that social support for this kind of application of genetics can be widespread. Equally, it is arguable that this gives us powerful reason to avoid the development or use of PGD technology for such purposes.

BRAIN IMAGING TECHNOLOGIES

Brain imaging technologies may be used to study the effects of drugs and the neurobiological consequences of chronic drug use. It may become possible to use imaging to identify former, current and potential addicts. Such screening has the potential added benefit of being more immediately accessible and less expensive than screening for drug antibodies. It doesn't necessarily raise any further ethical issues than those already raised in this report, although there may be concerns in other areas of neuroscience concerning medical intrusions into individual preferences, and the shaping of social and moral attitudes. The questions to focus on in the context of this report are whether an addict can competently enrol to take part in novel imaging research, since this is the most likely group to be targeted for research; and what the data will be used for, and by whom (bringing to the fore issues of privacy and confidentiality). For example, can characteristic patterns of brain activity in childhood and adolescence predict increased risk of addiction in later life?

We still have to ask whether such medical interventions will be used in the best interests of the patient. Will imaging be used to identify addicts against their will (similar to a lie detector test) and then used in coercive measures in the legal or employment fields (with questionable success) or for insurance purposes? And will it be used to identify those at risk of addiction and to justify non-consensual or intrusive measures, or to selectively abort foetuses that have certain brain signatures in utero? Notwithstanding all these ethical questions, it is still likely that neural imaging will provide further information about the mechanisms of drug action, the mechanisms of addiction, and the impact of long-term drug use, which will be useful both in the diagnosis of disease and in the development and targeting of treatment. Questions will continue to be asked about whether in the future it might be possible to use imaging to excuse or explain drug-related behaviours in the court system, but much of the information provided by brain scanning will be insufficiently sensitive to provide detailed explanations of particular behaviours by most people. The impact of these technologies is likely to be felt most strongly in basic science and research into treatment.

TARGETED NEUROCHEMICAL INTERVENTIONS

Future developments in psychopharmacology may allow specific psychological changes to be induced by *targeted neurochemical interventions* in the chemical events that underlie cognition and emotion. The effects would not normally be hereditary - unless they altered the germ cells in some way - and would more often be temporary, or at least reversible (although this is an aspect that must be resolved). However, these drugs may alter what the human species can achieve, by making them 'better than well', and it is this aspect that we are interested in here. There are great pressures to succeed in everyday life. Passing an exam or completing a job have financial and career consequences that can encourage the individual to enhance their chances of success. Furthermore, once success is achieved, there is the added concern for the user of returning to normalcy, which could be an incentive to continued drug use.

This dependency on a drug for improved performance may at least prolong drug use and increase the chance of an addiction developing. Success may involve encouraging states of mind, including concentration, anxiety and mood, and metabolic alteration such as stimulation to reduce tiredness and bradycardia to induce calmness. Further potential targets for pharmacological enhancement would include memory, executive function, appetite and libido. If these treatments work, there are many potential users, and a concern that this new breed of drugs may have abuse potential. Research into the mechanisms of addiction will give a better idea of how realistic this fear of creating new addictions is, and how far this fear should outweigh the likely benefits of new drugs. It is already apparent that drugs that are developed to treat a specific medical condition are being used for purposes for which they were not intended. Most, if not all, medicinal drugs have unwanted side-effects, and this sometimes includes addiction, but these are outweighed by the benefits of treatment.

When the same drugs are used by those who are not ill for the purpose of enhancement above 'normal', this balance is not evident. The use of such drugs for this purpose is not necessarily novel. But future developments may mitigate or remove some of the adverse effects of drugs, such as addiction and tolerance, which could mean increased interest in taking such drugs, both for medical and non-medical reasons. One example is the use of drugs that are normally prescribed to treat *attention-deficit hyperactivity disorder* (ADHD), such as methylphenidate (Ritalin). It is evident that this drug is being used by young people to help them with exams. This has led to a continuing debate over the increase in the use of Ritalin in schools and misuse in university settings and in other adult groups.

There is circumstantial evidence of addiction in some users. Other drugs developed for the treatment of memory and cognition defects (such as Alzheimer's disease) may be used to improve cognitive abilities in individuals who are not significantly cognitively impaired. Critics of the wider use of such drugs argue that fuzzy diagnostic criteria are being used as an additional reason for '*treatment*'. But, again, one must question whether there is a significant gap between, for example, using memory-enhancing drugs to slow down normal deterioration to a '*disease*' state, and using it to enhance 'normal' capacities. Selective serotonin reuptake inhibitors, which are used to treat depression, are being used by people who are not depressed to alter their mood and personality. To continue the list, beta-blockers are used to remove nervousness in stressful situations; modafinil (Provigil), which is normally used for narcolepsy, can make users more alert; and Prozac and similar drugs can be used to make people less shy, less compulsive and more confident.

Neural Enhancement

The possible concerns from such use are two-fold. First, are there health harms that may be experienced by those who misuse such drugs? Second, are there adverse social impacts of the wider use of such drugs? Addiction may play a role in both of these concerns, since uncontrolled use may lead to problems with drugs similar to those that we already have experience with. Many drugs developed for medical use have side-effects, and these may be outweighed by the medical benefits of the drug. Are the benefits of enhanced cognitive ability to be measured in the same way? Furthermore, such neural enhancements modify a complex system, and may cause unforeseen longterm problems.

Enhancement in early life, for example, of memory, could exaggerate declining capacities later in life. Psychopharmacological enhancements to remove unwanted memories that may present serious debilitating effects in the individual, may be possible; but what would the consequences be for wanted memories or brain functioning? Or should we individually value memories, regardless of their content, as part of what we are? Many people would want to eliminate traumatic or embarrassing memories. It is far from clear how such technologies would be used, whether they would be beneficial or harmful, and what impact they would have on individuals and society. On the one hand, it may be useful to eliminate the memory of the horrors of war or sexual assault; on the other hand, it could be equally useful to eliminate the memory of some criminal activity so as to be able to lie more convincingly under oath (or brain scan!). The NCB has argued that we should be concerned about 'medicalising' human behaviour. As we saw earlier, the diagnosis of the risk of a type of disease related to drug misuse may lead to fatalistic behaviour; likewise, the novel reclassification of types of behaviour, such as '*shyness*', may lead to a desire to correct one's predicament by using enhancing drugs.

A great deal has been written on questions of what it is to be human and whether enhancement alters this in some way. Concerns have been expressed that such an approach diverts attention away from the value of life and its natural imperfections. It may alter our comprehension of our responsibilities and encourage us to think of our actions as chemical reactions in the brain, and regarding personal inadequacies or deviancy as imperfections to be fixed through chemical means. There are also questions of desert or merit. Is it justifiable to improve oneself through hard work, perhaps by exercise to improve blood flow to the brain, but not to achieve the same results by taking a drug? Some argue that efforts to improve the capacities of the human species do not raise significant problems, since industrialisation and other periods of intense progress have also had this effect. Furthermore, it is argued, we have for some time used numerous substances, such as alcohol, tobacco and caffeine, to alter mood and capacities. Enhancers could also instigate change for the better if properly managed.

People might achieve more by improving their concentration or cognition, and this may be compensated for in schools by changing the emphasis to understanding and problem solving over recall tests. Perhaps it could be argued that we should intervene in some aspects of use such as sports, where the use of enhancement drugs is termed '*cheating*'. But in this case, what is the real difference between sport and workplace or university competition? The drugs are designed in both cases to provide a competitive edge. Perhaps, then, we should control their use by some individuals, as we regulate the use of sports enhancement drugs. The most immediate candidates would be children, on the grounds that they may not fully understand the impact of drug misuse (not that we do, at least presently!), and that they have longer for the adverse effects to develop over time. Conversely, should we regulate the use of such drugs for some groups, for example, with those who work in situations where enhanced concentration may save lives, make action safer, or increase output?

If neurological enhancement can improve performance in certain activities, and this makes them safer (in pilots or long-distance drivers, for example), is there an ethical imperative for regulators to insist on their use? Perhaps tellingly, the new drugs of enhancement are already being seen in a negative way and there is resistance to their use in those who do not need them for some medical condition. The main concern seems to be that such drug use may expose individuals to unexpected consequences, and that such problems will alter the very nature of being – the mind.

Social Enhancement

A different kind of argument against enhancement is based on issues of justice and equality. This approach shifts our attention to the continuing inequalities in our society, and the potential for new forms of discrimination through unequal access to enhancements. One question for policy is whether we should prohibit the use of such drugs on the basis of the harms–benefit calculus, or concede that, while there may be adverse effects in such drug use, it is the individual's responsibility to take these drugs appropriately. There are others who could be forced into taking such drugs. Children may be a particular concern and we might need to question the wisdom of parents who think that giving such drugs to them is in their best interests. But it is possible that the haunting visions of the eugenics movements in the twentieth century drive such concern.

Why separate self-development through education (which itself may be biased towards different sections of the community) from the use of biotechnology? Another issue is a possible increase in social inequalities through enhancement technologies. It may be argued that such enhancement may be used to enable individuals to better deal with injustices and discomfort in the world, instead of doing something about them. Furthermore, public acceptance of enhancement may make it impossible to reject it, especially for, say, underperforming children or employees with dangerous jobs, or as a remedy for poor cognitive performance. These inequalities already exist but that is not a compelling reason to defend coercion. There are already medical interventions that are only universally available to those who can pay – for example, IVF treatment – and the only means to remove this inequality would be to make it universally available or prohibit it.

Herbal Drugs with Neurological Effects

A number of herbal drugs have been associated with neurological effects such as cognitive enhancement. Ginseng has been associated with memory improvement, while ginkgo biloba appears to improve the concentration of those who take it, although there are some questions concerning the experimental evidence. Research into the effects of these drugs is leading to developments in their medical use for treatment in dementia and other brain-compromising illnesses. These drugs can chemically alter the brain, and therefore behaviour, but they are not regulated in the same way as conventional pharmaceuticals. Despite the mixed evidence of their effectiveness, such drugs sell well, possibly because of folk belief about the greater safety of 'natural' neuroenhancements.

They are not formally regulated or monitored for detrimental effects, despite evidence that they have, among other undesirable effects, addictive elements such as craving, overdose and withdrawal, especially after long-term use. One means of controlling drug use is to allow potential users to make informed decisions once they are aware of the risks. But this option is not possible if such drugs are not included in standard pharmacological tests and regulation. This is evidently a problem both for potential users and the medical professional, where questions regarding adverse effects have yet to filter through.

Social Consequences

Drugs of various kinds have always been used in human societies for a wide variety of non-medical purposes – as stimulants, social disinhibitors, aids to relaxation, and so on. It would be inappropriate to think of the new drugs, or of new ways of using drugs, purely negatively. However, our focus here has been on problem drug use. While allowing for this, we note that one consequence of the introduction of a new drug or way of using drugs can be the evolution of new social forms of life around this new '*social technology*'. These new forms of life may be seen as positive by some and negative by others, who perhaps fear that these new forms of life are disruptive of existing ways of life or social values.

The current debate about enhancement technologies illustrates this tension very clearly, as does the debate about the non-medical merits and demerits of cannabis use (especially in the 1970s) and ecstasy (3,4- methylenedioxymeth-amphetamine (MDMA) use in the early 1990s. We have very little evidence or plausible predictive social science that would allow a prospective moral assessment of new drugs as social technologies, and we can't do much more than draw attention to this rather different perspective. We are cautiously confident, however, that core values such as autonomy, avoidance of drug- (or drug-use-) related harm to others, and avoidance of drug dependency will remain central to future assessments of drugs and their uses.

CONCLUDING REMARKS

Advances in treatment for addiction or mental health disorders are to be welcomed, whether through vaccinations or a greater understanding of genetics. However, the use of such advances must be balanced against certain key ethical questions. This report has highlighted the ethical questions that arise from prominent developments in neuroscience, drug use and addiction. Such developments include vaccination against addictive drugs, screening for genetic factors influencing response to drugs or disposition to addictive behaviour, brain imaging, and the use of drugs to enhance human performance. We have concentrated on areas of worry, risk and concern, since these (in reality or in perception) are likely to drive regulation and public debate, unless there is, for each technology, some overwhelming or exciting benefit which captures the public imagination. The discussions that follow this report should raise the question: Which (if any) of the technologies now being developed in neuroscience can legitimately be put into practice and with what ethical limitations?

The rapid development of new medical applications often raises specific anxieties, which in this report have been repeatedly highlighted as concerns for autonomy in individuals' capacities for self-determination (consent, confidentiality, and the like). This sometimes has the effect of over-emphasising the idea that progress inevitably (or uncontrollably) leads to negative consequences, often at the cost of individual rights and freedoms for the benefit of society. It is therefore important for analysis of the issues to demonstrate that science is not necessarily going to be used for malign or discriminatory purposes. Indeed, there are benefits for individual rights, such as improved autonomy through better treatments and management of conditions, as well as knowledge of one's genetic susceptibilities.

We have concentrated on drug-related (and more generally technology-related) harms, and specifically those related to the harms and risks of drug addiction, because our analysis is based on a liberal, human-rights-oriented framework of analysis, in which personal autonomy and avoidance of harm to (innocent) others are central. We have throughout had to generalise, either because experimental detail is lacking, or more often, because of a lack of space. This leaves a number of questions to be asked about specific drugs of misuse. First, we must assess the cost-effectiveness, and medical effectiveness of and, where necessary, compliance with, the interventions mentioned, in comparison with existing interventions and management strategies. Key to this ethical discussion are the different aspects of safety with regard to different neurochemical interventions. Second, the issue of coercive treatment (for example, of vaccination against addiction as an alternative to harsher penalties) could raise serious issues about the balance of individual autonomy and social harm or benefit. Even if such imposed treatment were effective, it might be considered a serious violation of more basic human rights.

The choice of such interventions by an individual fully informed of the risks and benefits is quite a different matter ethically, and offering a treatment should not be ruled out. It may be highly effective as a means of clinically managing addiction, since consent in light of situational constraints may still be valid. In most cases (except for clearly 'out of it' states or during periods of severe craving) an addict can still give real consent, ie weigh up the consequences of action or inaction. The difficulty would arise if an individual were forced to accept treatment, once addicted or to prevent addiction. It is unclear whether this would lead to demonstrable benefits for that person in the short term. There may be some longer-term benefits. For example, medical professionals may consider that a certain treatment would palliate an addiction, and this, in the long term, would be in the addict's interests. But would these long-term benefits justify measures to override autonomy from the perspective of best interests or public good? Under the present principles of medical ethics, the answer would probably be 'no', since the harms to self-determination would be considerable.

Similarly, violating individual liberty rights for long-term social goals would rarely be supported by current norms of medical ethics. But medical ethics is only one perspective here, as we have noted above: public interests in public safety and criminal justice have also to be considered, although these are, if anything, even more controversial and open to political and ethical disagreement. Third, as long as information on the health harms of certain drugs for enhancement is properly disseminated and understood (which at present is perhaps not the case), some would dispute whether we should try to legislate against their personal use, though it may need regulation as we have for medicines and alcohol. They might argue that it is difficult to see how society can justifiably concentrate on limiting the actions of an individual to temporarily improve their cognitive capacity through artificial means, when existing, and more pressing, social inequalities exist. And indeed, using neurological enhancement drugs may not differ substantially from other forms of 'mental' enhancement technology, such as education.

Furthermore, safe means of temporarily enhancing one's cognitive abilities, including a remote risk of causing an addiction, may allow a person to expand what they can achieve in everyday life. This may not necessarily be a bad thing, since society can adapt to provide appropriate measures of management (for example, by modification of existing tests and exams). It may also make some human activities safer, although questions regarding the justification for choosing chemical rather than social mechanisms (eg reduced working hours to minimise fatigue) to achieve this effect should be addressed. Finally, how is knowledge of a genetic susceptibility to addiction going to affect any one individual? These are issues for experimental and clinical physiology to address in detail.

Significantly, it will be difficult to inform each person with certainty what 'genetic susceptibility' actually amounts to; how their (change in) behaviour will alter the effects of a drug; and, indeed, how any drug will affect any one individual, given the different biological, chemical and neurological effects of the drug itself and its context. Such information, carefully managed and given to the patient, and responsibly used by that individual, could present significant advantages for our life choices. Equally, it could be burdensome to live with and hard to interpret for individuals and indeed for society. Careful evaluation is needed to study the social and personal impacts of new technologies, just as much as their technical and economic impacts.

Annotation

1. The functional study of neurology, including all component parts from the neurones, to the brain and central nervous system.
2. The autonomy of the addict enrolled in research may become of increased importance in discussions concerning their ability to consent to certain procedures, insofar as the thesis that '*addiction is a brain disease*' comes to dominate research and public discourse on addiction. Such an understanding of addiction may constrain researchers' ability to undertake certain types of research, and the means to justify it, on, for example, persons who are still actively addicted or who have '*recovered*' from addiction, and the ability for addicts to give free and informed consent.
3. Since the Human Rights Act 1998, it has become increasingly important to present issues of ethics in terms of legal rights. These rights are often vague expressions of intent, and it is the responsibility of the law courts to unpack their exact meaning for Government policy.
4. Watershed declarations, such as those at Nuremberg and Helsinki, stress that the rights of individuals must not be detrimentally subordinated to the advancement of science, medicine or society.
5. The use of such medical interventions will rest significantly on the demonstration of their safety and efficiency. This presupposes that research studies evaluating such treatments can be done, since if it is accepted that drug addicts lack the capacity to give consent to participate in research, then forward progress on these lines may be in doubt.
6. Relapse rates in the first year of treatment of almost all drugs of addiction is of the order of 70 per cent.
7. Either by '*mopping up*' the drug from the blood by binding a blocking antibody to it (the antibody-drug complex cannot enter the brain, although the vaccine is only effective as long as there is enough of the antibody to bind); or by a catalytic antibody which moves to the brain and binds to the cocaine, stopping its effect on the reward system. The use of such strategies on actively addicted people may be better termed as simple treatment, since the preventative aspect is lost.
8. The realistic use of vaccines may also depend on the quantity of the drug typically taken by an addict. For example, an alcohol vaccine may not work because of the sheer quantities typically taken (measured in grams of alcohol), because a vaccine will typically only mop up drug

molecules on the basis of milligram doses. So they may be effective on drugs taken in smaller amounts, such as heroin, cocaine and nicotine.

9. The most effective way of helping tobacco users quit involves a combination of pharmacotherapy with advice and behavioural support.
10. In the case of HIV, the possible stigmatising effects of the virus have led to mechanisms to preserve individuals' status. Express consent is required for testing for its presence in blood; and clinics that can preserve anonymity can be used by those at risk of exposure.
11. Nutt and Lingford-Hughes put relapse at about 70 per cent or more in the first years for the treatment of almost all drugs of addiction.
12. The National Institute for Health and Clinical Excellence (NICE) guidance for the prescription of anti-smoking drugs (nicotine replacement therapy) and bupropion (Zyban) recommends that repeat prescriptions for these drugs should only be given to addicts that show a commitment to quitting, with those who fail to demonstrate such a commitment having to wait for six months before trying again.
13. By tackling the problem of addiction, drug misuse will still be harmful to the individual. Other health harms, including death from experimenting with combinations of psychoactive substances, which is a significant and increasing cause of premature death, will remain a substantial concern.
14. There is anecdotal evidence that heroin dependence could be '*cured*' within 24 hours using general anaesthesia and naltrexone. Similarly optimistic views are expressed in *The Heroin Century*.
15. The question is how subtle coercion, which may be justified in limited circumstances, differs from that which is aggressive to the point of being '*no option*'. It is also important to note where the coercion originates, since, for example, that from a concerned family member or friend may differ in nature to that from the paternalistic intentions of the state.
16. We are only talking about criminal activities in which the harm caused is foreseeable and is a controllable effect of the crime, ie the addict knows or has good reason to believe that, through their criminal actions, it is likely that it will lead to the harms in question, and therefore it is within their power to prevent, or lessen the probability of their occurrence. A further distinction to make is that the crime is, more often than not, directly associated with the outcome (unlike drug use and possible criminal activity), in that there is often not an intervening action.
17. So-called '*victimless crime*'. Indeed, a libertarian may argue that the freedom of the individual to pursue pleasure is a right that should not be challenged unless it conflicts with the well-being of other persons. Of course, it could be argued, as has been, that there is no such thing as victimless crime.
18. The '*alternative*' here may be framed as '*rehabilitative*' or '*treatment-oriented*', as opposed to being '*punishment*'. It may also be that the vaccine is '*offered*' to offenders who are convicted of drug- or alcohol-related crimes even where there is no evidence for addiction - for example, those convicted of drunk or drugged driving. Here, the emphasis is more clearly on retributive justice (punishment) and prevention of reoffending than on treatment for an illness. This may

increase the stigma attached to use of the vaccine if it is seen mainly as a tool of criminal justice.

19. The law does not recognise that a prison environment may itself undermine an individual's freedom to give or refuse consent. Furthermore, in the field of medical law, consent to treatment from a competent adult obtained by coercion is invalid.
20. It may also be used to protect inmates or prevent drug use by inmates in prisons, where drug use is a significant problem. It is possible that vaccines could be used in institutions such as the armed forces, where problem drug use is a significant concern relative to the effectiveness of personnel on duty.
21. Assuming '*criminals*' are not in control of their '*drug*' consumption, therefore it should be '*treated*', and not punished. But then, could criminal activities associated with drug consumption be considered as out of their control? Perhaps the distinction can be made that the consumption of a drug and the drug-related crime cannot be allowed to continue because of the harm caused, whereas sole drug consumption does not cause harm (there is no victim) except to the individual.
22. Unless it is born alive, thus there is no civil action available to the foetus in the UK for its negligent death (while *in utero*, it remains a 'person-in-waiting').
23. Parents already make choices 'on behalf' of their children that will affect their lives as adults, such as diet and education.
24. Although the increased intake to induce the 'high' may also be a financial deterrent to users.
25. For example, with regard to addiction to tobacco, genes have been associated with increased likelihood of smoking, smoking more cigarettes, and a lower likelihood of stopping smoking.
26. Genes have been associated with possible candidate diseases caused by tobacco, such as heart disease and cancer, and these have been associated with a smoking addiction. Additionally, there are presently a number of genes associated with specific forms of cancer. In some cases, changes in behaviour and medical interventions have been shown to reduce the chances of disease-related mortality.
27. It has been argued that genotyping for known conditions is only justified when there is an effective intervention to prevent the disorder in those who possess the susceptibility genes. At present, no effective interventions exist in the form of pharmacotherapy. However, a potential addiction or the risk of a disease caused by addiction may be avoided if the individual alters their behaviour or avoids exposure to a potentially addictive drug. This presents problems when the individual is in a position that means they can't avoid the drug, such as living in a deprived area where drug misuse may be more evident.
28. A study has shown the prevalence of both of these attitudes in nicotine use. Although individuals were more amenable to effective cessation methods, they put less importance on the use of willpower. This underlines concerns that genetic risks are perceived as immutable, weakening the belief that changing behaviour will reduce risks.
29. An epidemiological 'risk' may be associated either directly or indirectly with addiction or drug-related diseases, but 'fixed' at birth (for example, sex or ethnicity), or acquired during life, for

example, exposure to the drug itself. In the case of addiction research, it will be a mixture of both elements that will be important, because a person can't become addicted to something they have neither taken nor been exposed to. This information would be of great importance in medical interventions, drug prescription and addiction.

30. French *et al.* (2004) analyse in detail the pros and cons of workplace testing in pre-employment screening and post-employment surveillance.
31. Slowing of the heart rate.
32. Including less serious conditions, for example, mild depression and various eating disorders. This also may lead to increased medicalisation of conditions not previously seen as needing 'treatment'.
33. But one must be sceptical of such claims of '*phantom epidemics*', more often than not confined to the popular media.
34. It should be borne in mind that using neurological enhancement drugs might cause collateral damage to the brain. While, for example, we enhance memory, a side-effect may be impairment to, or trade-off with, another capacity.
35. And, again, to highlight the potential contradictions – marijuana is presently seen as harmful (hence the legalisation of its use) while there are potential medical benefits.
36. It is unlikely that all psychoactive drugs will be entirely safe, and they will always present some risk to certain users. Potential addiction is one such risk.
37. Indeed, a ban on such use may itself be more coercive in legally stopping others from enhancement, than the effect of the (possible) gradual social acceptance of drug enhancements.

13

Prospects of Psychoactive Drugs

The psychoactive substances of the future and responses to them will be the inheritors of the past. The past is not '*another country*'. It is predictive of the issues and responses of the future and their implications. The future may be marked by greater consumerism and individualism and a more hedonistic culture, with rising consumption of drugs and alcohol, or by greater puritanism in society. Both trends are visible and one may counterbalance the other. There will always be new drugs that emerge and are considered to be problematic and addictive. These substances, like morphine and cocaine in the nineteenth century, are often uncritically welcomed and are used as '*cures*' for addiction or other medical conditions. These new problem drugs are likely to emerge through pharmaceutical innovation. How their use is viewed will depend on a range of factors far removed from the pharmacological effects of the substance itself. Such factors, their value and their implications should be reviewed as part of any risk assessment.

The simple use of the concept of addiction to categorise these complex interactions may not always be helpful. Drug effects are highly dependent on individual circumstances and expectations. In the past, there has been an absence of concern about drug effects that would now be termed addiction. Lay assessment of drug effects and the balancing of different drugs is possible without formal professional intervention. Society's assessment of drug effects is conditioned by who is using the drug and the social threat such use is seen to pose. Policy alone is not the answer, but must be carefully timed to fit with social acceptability and culture. There will be conflicting trends in policy, with liberal and punitive approaches coexisting and balances between them realigning over time. Stages of policy making may occur where different policies are appropriate, given different points in a process of cultural change. Such processes will interact with culture, and culture and policy will reinforce one another.

There is cultural change over the last half century towards the reduced use of some legal substances such as tobacco, counter-balanced by a rise in other substance use. A balance of interests, for example, involving pressure groups and industry, should be retained in policy making in order to avoid '*over-destabilising*' culture. Restrictive policies can work, but they must fit with popular support and have support from economic interests. Clear scientific and media messages are also important. The availability of replacement substances, under medical or lay control, plays a role in containing harmful use. It is important to retain substances seen as problematic or addictive in the future within a normal market rather than a black-market model, which is much more difficult to regulate.

It is important to define what the interaction between policy and culture is intended to achieve. Harm can be amplified by greater restriction of those who continue to use. Cultural change can begin locally and can transfer to a national scale. Transferring models from one national culture into a different one is often inappropriate. Different societies have responded differently to psychoactive substances and the nature of those responses has changed over time. History offers us plenty of examples of the arrival of new substances and how their arrival was dealt with. It also offers us examples of successful containment of '*epidemics*' of drug use, of the rise and fall of particular substances, and of changes in how the use of those substances has been conceptualised. All these changes can be analysed and explained.

Those explanations, and the shifting historical patterns of substance use and regulation, offer food for thought for an exercise which is aiming to illuminate possible future patterns. They offer an insight into how problems have been conceptualised and responded to, and what different mixes of regulation and culture have produced. For this review we concentrate on three main substances – those regulated in western societies as illicit drugs (including the opiates, cannabis and cocaine); alcohol; and tobacco. They offer a range of issues for consideration which can then be applied to possible future developments. We range widely in terms of time up to the near present and bring together historical examples of substances that are often considered separately.

OBJECTIVES OF STUDY

Undoubtedly there will be new psychoactive substances in the future. There are plenty of examples of new drugs (morphine, codeine and cocaine in the late nineteenth century, LSD in the 1940s, Valium in the 1960s, ecstasy in the 1990s) emerging from old ones through technological innovation. Throughout history, drug use has ebbed and flowed and different substances have been perceived as problems at different stages in time. These processes will continue in the future. It is important to remember at the outset that new drugs, or new methods of use, or the involvement of different types of people seen as social threats, have never failed to arouse disproportionate attention from social and medical commentators. Some forms of drug use may arouse particular fear, while other allied forms arouse little comment, in part because the consumers are seen as less threatening.

The association of drugs and race, for example, has a long history, which we discuss below. Such issues raise questions about who is defining a particular drug-using phenomenon and why. Concepts associated with drug use have changed and will continue to change over time. They are not timeless and value-free. The concept of addiction also has a history, changing both its compass and its given significance, according to the concerns of the society of the time. Our point is that attitudes and responses towards psychoactive substances have been socially constructed. They are not, and never have been, direct reactions to unambiguously medical or pharmacological problems.

In this section, we analyse some of the issues which have led to changing attitudes towards particular substances and how they have come to be seen as threats. We make a dual argument. On the one hand, it is clear that there were problems associated with the use of some of these substances; but, on the other hand, responses to those problems have been dependent on a range of factors far removed from the effects of the drugs themselves. Our key themes are:

1. Changing cultures of use
2. The role of technology
3. Professional interests
4. Scientific theories and the concept of addiction
5. Economic interests
6. Internationalism.

Modernization of Cultures

Culture and its change are an important dynamic. An apparent rise in consumption of a psychoactive substance, or a connection between its use and groups seen as threats to social order, can make the practice appear as a social problem. In the eighteenth and early nineteenth centuries, drink was built into the fabric of English social life. It played a part in nearly every public and private ceremony, commercial bargain and craft ritual. Paying wages in pubs was common. But such practices were less welcome in the context of a maturing industrialised economy that required more disciplined and time-aware workers. Likewise, the use of opiates, which was also commonplace for both medical and non-medical reasons (the distinctions we would draw were absent then) was seen as more problematic when associated with the industrial working class, or with the Chinese in the docklands area of London in the late nineteenth century. Who is seen as using drugs and how threatening those groups appear to be in the context of the time are important in defining which drugs are a '*problem*'. There can also be the opposite effect, a lack of perceived threat when the problematic aspects of legal substances are played down to the detriment of users and families.

The worldwide use of Valium, mainly by women, was a case in point. Current discussions of the balance of concern, and allocation of policy time and resources between the problems of illicit drug use and problems associated with legal substances such as alcohol and tobacco, indicate the strength of the 'fear factor' and its importance in looking to the future. Problem definitions themselves have an impact on cultures and help to frame a process of cultural destabilisation. Opiates in the nineteenth century provide one example of this. Today there is much discussion of the '*normalisation*' of drug use, but it is important to remember that, in the nineteenth century, using opiates was far more normal for the vast majority of the population than it is today. In more recent times, during the last 50 years, the culture of smoking in the UK and other countries such as the US has changed. In the late 1950s the Cabinet Secretary of the day, Norman Brook, was horrified that it might one day be considered inappropriate for the Prime Minister to smoke in public. Nowadays it would be unthinkable that Tony Blair would do so, or indeed would smoke at all. Smoking has moved from being a cross-class activity in the 1950s to one associated in policy terms today with lower social classes and with teenage mothers in particular. Such processes of cultural change contribute to the definition of a problem.

Technical Changes

Technical change can also lead to sudden changes in patterns of use. For the opiates, the advent of the alkaloids (morphine and codeine in the early nineteenth century) and of hypodermic medication led initially to the restriction of the mass market and a more medicalised model. Initially, medical enthusiasm for the new mode of injection was great. Dr Francis Anstie's enthusiastic welcome for the hypodermic injection of morphia in 1868 has contemporary echoes in the enthusiastic welcome for 'safer' drugs with no 'addiction potential' today: ... it is certainly the fact that there is far less tendency with hypodermic than with gastric medication to rapid and large increase of the dose, when morphia is used for a long time together. But because hypodermic administration was more under medical control, the usage of this form of the drug was more visible and fears of an epidemic of morphine injection consequently greater. For its time, the hypodermic administration of medication was very near the leading edge of medical innovation. But medicine was not modernising in a vacuum.

Medical changes such as the increasing refinement of opium and the invention of the syringe were embedded in a society that was experiencing a host of other technological '*revolutions*'. The addiction concept thus came to embody not simply concerns over the risks attendant in the hypodermic injection of morphine, but also much broader fears that humans might somehow become the slaves of their own technological innovation. For alcohol, technological change meant an amplification of the mass market and the ability to produce and market a standardised product. The advent of flue-cured tobacco and the Bonsack machine for the mass-produced cigarette led to market expansion. The chewing of coca leaves in South America caused little comment because there the substance was used as an aid to work and to help contain social problems.

The isolation of alkaloid cocaine in the late nineteenth century and its importation into western societies was initially also seen within the '*aid to work*' model. But medical enthusiasm for the new drug again led inexorably to fears of epidemic usage. So technological development can have a dual impact. It can expand usage of a particular substance, or it can be met with an over-optimistic welcome followed by a subsequent over-reaction to an apparent epidemic spread, which contributes to our process of 'problem definition'. Technology impacts in other ways too, in the development of new techniques for the production and dissemination of knowledge, which is discussed below.

Professional Influence

Professional interests also play a significant role in the process. In the nineteenth century, opium use did present problems. A large number of infants and adults died from opium overdoses, and opium poisoning was frequent because of the uncertain standardisation of the drug. But those who led concern over the use of the drug, in particular the public health movement and the medical and pharmacy professions, the latter leading the initial move for legal restrictions on its sale, also had professional motivations in view. The 1868 Pharmacy Act, for example, aimed to place sale of the drug (until then available over the counter) under pharmaceutical control. But its main aim was not to restrict sales so much that the profits of pharmacists were reduced. Britain had a system of pharmaceutical regulation of what would later be called 'dangerous drugs' at least until the First World War. As we will see later in this chapter, that system changed to a medico-penal one in the 1920s, a reflection of the influence of the medical profession, which had been establishing its professional authority since the 1850s.

The role of the American medical profession was weaker in comparison, only establishing its professional status in the early twentieth century. This difference affected approaches to drug control in both countries. A more recent example of this process of professional influence in defining the nature of a problem comes with the changes in drug policy in Britain in the 1960s. Until then, drug addiction had been dealt with by a range of professional medical personnel, including those who ran private nursing homes and practices, general practitioners and NHS consultants. Much drug addiction (as it was then called) was dealt with within general practice. Drug users were typically middle-aged and middle class, often with some medical connection themselves. The expansion of drug use in the 1960s and the use of drugs by a younger and more hedonistic group led to two committees of enquiry and to the ultimate replacement in the late 1960s of this earlier system by one based in hospital drug dependence units and run by consultant NHS psychiatrists. By the late 1970s and early 1980s these specialists, based primarily in London, had introduced abstinence-based regimes. This period is still controversial, but it has been cogently argued that such professional interests, in the context of other changes, led to an amplification of the problem and helped create a drug '*black market*'.

Modern Theories

Scientific *modern theories* can also be seen in this dual light – of helping to explain and helping to amplify a problem. It is important to note that the addiction concept is animated by the play of two diametrically opposed meanings embedded in the definition of the word 'addiction' itself. These two senses of the word differ in how they assign responsibility for the condition's origin. The word 'addict' first appeared in English around 1529 as a legal adjective derived from the Latin *addictus* which, according to the *Oxford English Dictionary* (OED), meant to be assigned by decree, made over, bound, or devoted. In this sense, the word described the state of someone who was formally bound (to another) by a court of law. This definition was set against a second, reflexive sense that emphasised the volition of the defendant, rather than the strict sense of judicial assignment pronounced in the legal term.

It thus meant 'to bind, attach, or devote oneself as a servant, disciple, or adherent (to any person or cause)' and also 'to devote, give up, or apply habitually to a practice' (OED 'addict'). Both of these meanings of the word are at play in the noun '(drug) addict'. 'Modern' theories of addiction go back only into the nineteenth or late eighteenth century. Their emergence used to be seen simply as '*medical progress*', medical science gradually explaining behaviour which had previously been inexplicable. But historians have developed a more complex understanding of such developments which in turn throws light on the use of such concepts in contemporary policy. Until the mid- and late-nineteenth century, the belief was strong in the medical profession that drink was good for you, and many medical preparations contained alcohol.

Pharmacists often held alcohol licences. Drink was often prescribed, as hospital and infirmary records tell us. Regular drinking and opium-taking were seen as 'habits', not as addictions. But at the same time, medical opinion hostile to alcohol was emerging – dating back at least to 1804 when Thomas Trotter published his *Essay, medical, philosophical and chemical on drunkenness.* Trotter, like Benjamin Rush in America, called the habit of drunkenness a disease, which should be managed by the discerning physician. The US sociologist Harry Levine has ascribed these developments within the context of a broader shift in notions of the relationship between will and desire at the turn of the nineteenth century.

Perhaps more importantly, Levine's work shows that the disease concept of habitual alcohol use, which he identifies as synonymous with the addiction concept, relied on broader social and cultural beliefs that had little to do with the pharmacological or social effects of drug use. In Britain, the historian Roy Porter pointed out that the components of what were later to be determined addiction were around in the eighteenth century. But it was only later that structural changes in society made them significant. He identifies the rise of Evangelism and philanthropic lobbies, alliances with the nascent temperance movement, and the medicalisation of drunkards into alcoholics. Porter's work rightly stresses the importance of structural factors in making habitual use a significant concept. This is often forgotten by commentators who seek to trace discussion of addiction back into the sixteenth century or before. It is not that people did not discuss these ideas then, but the issues failed to attain social and medical significance.

Why people choose to use such categories at one point in time and not at another is a central concern for our present argument. At the end of the nineteenth century, these beliefs coalesced into a scientific specialism. In England, the Society for the Study and Cure of Inebriety was established in 1884. Such medical societies were established in other countries as well, including the United States. Drinking was a disease, initially the disease of inebriety and later that of addiction. This was, by the time of the First World War according to Sir William Collins, president of the Society: a disease of the will – if one may couple terms derived from the opposite poles of the material and the volitional – and assuredly a disease in which the individual possessed has in many instances a most essential cooperative influence in his own worsement or betterment. Collins and his supporters were both doctors and temperance supporters. They hoped that their work would put the temperance cause on more '*scientific*', and therefore firmer, ground. It is important to remember that their theories applied to the opiates as well as to alcohol. Ideas about addiction changed over time. In the late nineteenth century, under eugenic influence, inebriety and addiction were seen as a hereditary condition.

The Lamarckian idea of the '*inheritance of acquired characteristics*' underpinned much discussion. The focus at this time was on the role of women and the impact on children through the transmission of the characteristics of drinking. These debates are revived, unconsciously it seems, in scientists' contemporary discussions of the transmission of genetic characteristics. The '*alcoholic family*' was a matter of concern in the late nineteenth and early twentieth centuries. It is important to locate such ideas and concepts in the context of the time. They were of particular prominence because of concern about Britain's future as an imperial power in the wake of the failures of the Boer War and the lack of physical fitness of the working-class population which had been revealed. There was concern about the '*future of the race*'. Such concepts are rooted in political imperatives and situations and often fulfil a political role. Such formative influences should also be considered in the present. Such hereditary ideas were seen as outmoded after the First World War. 'Addict' became the dominant name for an habitual drug user, or, someone who suffered from an 'addiction', at around this time.

Again this change reflected structural factors rather than '*medical progress*'. The fear of technology was amplified by the horrors of the First World War. Psychological emphases grew more common after the wartime development of interest in the neuroses. Their popularity was also related to a change in positioning of the psychiatric profession, which was moving out of the asylum to seek a middle-class clientele. Such theories assumed their greatest importance after the Second World War. In the 1960s the concept of dependence was used to bridge the categories of addiction (with physical connotations) and habituation (mind). Theories of the mind and body were also linked in the practice of giving up. Such categories were reinforced through their elaboration and dissemination at the international level through the authority of the World Health Organisation, established in 1948.

Psychologists disputed individual and collective psychology at this period and wondered whether there was such a thing as an '*addictive personality*'. Today concepts of mind and body are coming together in a different way through the discipline of psychopharmacology. Addiction is thus not a timeless concept but has been the outgrowth of changing interest groups and structures within societies. Some commentators have seen the function of the definition as one of medical surveillance. Such analyses are illustrated by the example of substances not initially categorised as addictive. Tobacco, for example, is a latecomer in this respect and the concept of addiction has only been significant (in the sense that Porter discussed addiction to alcohol) since the 1980s or 1990s. This can be related to a number of factors quite external to '*scientific progress*'. Among them are: the different historical route taken by the scientific investigation of tobacco, through public health epidemiology rather than psychiatry; the closer linkages with primary care and psychology in the 1970s; the distance of such networks from psychopharmacology; the differing role of the law; and different relationships with the industry. Again, structural and professional rationales determine the nature and timing of concepts.

More recently, there have been moves to extend the concept of addiction to include other activities such as shopping and eating and to see it as a feature of social dislocation within free-market economies. The increasing elasticity and diversity of the concept may make it less useful in the future. Increasingly in the last few decades, policy has been framed within what is termed an '*evidence-based*' format. As well as considering the rise, acceptance and expansion of addiction as a concept, we should also consider the role and diffusion of '*new knowledge*' more generally about the impact of psychoactive drugs. For example, the postwar rise of risk-factor epidemiology (initially significant in relation to smoking and lung cancer) saw this method become part of the '*toolkit*' of public health. Such methods have influenced the development of measures of harm (for example, units of alcohol consumption) and facilitated the categorisation of populations into categories for intervention.

Commercial Values

Economic positioning is an important variable in deciding what counts as a problem and what does not. The historian David Courtwright points out that, at the turn of the twentieth century, the economic embedding of alcohol was considerable. The alcohol industry had a size and fiscal importance in western nations that dominated economic and diplomatic affairs. The French alcohol industry affected the livelihoods of 4.5-5 million people, or roughly 13 per cent of the French population. At the same time, alcohol taxes were a bedrock of western government finance. The same was true for many colonial governments in Africa and Asia.

Industrial interests, from the early twentieth century at least, were allied with political ones - the drink trade's connection with the Conservatives in Britain began after the People's Budget of 1909. There was an emergent difference with the opiates. By the end of the nineteenth century, the pharmaceutical industry was becoming established. But in general, production and trade in opium products and coca was more confined. Poor nations and colonies in southeast Asia grew most of the opium. Peru and Java accounted for most of the coca. A handful of industrial nations manufactured morphine and cocaine. Germany was the world's major producer of cocaine, and Britain the main manufacturer of morphine. Both resisted international regulation just before the First World War, but through the Board of Trade rather than any political alliance.

The economic and political embedding of cocaine was weak in comparison with that of alcohol and it was an easier candidate for restriction. Economics affects the ways in which the use of different drugs is perceived. It also affects the ways in which the usage of different drugs spreads globally. Courtwright again speculates on why some substances have attained global status as items of consumption and as '*problems*', while others have not. He traces a continuum of spread from local to regional to hemispheric and to worldwide use. Tobacco has had the most dramatic historical movement along the continuum over a period of centuries. But other substances have not made the transition. A world of betel juice and kava bars is now unlikely. Plant drugs now have to compete with chemicals. In fact the stories of morphine and cocaine provide some of the earliest examples of the transition of plant to manufactured medicine.

The more recent history of khat illustrates how a lack of technological innovation can impede the wider diffusion of a substance. The drug has to be imported weekly into the UK as it quickly loses its potency. The introduction of synthetic drugs by multinational pharmaceutical companies provides a major economic incentive in the market today and these are inevitably seen as 'clean' alternatives to plants. Some will also translate into underground usage. Some of the earlier drugs in turn are shifting their markets into the developing world. Tobacco, heroin and alcohol all provide examples of this reverse process.

Spreading Worldwide

As the above discussion suggests, the perception of any given psychoactive substance as a problem has also been affected by over 500 years of globalisation. Internationalism remains a key feature of anti-substance alliances. The first international alcoholism congress was held in Paris in 1878. In 1906 the first international association on the subject was set up and located in Lausanne, where it still sits, now called the *International Council on Alcohol and Addiction* (ICAA) But alcohol was never a serious candidate for international regulation. The closest approximation to international effort was in the African-based regional control arrangement arrived at between the parties to the General Brussels Act of 1889-90 and included in the anti-slavery provisions of the Act. Paradoxically it was the opiates - with the focus of anti-opium effort being overseas - which ended up with an international control system that has dominated and helped to determine systems of domestic regulation.

The story is one of those strange historical conjunctures whereby an earlier draft regional system was transmuted through American efforts, in their turn prompted both by missionary concerns and strategic imperatives, into a nascent worldwide system before the First World War. Germany and Britain resisted, defending their trade interests in the manufacture of cocaine and morphine. But the post-war settlement saw these export controls imported into the peace settlement under the supervision of the League of Nations. In the interwar years, the international system operated simply as a trading control mechanism and defined the boundaries between medical and nonmedical use. As historians such as William McAllister (1999) have recently shown, that system of control of trade changed after the Second World War, again under US influence, into a strongly prohibitory regime whose impact continues to be felt in smuggling, illicit trade and domestic drug-control legislation. Control, as it changed, helped to create and amplify the problem. The attraction of the profits to be made in the illicit drug trade have distorted the agricultural economies of countries such as Mexico, Peru and Afghanistan, where growing crops that are restricted in the rich countries of the first world has advantages. US drug policy has increasingly been marked by a willingness to intervene in the affairs of other countries and to demand the support of other 'drug-consuming' nations in achieving its goals. The involvement of the CIA in drug running in Vietnam or, more recently, US support for the Taliban's ban on opium cultivation in Afghanistan, reflects a process of globalisation and international control that is now nearly a century old.

SET AND SETTING HYPOTHESIS

The American psychiatrist Norman Zinberg (1984) famously stated that the effect of a particular drug was dependent on '*set and setting*', ie the situation in which the drug was taken, the expectations of the user and so on. There were no intrinsic effects of substances, but much depended on the social situation of the user. To this, we add the historical dimension. The effects of particular drugs and the ways in which those effects are talked about will vary according to positioning in time. If we look at opium as an example, this becomes very clear. As we have seen, the concept of addiction was not an important part of discussions of drug use in the early nineteenth century. Opium had the same pharmacological properties then as now, and what we would now call its addictive properties were discussed in great detail by Thomas De Quincey in his *Confessions of an English opium eater* in 1821. But this was met with little concern and aroused relatively little comment at the time.

The setting within which such comments were published was not one where the elements of '*problem definition*' we have discussed above had yet coalesced. De Quincey's writings were more significant in this way later on in the century. In the nineteenth century the concern was for balancing good and bad effects. The overuse of drugs was seen as a bad habit, rather like overeating, and longterm use was of little concern. Dean Isaac Milner told the reformer of the slave trade, William Wilberforce, who had originally started taking opium in 1788 to relieve a stomach ulcer: Be not afraid of the habit of such medicine, the habit of growling guts is infinitely worse. There is nothing injurious to the constitution in the medicines and if you use them all your life there is no great harm. But paroxysms of laxity and pain leave permanent evil. The distinction between what we would now term medical and non-medical usage and effects was not established at this stage.

De Quincey, who used opium for its power 'over the grander and more shadowy world of dreams' had initially taken it to relieve the curse of toothache. Such origins of use and for continuation became a matter of debate between De Quincey and Samuel Taylor Coleridge. De Quincey claimed after Coleridge's death that the poet had taken opium not for medical reasons but as a source of luxurious sensations. That these arguments could take place indicates that the historical setting of particular effects and how they are defined is important. Overtly recreational use was, at that time, largely confined to a small circle of Romantic writers and poets and generally disapproved of by respectable society, although no more than overindulgence in alcohol. In the public discussions of opiate use that did take place, for example, in the public health movement and the public enquiries of the 1840s and 1850s, the effects of opiates were often compared with those of alcohol.

The passivity induced by opium was, it was argued, preferable to the violence and lack of control of the drunkard. Such views were related to the fears of the industrial working class brought about by urbanisation and the separation of classes. But, as well as the public discussion of different effects and their political rationale, there is evidence of self-moderation of consumption, of lay knowledge of effects and how to moderate them. This emerges in some accounts of opiate use in the Fenland area, where usage was particularly high in the nineteenth century because of the marshy and malarial nature of the land. A Dr Hunter, reporting to Sir John Simon of the Privy Council Office in 1863, wrote: A man who is setting about a hard job takes his pill as a preliminary, and many never take their beer without dropping a piece of opium into it. To meet the popular taste, but to the extreme inconvenience of strangers, narcotic agents are put into the beer by the brewers or sellers. The perceived effects of drugs also varied according to who was seen to be using them. We have already discussed the connection between the effects of drug use and the industrial working class.

Beliefs about drug effects also expressed the racial fears of the late nineteenth century. In the turn-of-the-century US, discussions of cocaine fitted strongly with formulations of racial difference. The widespread belief that cocaine use was a distinctive element in many African-American communities has been challenged by historians, but thinking about the figure of the black cocaine user within its wider cultural context helps us to understand Dr Christopher Koch's 1914 claim that 'most of the attacks upon white women of the south... are the direct result of a cocain-crazed Negro brain'. Writers such as Koch were alarmed by the perception of a threat borne by black emancipation. Writers found in cocaine use a powerful figure for what they believed was the threat held by the new social reality of black freedom - the emancipation of a predatory 'brute' into free, white society. Such views were the expression of wider fears and tensions in American society of the period. In Britain, fears about race were expressed differently, through the image of the threatening Chinese opium smoker in dockland areas, most notably the fog-shrouded East End of London.

Such views, like the American perceptions, have been critiqued by historians and others have seen in them an orientalist discourse which had its origins with De Quincey. They were part of the expression of contemporary fears of the '*residuum*' that might overwhelm society and also of the potential domestic legacy of Britain's empire. The racial issue and its connection with drugs was formulated and expressed differently in different national situations. As well as race, gender and childhood have also been important factors in perceptions of drug effect. In the late nineteenth century, it was women who, in Britain and the US, were seen as particularly prone to the non-medical hypodermic injection of morphia. Women as mothers were also of concern for the movement against alcohol.

Juvenile smoking was associated with the menace of the lower-class 'hooligan' who threatened the future of the nation. Such fears of drug effects have their contemporary echoes in the media condemnation of '*laddette*' culture or the fears of public disorder by young people under the influence of alcohol. The legal setting has also influenced the effects of using drugs. It has helped to produce the sense that drug use somehow equates with social rebellion, and this has proved to be attractive to many people who feel a need to challenge what they see as the conventional boundaries of 'normal' society. Ironically, the vehemence of the antidrug crusaders has fuelled the drug use which has typified various countercultural movements, particularly in the twentieth century. Narcotics have played a similar counter-cultural role ever since De Quincey suggested that drugs might offer a pathway out of the accepted intellectual conventions of his day.

This idea has gained wider appeal as drugs drew ever more negative publicity in the second half of the twentieth century. Put another way, when Nancy Reagan exhorted American youth to 'just say no' to drugs, there was a sizeable minority who were eager to say 'yes, please', simply because Nancy Reagan said do the opposite. Fashionable, countercultural drug use has flourished in the setting produced by antidrug programmes that have demonised drug use as immoral or antisocial and it remains attractive to many people today, something most recently demonstrated in the so-called '*heroin chic*' fashion photography of the 1990s. In Britain, the anti-heroin campaign posters of the mid-1980s made their way onto the bedroom walls of teenagers in Liverpool - the model became an icon rather than a warning.

MANAGEMENT OF PSYCHOACTIVE DRUGS

History provides many examples of the mechanisms that have been used to manage psychoactive substances. Societies and governments have responded to drug use in different ways. In this section, we briefly survey a number of case studies in which use of substances has been responded to by different mechanisms. We focus primarily on those used by western governments, although we are aware that non-western and colonial governments have often used different forms of regulation from those in place at home. Our case studies are:

1. Alcohol controls during the First World War
2. Prohibition in the US
3. Drug policy in the US and UK in the 1920s
4. US soldiers returning from the Vietnam War in the 1970s
5. Drink driving in the UK
6. Decline in tobacco consumption since the 1970s

7. The rise, fall and rise of cocaine use in the twentieth century
8. Alcohol policy in Russia in the 1980s and 1990s.

Temperance Movement for the Use of Alcohol

Although the 19th-century temperance movement had long called for controls on the sale of alcohol, a comprehensive strategy was adopted only during the First World War under the impact of wartime conditions. Munitions-makers and shipbuilders had complained that drunkenness was a cause of low productivity. The government set up the Central Control Board and this imposed a wide range of restrictions in 1915. These included reducing the hours of sale of alcohol (the afternoon closure), lowering the alcohol strength of some drinks and outlawing the sale of alcohol on credit. In some areas, for example in Gretna, the trade was brought under state control.

Beer and spirit production was reduced by half during the last years of the war. This was at a time when a wartime spirit prevailed and greater restrictions on personal freedom proved acceptable. The impact of these changes is measurable through the decline in deaths from cirrhosis of the liver. These dropped dramatically after the restrictions were introduced, as did overall alcohol consumption and convictions for drunkenness among civilians in areas affected by the Board's controls and, indeed, outside them. Other factors could also have affected these outcomes. Levels of drinking had been falling slowly since 1900 and remained low until after the Second World War. The depression of the 1930s and expanding alternative leisure activities also had an impact. But the restrictions on production and increases in price were clearly important. The Central Control Board did not long survive the war but its restrictions remained in place into the 1960s and consumption remained low.

Strict Action

The case study of US prohibition is often used to support an argument that stringent restriction of psychoactive substances doesn't work and leads inevitably to the creation of a criminal black market. The historians' assessment of the '*prohibition experiment*' makes a different case. The 1920 Volstead Act was the outgrowth of earlier localised experiments in individual 'dry' states. It represented the interests of the older rural Protestant interests in American society, supported by industry and by women's campaigns. Prohibition was never a total ban and a complicated licensing system provided many loopholes. Sources of illicit alcohol soon appeared in the industrialised cities where the culture of drinking was much more entrenched. But the effects of prohibition were remarkable. In the early 1920s, consumption dropped to about 30 per cent of pre-prohibition levels. In areas where there was public support for the laws, they were effectively enforced and patterns and forms of drinking changed.

The old culture of the saloon with its heavy drinking ended, and new groups of drinkers such as women emerged. Spirit drinking rose. Certainly prohibition did see the growth of a criminal black market and illegal speakeasies, but, by 1927, drinking was still only at two-thirds the level it had reached before the war. Prohibition came to an end not because it failed, but because of the advent of the Great Depression in 1929. The prospect of taxable revenues from alcohol was too much for the government to resist, and restriction appeared also to be outmoded.

Prohibition Policies

Prohibition also had its impact on the development of drug control policy in the US and the UK in the 1920s. Here, the contrasting policy routes taken by the two countries are often used to point to a '*lesson from history*'. American-style prohibition of drugs led to an amplification of the black market, while the British liberal and medical control system resulted in a 40-year 'calm' when drug problems were minimal. Again, the work of historians doesn't completely support these interpretations. Controls were introduced just before the First World War in the US and just after the war in the UK.

The operative factor was the existence of a new system of international control of drugs, given force by its incorporation in the Versailles peace settlement after the war. This international system was the outgrowth of US missionary and strategic concern about opiate use in the Far East. Controls at national level in the US came through the Harrison Narcotics Act of 1914, interpreted through legal decisions to implement a policy where maintenance on opiate and other drugs through a medical practitioner was forbidden.

A large-scale criminal black market was a result and drug users were confined, often against their will, in treatment centres which were like prisons. As Courtwright has commented, by 1940, the US addict population had been transformed: 'The secretive, female, morphine addict had given way to the hustling, mainlining male junkie'. But, Courtwright argues, the policy changes of the post-war years can't be given sole responsibility for this change. The decline in medically induced addiction had been marked before the 1914-18 war and the pattern of addiction was shifting. Legal changes had not been solely, or even primarily, responsible for the transformation of the addict population, and government policy simply succeeded in making a bad situation worse. In Britain, the opposite occurred, at least ostensibly.

The British medical profession was more powerful earlier than that of the US and was far more entrenched in the treatment of addiction. It stood firm against Home Office moves to introduce a US-style system which could see addiction stamped out. Home Office officials were also uncertain about the medical rationale for control practices and had to turn to the profession for validation of their policy approach.

The Rolleston Committee on morphine and heroin addiction (1924-26) is often seen as having introduced a 'British system' of medical maintenance prescribing, which prevented the development of a black market and was widely commended by US observers of drug policy in the 1960s as a suitable model to follow. The reality, again, was rather different. The 'British system' was continued in the 1920s because addict numbers were low and were mainly middle class and iatrogenic. The system reflected and did not create this addict profile.

The 'British system' was the result and not the cause of Britain's lack of an addict problem for the next 40 years. The system was less medical than it seemed. What had been established was a mixed criminal justice and medical system of control in which the balance of power could alter as the pattern of addiction itself shifted, as it did in the 1960s. These two examples show us that policy changes do not have the immediate impact on patterns of drug use which commentators often suppose. In both the US and the UK, policy changes fitted into already existing patterns of changing drug use, both reflecting and reinforcing them.

Alarming Use of Drugs

The story of the epidemic of drug use among US soldiers in Vietnam which disappeared, or was replaced by less harmful forms of drug use, on their return to the US, is widely known through the work of Lee Robins. Wartime availability of cheap heroin in smokable or 'snuff' form led to a huge demand for the drug in the war zone. But, of those who returned home, within a year, only 5 per cent of those who had been addicted to opiates in Vietnam were still addicted. Nearly half the men who considered themselves addicted in Vietnam tried opiates after their return, but only 6 per cent became addicted again, and few received any treatment for their drug use. Many changed to safer forms of drug use on returning home, most notably the widespread use of cannabis. The story of the Vietnam veterans illustrates the importance of the 'set and setting'. It demonstrates the significance of environmental impacts on patterns of drug use and of cultural norms and shows that many drug users are able to moderate patterns of use without formal interventions.

Use at Public Places

The arrival of restrictions on drinking and driving in Britain in the 1960s and 1970s provides an example of the complexity of interactions within policy and with culture. As Greenaway has pointed out, there was a background of clear and mounting scientific evidence, backed up by policy change in other countries. Public and media opinion pressed for action, although politicians remained cautious for some while. Sections of the drink industry were supportive since they had much to gain from improved standards in this area. The policy area was defined as road safety, where the tendency was for greater regulation, rather than drink, where the trend was for liberalisation. Pressure group input was important. Regulation (although not random testing) was accompanied by large expenditure on anti-drink-driving publicity and by a significant decline in road casualties as a result of drink. Drink driving shows how regulation and culture can interact, with one feeding off the other. It also shows the importance, for an intervention, of how the area of concern is defined. Its definition as road safety rather than drink was at a time when society favoured a liberal approach to drinking.

Substantial Reduction in Tobacco Consumption

There has been a substantial reduction in the proportion of smokers in the UK population, from 51 per cent of men and 41 per cent of women in 1974 to 28 per cent and 26 per cent respectively in 1998. The number declined steadily throughout the 1970s and 1980s, levelling out in the 1990s. However, figures for the second half of the 1990s showed smoking falling again among both men and women. A clear class gradient in smoking has developed since the 1970s, when smoking was a cross-class activity. In 1998, men who lived in 'unskilled manual' households were nearly three times as likely to smoke as those who lived in professional households.

The trend has been particularly marked for women, where smoking and lone teenage mothers have been closely associated. The reasons for this decline are various. Government policy throughout the 1970s and into the 1980s pursued a twin-track approach. It worked with the tobacco and pharmaceutical industries to modify the product (safer smoking and tobacco substitutes), and its marketing (packet labelling) and to produce cleaner, more medicalised, supplies (nicotine replacement). It also helped to fund anti-tobacco pressure-group activity (for example, by Action on Smoking and Health (ASH)) which gained a high media profile, and by raising taxes and mounting health education campaigns. Government pursued a policy balancing act which was successful in reducing the consumption of tobacco. In more recent times, prohibition has come onto the policy agenda through the advertising ban and smoking bans in pubs and open spaces. Government has felt able to introduce more stringent regulation as the normality of smoking and its cultural embedding has declined. This in turn will contribute to further decline and to the confirmation of a hard core of users, now increasingly defined as '*addicts*'. These developments have been paralleled by an interesting counter-process, the apparent liberalisation (although this has been disputed) of the law on cannabis and the rise of cannabis smoking as a more culturally acceptable activity.

Elasticity in the Psychoactive Drugs

For our last two examples, we turn from '*success stories*' to those where drug use has not been contained. Cocaine proves one example. As Musto (1999) has pointed out cheap cocaine fed a global epidemic which lasted from the 1890s to the mid-1920s, with peaks in different nations at different times. Medical and semi-medical usage was common at first, followed by a spread to the underworld where opium smoking and drinking were already common. The first two decades of the twentieth century were full of reports of increased cocaine use in the '*demi-monde*'. Then in the late 1920s the epidemic subsided and world exports began to decline. As was the case with the opiates, the drug was scarce in Europe and the US at the outbreak of the Second World War. This is partly attributable to a generational learning pattern amongst physicians, but the drying up of supplies was also important. The rise of cocaine use from the late 1960s was associated with more abundant supply but was initially a matter only for middle-class whites in the US.

The arrival of crack cocaine in 1985 saw cocaine acquire a reputation as a dangerous and addictive substance. Patterns of usage were important in this perception. Crack contributed to the problems of poor inner-city areas. But its usage by those already feared - poor immigrant and ethnic groups - contributed to its image. The subsequent '*War on Drugs*' policy followed by the Reagan Government in the 1980s saw a move away from treatment towards criminal sanctions, law enforcement and international control efforts.

The number of Americans using drugs continued to decline in the 1980s and 1990s but the harm experienced by those who did use and by the rest of society was amplified. The supply of drugs remained plentiful and producer countries expanded their crops, while problem drug use increased in the US, as did the violence associated with the crack trade. Alcohol consumption increased in the then Soviet Union from the 1960s. Interventions had little effect. In 1984, Mikhail Gorbachev introduced restrictions to limit the availability of alcohol and launched a mass campaign against drink.

The campaign failed but it achieved some striking early success between 1984 and 1987. Fewer people died from alcohol-related causes, and life expectancy increased sharply for both sexes during that period. But these successes were only temporary, and drinking and drink-related harm soon began to rise. As with US prohibition, the restrictions underpinned new patterns of drinking, with a rise in spirit consumption. The already-existing Russian black market was capable of supply and the later stages of the campaign coincided with a freeing up of the economy which reduced the state's ability to control supply. The price of alcohol relative to other products fell, and alcohol became an item of barter in the developing black economy.

CONCLUDING REMARKS

The case study examples lead us to some general conclusions about the means by which substance use can be reduced or managed. Change can come about through policies and interventions specifically directed at the perceived problem, but these must be carefully timed and must attune with social, economic and cultural change. The belief that policies can deal with problems is common, in particular with discussions of drug policy which use historical examples. But policy alone has relatively little impact. The prevalence of drug use often converges over time in countries with widely differing policies. The cultural context is of vital importance in reinforcing, or legitimating, forms of regulation. No British politician would have considered introducing smoking bans in the 1950s because of the political implications of introducing such a culturally and electorally unpopular measure. The example of Russian alcohol policy shows what happens to a policy without social and economic support.

The drug policies credited with bringing about diverse effects in the US and UK in the 1920s were, in fact, building on already existing patterns of use. We may be dealing with cycles of drug use which relate to broader issues beyond the impact of formal policies. The recent decline in tobacco use and rise in the use of cannabis may be a local example of this. In more distant history, similar processes were at work. As opiate use declined in China at the end of the nineteenth century, tobacco use began to rise. Prohibition or strict regulation can work, at least initially, as the examples of US prohibition and the short-term results of Russian alcohol policy show. How one defines success in these circumstances is, of course, important. A total ban may reduce overall numbers of users but increase harms for those persisting in use, as did US policy on cocaine and other drugs in the 1980s and 1990s.

The balance between different forms of intervention, between health and criminal justice approaches and the differing emphasis over time on one or the other seems to be the most common form of policy choice. Culture and its change is a vital component in the equation. Policy, of course, reinforces culture but there are other engines of change. The combination of science, the media and health activism has certainly been important in the British context for smoking, bringing a '*tipping point*' in which smoking has been increasingly seen as a deviant rather than normal activity. Whether that is a desirable result should also be part of the discussion of '*success*'.

Destabilising and redefining use and the user brings its own harms and potential dangers. Such processes of cultural change have typically been initiated by the middle class and framed with working-class consumers in mind, of which smoking offers the most recent example. There is often little understanding of working-class cultural norms, or they are seen as unacceptable. The timing of interventions in relation to culture is also important, as the examples of Vietnam, drink driving and smoking demonstrate.

Interventions of different types can be made, either before drugs are well established with extensive marketing networks, or, alternatively, when the process of cultural destabilisation is underway and resistance is likely to be less. And such sanctions, as the example of alcohol control in the First World War shows, must have the support of economic interests. One can also see a staging of different forms of intervention dependent on the stages of culture reached. Where a drug is regulated, demand can more easily be influenced through taxation and the control of distribution. The case studies also underline the importance of industry partnerships in these forms of regulation.

Although public health interests are now often hostile to involvement with industry in harm-reduction strategies (there are current examples from the tobacco and alcohol fields), such strategies have brought public health and industry interests together in the past and have been seen to have advantages. The availability of substitute drugs also affects the impact of policy. Cannabis use by Vietnam soldiers in the 1970s and the current availability of nicotine replacement therapy for smokers have both provided alternatives for users.

14

Addiction of Common Drinks

COCA

The coca bush is native to the Andes, where it has been harvested since ancient Inca days. Use in that era has been confirmed through analysis of hair from ancient corpses and from examination of artwork. At first only upper-class Incas and select individuals were permitted to use coca, but usage spread to Inca society as a whole after the Spanish conquest. In modern times the plant has been cultivated in India and Sri Lanka as well as Formosa, Indonesia, and Malaysia. During the 1960s Malaysia was the world's primary source of coca. Its leaves are the natural product from which *cocaine* is refined, and blood measurements confirm that coca users receive cocaine from the leaves. Cocaine content is commonly 0.6% to 1.5% of the leaves by weight. Cocaine is not the only drug component of coca, but relatively little exploration has been made of other components. Some investigators suspect that these other drugs are more important than cocaine in producing coca's effects. Traditionally coca has enjoyed wide use and social acceptance in the Andes, although leaf chewing (as opposed to taking coca in tea) is associated with lower social classes. Persons from middle and higher social classes who do not engage in physical labor may use coca recreationally. Short-term mental effects of coca leaves have been likened to those of coffee. Usage is much more common among persons living at high altitudes than at sea level.

Coca can be a social lubricant, much in the way that *khat* is traditionally utilized. As with wine, coca leaves produced in different regions under different conditions can have flavor characteristics making some varieties more sought after than others even if the drug content is virtually identical. Veins are stripped from the leaves, which typically are then chewed or sucked upon in a small quantity for hours, much like chewing gum. Coca is a local anesthetic and numbs the mouth when a person chews coca leaves.

Lime (the mineral, not the fruit) may be added to improve the body's absorption of cocaine from the leaves. Habitual leaf chewers routinely receive a daily cocaine dose of 0.25 grams spread over several hours. Coca tea or the leaves themselves are used to aid digestion, reduce gastrointestinal colic, fight asthma, soothe vocal cords (laryngitis), relieve stress and elevate mood, alleviate cold and thirst, produce sweat, and fight motion sickness. Milk with coca has been used against colic and diarrhea in babies. Coca cigarettes have been used to treat both asthma and colds. Proposals have been made to produce coca lozenges and chewing gum to make the natural product available in formats more familiar to persons coming from a European heritage. Habitual coca chewing interferes with the body's insulin and thereby tends to raise blood sugar levels. That finding should interest diabetic chewers but is also relevant to persons using leaves at high altitudes (common in the Andes) because the reduced oxygen supply at upper elevations tends to lower blood sugar.

Habitual coca use apparently compensates for that effect. Blood sugar tends to decline during exercise; coca can prevent that decline, and scientists suspect that coca can increase the body's effective use of blood sugar during exercise. Coca leaves improve a person's access to stored energy sources in the human body and thereby increase capacity for physical labor. During the nineteenth century such qualities attracted notice in Europe, but suggestions that coca be used in industrial and military labor were apparently ignored. In the 1970s a sample of Argentine miners showed that 65% chewed coca leaves every day, and another 14% used leaves less often. The more physical power and endurance required for a particular job, the more likely that a miner used coca leaves each day. One study of coca found the typical stimulant actions of raising pulse rate and blood pressure but also found that coca had a more unusual effect of decreasing the volume of blood plasma–a condition normally associated with bleeding or with not drinking enough fluids. This condition interferes with proper blood circulation during physical exercise, but centuries of experience suggest that coca-chewing laborers get along well enough nonetheless.

The study just referred to involved habitual chewers of coca; occasional chewers may not experience the same results. For example, in another study scientists concluded that regular chewers had more access to energy stored in body fat but that occasional chewers did not and also concluded that occasional chewers would not show the same improved endurance during physical exercise that is demonstrated by habitual chewers. One investigator has concluded that coca improves endurance but does not otherwise help physical labor (a person cannot lift more or run faster). Like many stimulants, coca reduces feelings of hunger and can thereby reduce food intake. The appetite suppressant effect is slight, however, simply helping a person to get by more comfortably when food is scarce. Coca chewers exhibit robust appetites when victuals are plentiful, and coca preparations can even be an element of meal-taking. Some research suggests that under low atmospheric pressure coca can improve the body's metabolism of carbohydrates and thereby improve nutrition of users. Coca itself is a good source of vitamins, iron, and phosphorus; and lime used with coca can provide almost a gram of calcium per day–important supplements where diet is often deficient in such factors. Malnourishment exhibited by rural coca chewers seems related more to poor food supply than to use of coca. Rat experiments, however, indicate that coca slows growth rate if used in mountainous altitudes.

History

Although individuals have to work hard to abuse coca enough to create problems, persons who succeed at that task can experience the kinds of hallucinations and other mental afflictions associated with stimulant abuse. That outcome is very unusual among persons who use coca in traditional ways. Contrary to what one would expect from a stimulant, scientific tests show that coca (like khat) retards reaction time and increases errors in work performance.

Long-term mental effects of habitual coca usage decrease thinking abilities in ways that are seldom noticeable in rural village life but that are clearly documented through scientific tests. Such decline would put persons at a disadvantage in coping with modern urban conditions, on or off the job. The decline is presumed to be evidence of organic brain damage. Coca depresses the immune system, presumably making users more susceptible to disease. Coca chewing is suspected of promoting spread of cholera, not from coca itself but from lime or other alkali substances chewed with the leaves, having the result of lowering the stomach's acid content and thereby providing an excellent environment for growth of microscopic cholera organisms. Coca is suspected of causing liver damage. Archeologists examining ancient human remains have concluded that coca chewing may cause tooth decay and loss. In contrast, modern-day chewers claim that the habit promotes dental health and makes users less susceptible to disease in general. Coca's effects on general physical health remain uncertain. Differing backgrounds of users and abstainers hinder efforts to measure effects; factors other than coca may be affecting health.

Tolerance and Dependence

Tolerance to the drug effect (resulting in a need to keep increasing the dose) is not observed among coca chewers; lack of that classic indication of addiction is evidence of coca's low addictive potential. Heavy coca users may exhibit mild signs of physical dependence with the drug if they stop using it, but any such transitory illness is too slight to be a factor in choosing to continue using the drug.

Combined effects

Not enough scientific information on the natural product to report.

In cancer

Mouths of habitual chewers show tissue abnormalities but no pre-cancerous conditions. Nonetheless, coca is considered a probable cause of cancer.

During pregnancy

Coca is considered a probable cause of fetal injury.

Additional information

In the 1970s a lift of all legal controls over coca was proposed on the theory that the natural product was far less harmful than pharmaceutical stimulants and might be just as attractive to persons who were damaging themselves through stimulant abuse. The proposal was rejected. Cocaine and ecgonine can be removed from coca leaves, and such "*decocainized*" leaves are legal to possess without a prescription. Not all leaves marketed as decocainized have undergone such treatment, and urine tests of persons drinking tea steeped from such leaves may be positive for cocaine use.

Coca paste

This substance is made from coca leaves but has the same sort of relationship to them that 100 proof alcohol has to 3.2 beer. Paste is yielded midway in the process of refining cocaine from coca; by volume the paste is anywhere from about 40% to 90% cocaine, thus as potent as typical street varieties of cocaine itself. Some persons desiring cocaine sensations prefer paste, which can be smoked without the heat destroying cocaine's drug effects. Such persons routinely add the paste to tobacco or *marijuana*. Because of coca paste's high potency, a user basically receives the same impact as with using cocaine. Descriptions of coca paste smoking and crack smoking are similar: euphoria, insomnia, compulsive use. Adverse effects duplicate those of cocaine (even including "*coke bugs*"–a hallucination of vermin crawling under the skin). Using coca paste is very different from using coca leaves.

The difference can be inferred from a 1996 report that combined surveys involving over 24,000 persons in several countries of the Andes region. In 1965 about 13% of Peru's population apparently used coca leaves, but only 0.8% to 3% of persons in the 1996 multicountry survey reported "*lifetime*" coca paste or cocaine use–meaning people had tried them at least once; the percentage of regular users would be lower yet. Direct comparison of those 1965 and 1996 figures would be invalid because of difference in dates and survey designs, but the huge gap between their percentages is consistent with a general rule that the stronger a drug is, the less popular it is. Also, the kinds of persons who use coca leaves and coca paste differ. The 1965 figure included over half the laborers in Peru's countryside. The 1996 survey found coca paste and cocaine to be used predominantly by educated middle-class urbanites. A 1992 study of hospitalized coca paste smokers also found most to be middle class, but almost as many were from a low-income background.

CAFFEINE

This drug is responsible for the stimulating jolt that coffee drinkers get. Many drinkers would probably be surprised to see caffeine listed as an ingredient in medicines they take. Caffeine is so widely used (typically in coffee, tea, soda, and chocolate) that it is scarcely considered a drug. Yet an overdose can be fatal. Caffeine makes people more alert, and experimentation finds that it can help persons function more effectively during sleep deprivation. Caffeine is commonly used in the workplace to increase employees' energy and output. Laboratory measurements indicate that a single dose of 250 mg to 400 mg at the beginning of a night work shift is more effective than several smaller doses spread out during the work period. Some studies find that caffeine helps extroverts perform simple physical assignments but over-stimulates introverts and thereby worsens their performance.

Scientific measurements prove that caffeine, by itself or in combination with *ephedrine*, improves athletic performance. Like some amphetamine class stimulants, caffeine has reduced hyperactivity in children when they received 600 mg daily, but it has been found ineffective for *attention deficit hyperactivity disorder* (ADHD). Studies demonstrate caffeine can mildly help asthma sufferers. Theophylline, a drug commonly used to widen airways and help asthmatics breathe, is related to caffeine. Caffeine is a standard drug to help premature infants that have interruptions in breathing. Rat experiments indicate that caffeine can promote weight loss. In humans a combination of caffeine and ephedrine has been used for that purpose. A study found that caffeine increased women's energy outlay and body temperature but that the temperature change correlated with smaller weight and waistlines only in younger women.

Caffeine has been suspected of promoting osteoporosis, a disease causing loss of bone density in older women; studies controlling other factors (such as cigarettes and drugs promoting calcium loss) found that caffeine had no tendency to reduce bone density, but one study published in 2000 and tracking almost 35,000 postmenopausal women found a slight correlation of caffeine usage to broken bones–a correlation implying loss of density. Coffee drinking is associated with loss of density.

In contrast, examination of over 1,200 older women in England showed that tea drinkers were less likely to have osteoporosis, leading investigators to wonder if something in tea, other than caffeine, affects bone density. Investigators examining caffeine consumption in a group of 8,000 men who were tracked for three decades discovered that the more caffeine someone ingested over the years, the less likely the person was to come down with Parkinson's disease. In a group of 46,000 men tracked for a decade, increased consumption of caffeinated coffee was linked to decreased likelihood of having gallstones; consumption of decaffeinated coffee did not have such a link.

As one analyst pointed out, such associations are interesting but do not prove cause and effect; for example, perhaps some physical aspect leading to Parkinson's disease also makes caffeine beverages unappealing–thus persons without the disease would consume more caffeine beverages than sufferers do, but that consumption would not mean that caffeine prevents the affliction. Mice experiments demonstrate that if caffeine is administered in the right amount and at the right time before exposure to radiation, the drug will allow mice to survive otherwise lethal amounts of radiation.

History

In the 1990s, 20% of the U.S. population was believed to be using over 700 mg of caffeine each day. That is enough to produce behavior mimicking an anxiety neurosis. If a person only ingests caffeine through pharmaceutical preparations or food products with labels listing caffeine amounts, intake can easily be measured. The amount in a restaurant cup of coffee is more difficult to measure; one rule of thumb says 100 mg to 150 mg. The substance can accelerate pulse rate; it can also make people more peevish and jumpy (even promote panic attacks) and interfere with getting good sleep. Caffeine can cause heartburn and increase urine output. Experimentation has confirmed that the drug's tendency to promote loss of body fluid will dry the vocal cords and affect voice quality. Analysis of over 30 years of reports about caffeine and blood pressure found that the drug reliably increased blood pressure when persons began using it but that the effect did not persist in all users.

Investigators measuring blood pressure among medical students found caffeine raising the readings far enough that anyone at risk for high blood pressure should avoid the drug during times of stress. Persons with coronary artery disease may be at significant risk for sudden cardiac arrest if they drink more than 10 cups of coffee a day. Excessive doses can dangerously reduce blood potassium levels, damage muscles, produce extremely rapid heartbeat, and cause delirium and seizures. As the twenty-first century began a case report associated caffeine with retina damage in several persons.

Tolerance and Dependence

Coffee was formerly treated as an illicit drug. Centuries ago possession was a death penalty offense in Spain and the Near East. In the early twentieth century one standard medical textbook warned of coffee addiction peril, and another medical volume described coffee as a gateway to opiate addiction. In a modern study volunteers showed no particular desire for caffeine but did find that 300 mg mimicked some effects of *dextroamphetamine.* Although caffeine is not a scheduled substance, users can develop a physical dependence on the drug that results in withdrawal symptoms including weariness and headache. Such symptoms are not inevitable nor are they necessarily troubling to persons experiencing them. In 2000 an international panel of experts convened by European drug regulation agencies described caffeine's potential for dependence as low. Tolerance can develop to some of caffeine's effects.

Combined effects

Caffeine itself can reduce headache, and an experiment involving hundreds of participants showed caffeine to substantially improve ibuprofen's ability to relieve headache. Phenylpropanolamine is a drug commonly found in remedies for colds. It seems to increase caffeine levels in a person using both drugs, and together the two can produce mood elevation, hyperactivity and manic behavior, confusion, high blood pressure, and stroke. In rat experiments, caffeine boosts the effects of *cocaine* and amphetamine, enough to transform normally tolerable doses into fatal ones. Human observations show that cocaine users tend to take lower doses of that drug if they also use caffeine. Perhaps the most common drug taken with caffeine is *nicotine.* Animal experiments find that interactions of that combination may make cigarette smoking more pleasurable.

Cigarette smoking increases the body's rate of metabolizing caffeine, which decreases the influence from a given amount of caffeine; British researchers found that smokers tend to use more caffeine than nonsmokers. Birth control pills can double the time that a given amount of caffeine lasts in the body. The drug can reduce drowsiness produced by *pentobarbital,* and it can reduce *diazepam's* interference with cognitive function. Caffeine is a traditional remedy for alcohol intoxication, but in fact it does not speed alcohol's elimination from the body, although caffeine's stimulant properties may help a drunken person function better. Estrogen replacement therapy appears to interfere with women's ability to metabolize caffeine.

In cancer

Caffeine does not seem to produce cancer in animal experiments. Indeed, green and black tea reduce development of cancer in mice, an effect in which caffeine is believed to play a part. In humans, however, caffeine is suspected of promoting premenopausal ovarian cancer and also cancer of the pancreas and bladder.

During pregnancy

Experiments examining caffeine's influence on pregnancy yield conflicting results, which may indicate the question is particularly complex or may simply mean that caffeine is an "invalid variable" having no effect. Here are examples. A 1994 survey of 259 women in the Netherlands found that caffeine raised the likelihood of becoming pregnant, but when the same subject was examined in the 1980s among almost 2,000 women in Connecticut, caffeine was found to suppress fertility.

During the 1990s a study of farm couples in Canada found that caffeine had no effect on fertility but that coffee-drinking women and tea-drinking men had lower birthrates, suggesting involvement of something besides caffeine in the natural products' effect. Rat experimentation shows that caffeine reduces female fertility, produces smaller than usual offspring, and may affect brain development. Research in Yugoslavia indicates that pregnant women who do not smoke cigarettes but do take more than 71 mg of caffeine daily have smaller infants. A British study found the opposite; slightly smaller infants came from cigarette-smoking women who used 1,000 mg or more of caffeine a week, but the effect was not seen in nonsmokers. Still other studies find no connection between caffeine and either birthweight or prematurity.

Caffeine affects vital signs in a human fetus even when the dose is so low as to have no influence on the pregnant woman. Question has arisen about whether caffeine promotes spontaneous abortion; a study published in 1994 found 140 mg to 280 mg a day to pose a significant risk; a rigorous study published in 1999 was unable to find such a hazard among moderate caffeine users; and a study published in 1993 saw caffeine as reducing the incidence of spontaneous abortion. Still another study found that women who drank decaffeinated coffee were even more likely to experience a spontaneous abortion than women who drank caffeinated coffee. Research in New Zealand indicated that sudden infant death syndrome (SIDS) was more likely if a woman had ingested more than 400 mg of caffeine daily during pregnancy, but an examination of SIDS in Scandinavia found no correlation with caffeine use during or after pregnancy. Instant coffee can damage DNA, but implications for general health or birth defects are unclear.

Testing caffeine on mice produced birth defects in limbs, and tests on chicken embryos produced heart deformities. Chicken embryos, however, are so sensitive to various chemicals that such results are not considered a warning of human danger. Indeed, a substantial body of research indicates that caffeine causes no human birth defects. Evidence does exist that caffeine can increase the likelihood of birth defects caused by *alcohol* and tobacco. An Egyptian study found that caffeine increases alcohol birth defects in rats. A statistical study showed that women who use more than 300 mg of caffeine daily around the time of conception and who do not smoke are less likely to have infants with Down syndrome. Given all the uncertainties, pregnant women are advised to use caffeine "*moderately*"—no more than 200 mg to 300 mg daily (150 mg or less is considered "*minimal*"). Caffeine increases milk production in nursing mothers and passes into the milk but appears unharmful to infants if the women are moderate users. On occasions when mothers use a lot of caffeine, however, their nursing infants may be fussier and have more trouble sleeping. A dose lasts longer in infants than in older persons.

ALCOHOLIC ADDICTION

Various kinds of alcohol exist; many are poisonous, such as wood alcohol, and are not intended for drinking. In former times physicians prescribed alcohol as a treatment for assorted afflictions. That usage is largely outmoded, but alcohol is still a common ingredient in cough remedies, is applied to skin as a disinfectant, and is a component of some injectable solutions. Alcohol is used as a partial antidote for methanol poisoning associated with inhalant abuse, and it is used to combat glycol poisoning. In some medical procedures alcohol is administered to create a form of anesthesia called a nerve block. Around 1970 at one hospital a combination of alcohol and *chloral hydrate* was routinely given to make pregnant women unconscious during labor, although alcohol is also a treatment for stopping premature labor.

The substance can be a sleep aid (a "nightcap" drink), but researchers have found that it interferes with dreaming, which is an important component of sleep. Tolerance can develop to sleep-inducing actions, and dream content can become disturbing for awhile after usage stops. Some persons drink alcohol to cope with stress. Evidence exists that moderate consumption of wine may help people live longer by reducing risk of heart disease and cancer.

History

As with many other drugs, moderate recreational use can be pleasurable while causing little harm. In contrast, excessive use can not only have psychological effects harmful to family and social functioning but also injure the stomach, liver, kidneys, heart, and brain. Rat experiments have demonstrated damage to the pancreas, damage that is boosted by cigarette smoke. A study found that men with spinal osteoporosis, a disease increasing the likelihood of broken bones, drink more alcohol than persons without the disease; this finding does not mean that alcohol causes the affliction, but it does indicate the need for further research.

Alcohol does not make peaceful individuals rageful, but it can lower inhibitions while leaving a person able to act out urges. Thus violent and criminal acts are commonly associated with alcohol intoxication. Impairment of mental and physical activity can occur during acute intoxication, making operation of dangerous machinery (such as automobiles) hazardous. During intoxication sensory perceptions are blunted, reducing awareness of tastes, smells, sounds, and pain. Lesser known problems are associated with alcohol. Male users experience a decline in *testosterone* levels, and females may experience menstrual difficulties. The nerve inflammation disease beriberi has been linked with alcoholism, and research has raised the possibility that alcoholism can worsen Alzheimer's disease.

Some studies report that drinkers have a slightly higher chance of developing cataracts, but a very large study involving 77,466 women found little, if any, relationship between the substance and cataracts. Experiments show that a drink of alcohol encourages more cigarette consumption and that persons who use both alcohol and *nicotine* tend to drink more when cigarettes are unavailable. Although alcohol can make a person feel hotter, that effect is superficial; the substance actually lowers body temperature, making alcohol counterproductive if a chilled person is trying to warm up; the substance should not be given to persons injured by exposure to cold temperatures. Alcoholics commonly have memory trouble, and small studies find that heavy-drinking nonalcoholics have impaired thinking skills even while sober.

Tolerance and Dependence

In addition to being a potent intoxicant, alcohol is one of the most addictive substances. Tolerance ("holding your liquor") and dependence may occur. Withdrawal can be dangerous, with death occurring (typically from difficulties leading to convulsions) despite intense medical supervision. The withdrawal syndrome is called delirium tremens. Its symptoms are similar to those of barbiturate withdrawal: weariness, nervousness, perspiration, tremors, vomiting, cramps, high blood pressure,

consumption by other persons, became politically powerful. Before the Civil War such groups had been able to get laws passed outlawing the sale of beverage alcohol in various communities and sometimes throughout entire states. Shortly after World War I this agitation culminated in the Eighteenth Amendment to the Constitution of the United States, giving the federal government power to ban manufacture and sale of beverage alcohol. Although purchase and consumption remained legal, the majority of Americans became so displeased with Prohibition that the Twenty-first Amendment to the Constitution was passed in the 1930s repealing the earlier one.

Index

A

A priori, 162
Activator of G protein signalling, 3, 16
Addiction, 68, 69, 214
Addiction-prone, 21
Addictive choice, 135
Addictive personality, 259
Addicts, 68, 267
Addictus, 257
Affective neuroscience, 25
Aid to work, 256
Alcohol, 276
Alcohol deprivation effect, 202
Alcohol-associated positive mood states, 211
Alcohol-induced motor incoordination, 157
Alcoholic family, 258
Amanita, 78
Amygdaloid complex, 43
Antinociceptive responses, 189
Anxiety, 157
Artificial selection, 169
Association, 278
Attention deficit hyperactivity disorder, 244, 273

B

Backcasting, 154
Behavioural tolerance, 5
Benign contagion, 146
Best reply, 152
Black market, 257
Bottom-up, 110
Bounded rationality, 30
Bounded willpower, 30

C

Cannabis, 79
Cannabis sativa, 188
Catastrophic, 149
Catha edulis, 80
Central nervous system, 117
Cheating, 245
Chloral hydrate, 277
Chlordiazepoxide, 278
Cloned enzyme donor immunoassay, 122
Cocaine, 270, 275, 278
Coke bugs, 273
Combination drug use, 34
Conceptualised addiction, 108
Conditioned place preference, 168
Conditioned stimulus, 167
Conditioned taste aversion, 168
Confessions of an english opium eater, 261
Congenic strains, 161, 162
Consenting addict, 231

Consuming, 167
Consumption, 166
Conventional, 149
Conventional behaviour, 149
Craving, 99, 178, 202
Criminals, 251
Cure, 230, 253, 250

D

Datura, 78
Decocainized, 272
Delay-independent, 33
Demi-monde, 267
Dependence, 214
Dependency syndrome, 69
Descending limb, 219
Dextroamphetamine, 275
Diagnostic and statistical manual, 94
Diagonal, 152
Diazepam, 275, 278
Disease, 239, 244
Dopamine transporter, 76
Drug, 251
Drug-preferring, 23
Drug-resistant, 23

E

Ecstasy, 25
Endophenotypes, 163
Enzyme immunoassay, 122
Enzyme multiplied immunoassay technique, 122
Enzyme-linked immunosorbent assay, 122
Ephedrine, 273
Epidemics, 147
Evidence-based, 259
Ex mortem, 222
Experimental psychology, 95

F

Faded out, 166
False positives, 181
Fetal alcohol syndrome, 279
First drink, 204
Flashbacks, 78
Fluorescence polarisation immunoassay, 122
Folk, 239
Framing, 138
Future of the race, 258

G

Gamma-hydroxybutyrate, 128
Gas chromatograph, 124
Gate, 40
Gateway effect, 137
Gene, 239, 241
Gene chasing, 239
General anxiety disorder, 239
Genomic medicine, 237
Gold standard, 124

H

Habit forming, 214
Hard drugs, 137
Harsh, 227
Heroin chic, 263
Heterodox, 149
High alcohol withdrawal, 160
High responders, 18
Homer, 15
Human rights, 229
Hypericum perforatum, 89

I

Immediate early gene, 223
Immunity, 236
In situ, 178, 209
In utero, 251
In vitro, 17
In vivo, 12, 17, 113, 126, 220, 221
Incentive salience, 204
Independence, 230
Infectiousness, 146
Inheritance of acquired characteristics, 258
Internet addiction, 70
Interval-specific congenic strains, 161
Invalid variable, 278

K
Khat, 270
Kindling, 208
Kinetic interaction of microparticles in solution, 122
Knockout, 172

L
Laddette, 263
Landing places, 154
Learning economies, 138
Lesson from history, 265
Lifestyle, 240
Lifetime, 273
Limited-access, 165
Liquid chromatograph, 124
Lobelia inflata, 89
Long sleep, 161
Lorazepam, 278
Loss of control, 27
Low responders, 18

M
Magic mushrooms, 78
Marijuana, 273
Mass spectrometry, 124
Mass spectrum, 125
Mechanisms of dependence, 103
Medical, 230, 258
Medical progress, 257, 259
Medicalisation, 239, 240
Metabolic tolerance, 5
Methadone, 278
Mind-altering, 236
Mind-manifesting, 78
Mind-wandering, 78
Miniature instrumentation, 128
Minimal, 276
Mistakes, 152
Mitragyna speciosa, 80
Moderately, 276
Modern theories, 257
Molecular mechanisms, 51
Molecular switch, 15
Mopping up, 249
Moral, 230
Morphine, 278
Myopia, 30

N
Neuroeconomics, 25
New knowledge, 259
Nicotine, 275, 277
Non-rational, 134
Normalisation, 241, 255
Nucleus accumbens, 38, 182

O
Objective, 147
Over-destabilising, 253
Over-represent, 140

P
Papaver somniferum, 75
Passiflora incarnata, 89
Pavlovian conditioning, 167
Peer pressure, 147
Pentobarbital, 275, 278
Phantom epidemics, 252
Pharmacodynamic tolerance, 5
Pleiotropy, 181
Polydrug abuse, 71
Polygenic, 181
Positron emission tomography, 103
Postmortem, 15, 17
Predictive genomic medicine, 237
Preimplantation genetic diagnosis, 242
Priming effect, 96, 204
Primrose path, 31
Pro-craving, 207
Pro-relapse, 207
Problem, 255, 260
Problem definition, 261
Prohibition experiment, 264
Pseudoendogenous, 128
Psilocybin, 79

Psychoactive drug, 117
Psychopharmacology, 11
Pueraria lobata, 89
Punishment, 250

Q

Quantitative trait loci, 159
Quick fix, 233

R

Rational, 136
Rational addiction, 150
Rational break, 104
Rational choice, 134
Rationally, 139
Real-life, 111
Recovered, 249
Rehabilitative, 250
Reinforcement, 157
Relaxing, 227
Reputation, 146
Revolutions, 256
Risk, 242

S

Salvia divinorum, 80
Sceletium tortuosum, 89
Scientific, 258
Scientific progress, 259
Second messengers, 2
Second-order, 99
Second-order schedules, 99
Seeking, 166, 167
Self-medicate, 5
Set and setting, 261
Short sleep, 161
Shyness, 245
Single nucleotide polymorphisms, 159
Social cognitive neuroscience, 25
Social lubricant, 74
Social technology, 247
Socially harmful, 242
Speedballs, 71
Spin, 126
Stable, 148
Standard inbred strains, 168
Stock of addiction, 146
Stress-induced sensitization, 57
Substance dependence, 68
Success stories, 267
Super-amphetamine, 33
Super-cocaine, 33
Supply chain, 137
Switched off, 231

T

Tabernanthe iboga, 89
Targeted neurochemical interventions, 243
Teaching signal, 8
Technology, 137
Testosterone, 277
Time-of-flight, 124
Tipping point, 268
Toolkit, 259
Top-down, 110
Transgenic, 172
Treatment, 241, 244
Treatment-oriented, 250

V

Vaccine, 236
Ventral tegmental area, 40, 60, 182
Victimless crime, 250
Visceral factors, 32

W

Wanting, 204
War on drugs, 267